Creditor Reporting System 2009

AID ACTIVITIES IN SUPPORT OF AGRICULTURE

Système de notification des pays créanciers 2009

ACTIVITÉS D'AIDE DANS LE SECTEUR DE L'AGRICULTURE

ORGANISATION FOR ECONOMIC CO-OPERATION AND DEVELOPMENT

The OECD is a unique forum where the governments of 30 democracies work together to address the economic, social and environmental challenges of globalisation. The OECD is also at the forefront of efforts to understand and to help governments respond to new developments and concerns, such as corporate governance, the information economy and the challenges of an ageing population. The Organisation provides a setting where governments can compare policy experiences, seek answers to common problems, identify good practice and work to co-ordinate domestic and international policies.

The OECD member countries are: Australia, Austria, Belgium, Canada, the Czech Republic, Denmark, Finland, France, Germany, Greece, Hungary, Iceland, Ireland, Italy, Japan, Korea, Luxembourg, Mexico, the Netherlands, New Zealand, Norway, Poland, Portugal, the Slovak Republic, Spain, Sweden, Switzerland, Turkey, the United Kingdom and the United States. The Commission of the European Communities takes part in the work of the OECD.

OECD Publishing disseminates widely the results of the Organisation's statistics gathering and research on economic, social and environmental issues, as well as the conventions, guidelines and standards agreed by its members.

ORGANISATION DE COOPÉRATION ET DE DÉVELOPPEMENT ÉCONOMIQUES

L'OCDE est un forum unique en son genre où les gouvernements de 30 démocraties œuvrent ensemble pour relever les défis économiques, sociaux et environnementaux que pose la mondialisation. L'OCDE est aussi à l'avant-garde des efforts entrepris pour comprendre les évolutions du monde actuel et les préoccupations qu'elles font naître. Elle aide les gouvernements à faire face à des situations nouvelles en examinant des thèmes tels que le gouvernement d'entreprise, l'économie de l'information et les défis posés par le vieillissement de la population. L'Organisation offre aux gouvernements un cadre leur permettant de comparer leurs expériences en matière de politiques, de chercher des réponses à des problèmes communs, d'identifier les bonnes pratiques et de travailler à la coordination des politiques nationales et internationales.

Les pays membres de l'OCDE sont : l'Allemagne, l'Australie, l'Autriche, la Belgique, le Canada, la Corée, le Danemark, l'Espagne, les États-Unis, la Finlande, la France, la Grèce, la Hongrie, l'Irlande, l'Islande, l'Italie, le Japon, le Luxembourg, le Mexique, la Norvège, la Nouvelle-Zélande, les Pays-Bas, la Pologne, le Portugal, la République slovaque, la République tchèque, le Royaume-Uni, la Suède, la Suisse et la Turquie. La Commission des Communautés européennes participe aux travaux de l'OCDE.

Les Éditions OCDE assurent une large diffusion aux travaux de l'Organisation. Ces derniers comprennent les résultats de l'activité de collecte de statistiques, les travaux de recherche menés sur des questions économiques, sociales et environnementales, ainsi que les conventions, les principes directeurs et les modèles développés par les pays membres.

This work is published on the responsibility of the Secretary-General of the OECD. The opinions expressed and arguments employed herein do not necessarily reflect the official views of the Organisation or of the governments of its member countries.

Cet ouvrage est publié sous la responsabilité du Secrétaire général de l'OCDE. Les opinions et les interprétations exprimées ne reflètent pas nécessairement les vues de l'OCDE ou des gouvernements de ses pays membres.

ISBN 978-92-64-07702-7 (print)
ISBN 978-92-64-07703-4 (PDF)

ISSN 1023-8875 (print)
ISSN 1609-6827 (online)

Corrigenda to OECD publications may be found on line at: *www.oecd.org/publishing/corrigenda*.
© OECD 2009

You can copy, download or print OECD content for your own use, and you can include excerpts from OECD publications, databases and multimedia products in your own documents, presentations, blogs, websites and teaching materials, provided that suitable acknowledgment of OECD as source and copyright owner is given. All requests for public or commercial use and translation rights should be submitted to *rights@oecd.org*. Requests for permission to photocopy portions of this material for public or commercial use shall be addressed directly to the Copyright Clearance Center (CCC) at *info@copyright.com* or the Centre français d'exploitation du droit de copie (CFC) at *contact@cfcopies.com*.

Foreword

This report examines data on aid in support of agriculture, with a focus on the period 2002-07. It is based on donors' reporting on Official Development Assistance (ODA) commitments and disbursements to the Creditor Reporting System (CRS) Aid Activity Database.

The report has been prepared by the Secretariat of the Development Assistance Committee (DAC) in collaboration with the DAC Working Party on Statistics.

The report contains **i) an analysis of aid to agriculture** including trends in donors' aid, geographical focus of flows, and a broader picture of donors' short and long-term interventions to address food security issues (food aid, rural development); **ii) individual donor profiles** with summary statistics on donors' bilateral aid to agriculture in the form of charts and tables, and **iii) the record of aid activities** for the agricultural sector, with amounts and descriptions, reported by bilateral donors and multilateral agencies in 2007.

Updated statistics on aid to agriculture will be regularly posted on the DAC website dedicated to agriculture at **www.oecd.org/dac/stats/agriculture**.

The complete CRS Aid Activity database contains records from 1973 onwards. It is available on the yearly *International Development Statistics* CD-ROM, and at **www.oecd.org/dac/stats/idsonline**.

Avant-propos

Le présent rapport passe au crible les statistiques sur l'aide à l'agriculture au cours de la période 2002-07. Il se fonde sur les engagements et versements d'Aide publique au développement (APD) déclarés par les donneurs dans la base de données sur les activités d'aide du Système de notification des pays créanciers (SNPC).

Ce rapport a été élaboré par le Secrétariat du Comité d'aide au développement (CAD) en collaboration avec le Groupe de travail du CAD sur les Statistiques.

Il contient **i) une analyse de l'aide à l'agriculture** présentant les tendances de l'aide des donneurs, le ciblage géographique des apports et un panorama des interventions à court et long terme des donneurs face aux enjeux de la sécurité alimentaire (aide alimentaire, développement rural), **ii) des profils individuels des donneurs** montrant des statistiques synthétiques sur l'aide bilatérale des donneurs à l'agriculture sous la forme de graphiques et de tableaux, et **iii) les données des activités d'aide** au secteur agricole, précisant leurs montants et descriptifs, notifiées par les donneurs bilatéraux et les agences multilatérales en 2007.

Les statistiques sur l'aide pour l'agriculture sont actualisées périodiquement sur le site Internet du CAD dédié à l'agriculture, à l'adresse **www.oecd.org/dac/stats/agriculture**.

La base de données sur les activités d'aide du SNPC est renseignée depuis 1973. Elle est disponible sur le CD-ROM annuel des *Statistiques sur le développement international* et à l'adresse **www.oecd.org/dac/stats/idsonline**.

Table of contents

INTRODUCTION: Basic aspects of the CRS Aid Activity database   p. 5

PART 1: Analysis of aid to agriculture   p. 7

PART 2: Bilateral aid to agriculture: donor profiles   p. 59

PART 3: List of aid activities in the agricultural sector, 2007   p. 85

Table des matières

INTRODUCTION :

Aspects fondamentaux de la base de données sur les activités d'aide du SNPC   p. 5

PARTIE 1 : Analyse de l'aide à l'agriculture   p. 33

PARTIE 2 : Aide bilatérale à l'agriculture : profils des donneurs   p. 59

PARTIE 3 : Liste des activités d'aide au secteur agricole, 2007   p. 85

INTRODUCTION: Basic aspects of the CRS Aid Activity database

(Creditor Reporting System)

The CRS Aid Activity database includes data on Official Development Assistance (ODA), and other official flows to developing countries.

The CRS was established in 1967, jointly by the OECD and the World Bank, with the aim of "supplying the participants with a regular flow of data on indebtedness and capital flows". Calculating capital flows and debt stock remain key functions of the system, but others have evolved in the course of years. In particular, the CRS aid activity database has become the internationally recognised source of data on the geographical and sectoral breakdown of aid and is widely used by governments, organisations and researchers active in the field of development. For DAC members the CRS serves as a tool for monitoring specific policy issues, supplementing the information collected at the aggregate level in the annual DAC Statistics.

> The CRS Aid Activity database comprises commitment and disbursement data on Official Development Assistance (ODA) activities in developing countries submitted by members of the DAC and multilateral institutions (135 000 – 155 000 transactions in recent years). Data in this publication refer to commitments, unless otherwise stated.

INTRODUCTION : caractéristiques fondamentales de la base de données sur les activités d'aide du SNPC

(Système de notification des pays créanciers)

La base de données sur les activités d'aide du SNPC contient des données sur l'Aide publique au développement (APD) et sur les autres apports du secteur public aux pays en développement notifiées par les membres du CAD et les institutions multilatérales.

Le SNPC a été établi en 1967 conjointement par l'OCDE et la Banque mondiale afin de fournir « régulièrement aux participants des renseignements sur l'endettement et les apports de moyens financiers ». Les statistiques sur les apports de ressources et le stock de la dette restent d'actualité, mais les fonctions du système ont évolué au cours des années. En particulier la base de données sur les activités d'aide du SNPC est devenue la référence internationale sur la répartition géographique et sectorielle de l'aide et est largement utilisée par les gouvernements, les organisations et les chercheurs actifs dans le domaine du développement. Pour les membres du CAD, le SNPC est un outil de suivi des questions politiques spécifiques, qui complète les informations agrégées notifiées dans les statistiques annuelles du CAD.

> La base de données sur les activités d'aide comprend les données relatives aux engagements et versements d'Aide publique au développement (APD) dans les pays en développement notifiés par les membres du CAD et les institutions multilatérales (de 135 000 à 155 000 transactions ces dernières années). Sauf mention contraire, les statistiques citées dans ce document concernent les engagements.

PART 1

Analysis of aid to agriculture

Focus on years 2002-07

www.oecd.org/dac/stats/agriculture

Key findings

Since the mid 1980s, **aid to agriculture has fallen by half**. In 2006-07, DAC member countries' bilateral annual aid commitments to agriculture amounted to **USD 3.8 billion**. Taking into account multilateral agencies' outflows, the total was **USD 6.2 billion**.

The share of aid to agriculture in DAC members' aid programmes has declined even more sharply: from 17% in the late 1980s to 6% in recent years, revealing a **clear relative neglect of the sector**.

Recent trends indicate a slowdown in the decline, and even the prospect of an upward trend: over the period 2002-07, bilateral aid to agriculture increased at an average annual rate of 5% (real terms).

Among DAC members, the largest donors in 2006-07 were **the United States (on average USD 932 million per year)**, Japan (**USD 821 million**) and France (**USD 451 million**).

Over the period 2002-07, aid flows to agriculture primarily targeted **Sub-Saharan Africa (31%) and South and Central Asia (23%)**. Least Developed Countries (LDCs) and Other Low Income Countries (Other LICs) received two-thirds of total aid to agriculture, indicating a clear focus on the poorest countries.

Total aid to agriculture and food security, including rural development, development food aid and emergency food aid, amounted to **USD 12 billion** per year in 2006-07, with **USD 7.6 billion** allocated to long-term interventions (agriculture and rural development), and the remaining **USD 4.4 billion** to short-term food aid (emergency and developmental).

Nearly half of all aid to agriculture is given in loan form, primarily by large multilateral organisations, including IDA, the regional development banks, and IFAD.

While **two-thirds of aid to agriculture goes to least developed and other low-income countries**, it is not well-targeted on the countries with the highest rates of malnutrition.

www.oecd.org/dac/stats/agriculture

Introduction

Context

This report examines data on aid flows in support of agriculture especially over the six year period 2002-07. It has been prepared by the Secretariat of the Development Assistance Committee (DAC) in collaboration with the DAC Working Party on Statistics.

Global food security is a perennial problem, and the recent food price crisis refocused the world's attention on this challenge. The food prices peaks reached in mid-2008 caused an additional 100 million people to fall back into poverty, and some 1 billion people are now hungry. Developing country agriculture provides the supply and incomes needed to ensure available and affordable food. Its contribution to poverty reduction and hunger is emphasized in the Millennium Development Goals (MDGs) through *MDG1. Eradicate extreme poverty and hunger* with a focus on hunger through *Target 1.C. Halve, by 2015, the proportion of people who suffer from hunger.*

Structure of the report

This report is structured in three parts. Part 1 is an analysis of **aid to agriculture** including trends in donors' aid, geographical focus of flows, and a broader picture of donors' short and long-term interventions to address food security issues (including food aid and rural development). Part 2 contains **individual donor profiles** with summary statistics on donors' bilateral aid to agriculture in the form of charts and tables. Part 3 is **the record of aid activities** for the agricultural sector, with amounts and descriptions, reported by bilateral donors and multilateral agencies in 2007.

Statistics shown include data up to 2007. Donors' commitments undertaken in response to the high food price crisis in 2008 and 2009 do not therefore appear in this report. Data on 2008 flows will become available at the end of 2009, and will be posted in the form of a factsheet on the DAC website dedicated to agriculture early 2010.

Source of the information

Statistics are based on donors' reporting on Official Development Assistance (ODA) commitments and disbursements to the Creditor Reporting System (CRS) Aid Activity Database. Data were confirmed by members, through a special request to them in Spring 2009.

Data that support this report are available from the DAC website

www.oecd.org/dac/stats/agriculture
www.oecd.org/dac/stats/data

I. Trends in aid to agriculture

I.1 Definitions and terminology

Coverage of "aid to agriculture"

The DAC statistical definition of aid to agriculture includes agricultural sector policy, planning and programmes, agricultural land and water resources, agricultural development and supply of inputs, crops and livestock production, agricultural services, agricultural education, training and research as well as institution capacity building and advice. Forestry and fishing are identified as separate sectors from 1996, and are shown as part of aid to agriculture in this report.

The definition therefore excludes rural development (classified as multi-sector aid), agro and forest-industries (industry), developmental food aid (general programme assistance) and emergency food aid (humanitarian assistance). These sectors are not taken into account in main statistics shown in this report, but, given their relevance to food security, they are included in section III, and shown separately in the donor profiles of Part 2.

The definitions of all sectors and sub-sectors are shown in the Annex.

Sector-allocable aid

In order to better reflect the sectoral focus of donors' programmes, when calculating the share of aid for agriculture in total bilateral aid, contributions not susceptible to allocation by sector are excluded from the denominator (general budget support, actions relating to debt, humanitarian aid, administrative costs and other internal transactions in the donor country).

DAC members' imputed multilateral contributions

DAC members, in addition to undertaking bilateral aid activities in the agriculture sector, also contribute to multilateral agencies active in the field of agriculture. In order to provide a full picture of the total ODA effort the donor makes in respect of the agricultural sector, data on DAC members' imputed multilateral aid for agriculture have been compiled.

The calculation of imputed multilateral contributions is done in two steps[1]: 1/ a 3-year average of the share of each multilateral agency's aid flows for agriculture in their total sector-allocable aid is calculated (core resources only); 2/ the share obtained in step 1 is multiplied by a member's contribution to the core resources of the agency concerned in the years concerned; the resulting amount is the imputed flow from that donor to the agricultural sector through the multilateral agency concerned. It can only ever be an approximation.

I.2 Bilateral and multilateral aid to agriculture

This section presents statistics on aid to agriculture by bilateral donors (DAC countries) and multilateral institutions.

---

[1] See also "The OECD methodology for calculating imputed multilateral ODA" under www.oecd.org/dac/stats/methodology.

Chart 1 below illustrates the long-term trend in aid to agriculture, and Chart 2 focuses on the last decade. Aid to agriculture has fallen by half since the mid 1980s, from USD 12 billion annually in 1984-85 to USD 6.2 billion annually in 2006-07 (bilateral and multilateral commitments).

Chart 1. Trends in ODA to agriculture

1971-2007, 5-year moving averages and annual figures, constant 2007 prices

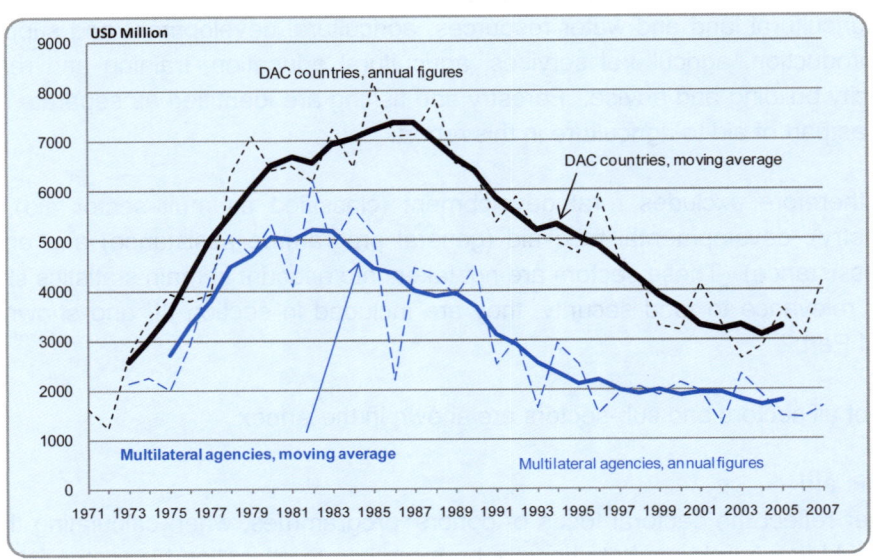

Chart 2 shows a slowdown in the decline over the last decade, and even the prospect of an upward trend: over the period 2002-07, bilateral aid to agriculture increased at an average annual rate of 5% (real terms). In 2006-07, bilateral aid to agriculture reached **USD 3.8 billion**.

The share of aid to agriculture in total bilateral sector-allocable aid is an indication of the extent to which donors' aid programmes focus on agriculture, and the priority this sector is being given. For DAC countries, the share has fallen from 13% to 6% over the last decade (1995-2007). It seems that the decline has now stopped, with the share stabilizing at around 6% since the early 2000s.

Chart 2. Trends in DAC countries' ODA to agriculture

1995-2007, 3-year moving averages, constant 2007 prices

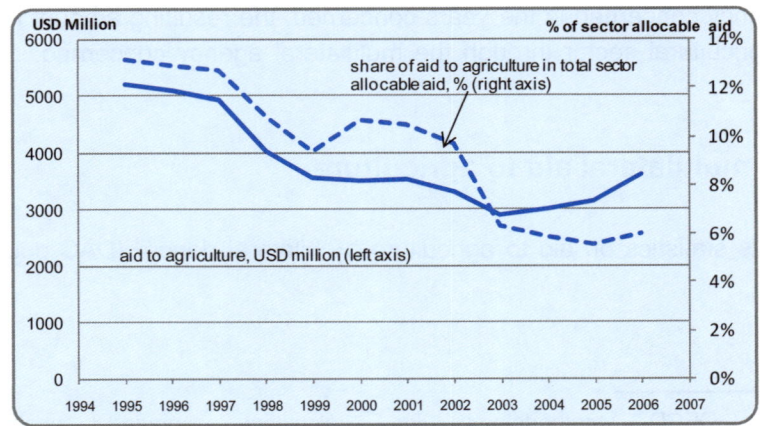

Table 1 below provides the breakdown of aid to agriculture by individual donors and multilateral institutions. Of DAC countries' combined aid for agriculture of USD 3.8 billion over 2006-2007, the United States is the main donor, accounting for 25% of the total, followed by Japan (22%) and France (12%). IDA is the largest multilateral donor to the agriculture sector, accounting for 45% of multilateral aid, followed by EU institutions (19%).

Greece, Luxembourg, New Zealand, Sweden and the United States more than doubled their aid commitments to the agricultural sector over the period 2002-2007; Austria, Finland, France, Ireland and Norway also increased their aid to agriculture significantly, and Denmark, Germany and Spain less markedly. Australia, Netherlands, Portugal, Switzerland and the United Kingdom reduced their aid to agriculture, while there was no significant change in the pattern of aid for Belgium, Canada, Italy, Japan, and the EU institutions.

A number of individual donors extend relatively high proportions of their aid to agricultural projects: Japan (10%), Belgium, Denmark, France and Switzerland (9% each), Finland and Sweden (8% each), and Ireland and Spain (7% each) are above the 6% DAC average for years 2006-07.

Table 1. Aid to agriculture by bilateral and multilateral donors
2002-07, annual average commitments and shares in total sector-allocable aid, constant 2007 prices

	Commitments, USD million			% of Donor Total			% All Donors		
	2002-03	2004-05	2006-07	2002-03	2004-05	2006-07	2002-03	2004-05	2006-07
Australia	114	84	82	10	7	5	2	2	1
Austria	7	9	12	2	4	4	0	0	0
Belgium	95	93	91	10	10	9	2	2	1
Canada	115	213	114	7	11	6	2	4	2
Denmark	67	204	83	8	13	9	1	4	1
Finland	22	48	34	6	11	8	0	1	1
France	255	184	451	7	4	9	5	4	7
Germany	239	229	317	5	5	5	5	5	5
Greece	2	1	5	1	1	3	0	0	0
Ireland	23	22	33	7	6	7	0	0	1
Italy	48	27	45	7	4	6	1	1	1
Japan	820	702	821	12	9	10	16	14	13
Luxembourg	4	11	9	3	8	6	0	0	0
Netherlands	226	148	121	10	6	3	4	3	2
New Zealand	5	9	11	5	6	5	0	0	0
Norway	83	115	117	6	8	6	2	2	2
Portugal	6	3	2	2	1	1	0	0	0
Spain	136	115	163	9	9	7	3	2	3
Sweden	44	79	144	3	5	8	1	2	2
Switzerland	85	70	69	13	9	9	2	1	1
United Kingdom	221	143	113	6	4	2	4	3	2
United States	337	669	932	3	4	5	7	13	15
Total DAC countries	**2953**	**3179**	**3768**	**7**	**6**	**6**	**58**	**63**	**61**
AfDF	306	207	245	23	16	15	6	4	4
AsDF	286	226	223	16	15	14	6	4	4
EC	471	288	444	7	3	4	9	6	7
FAO	219	213	203	100	100	100	4	4	3
IDA	725	718	1075	10	10	12	14	14	17
IDB Sp.Fund	48	42	0	8	10	0	1	1	0
IFAD	107	175	209	24	40	44	2	3	3
UNDP	0	7	1	-	2	0	0	0	0
Total Multilateral	**2161**	**1875**	**2400**	**12**	**9**	**10**	**42**	**37**	**39**
Total	**5114**	**5054**	**6169**	**8**	**7**	**7**	**100**	**100**	**100**

Notes:

General budget support, once integrated in developing countries' domestic budgets, will contribute to the development of the agricultural sector, but this contribution is not specified and not taken into account in the above figures (see also "Aid modalities" under section I.3).

Figures for **FAO** are Secretariat estimates based on the agency's ODA coefficient and core budget.

Belgium internally uses a multiple-purpose coding system that leads to higher figures: USD 122 million for 2002-03, USD 121 million for 2004-05, and USD 124 million for 2006-07.

The data show that **IFAD** does not extend the totality of its aid flows for agriculture (44% in 2006-07). Most of the remaining flows fall under rural development in the DAC sector classification, and are taken into account in a broader measure of aid to agriculture and food security (see section III).

DAC members' imputed multilateral contributions to agriculture

As explained in section I.1, DAC members' imputed multilateral contributions can be estimated (Table 2) and added to their bilateral contributions (Table 3).

Table 2 indicates that, out of DAC countries' total core contributions to multilateral agencies, **USD 1.4 billion** can be imputed to agriculture over 2006-07. IDA is the largest multilateral channel representing 37% of this amount, followed by the EU institutions with 24%.

The annual total for imputed multilateral contributions over 2006-07, USD 1.4 billion, does not match total multilateral outflows shown in Table 1 (USD 2.4 billion annually over 2006-07) because multilateral flows in a given year are not derived directly from donors' contributions in that year. In the case of international financial institutions (IFIs), lending in any one year is much greater than donors' contributions (grants), since it also draws on reflows of principal, interest receipts, and transfers of funds within the IFIs. There are also delays between receipt of funds by the multilateral agency and their disbursement to developing countries.

Table 2. DAC members' imputed multilateral contributions to agriculture

annual average 2006-07, USD million, constant 2007 prices

	through EU institutions	through AfDF	through AsDF	through IDA	through IDB	through FAO	through IFAD	through UNDP	Total imputed multilateral contributions
Australia	0.0	0.0	5.6	13.9	0.0	3.7	0.7	0.1	24.0
Austria	8.0	4.1	1.2	9.1	0.0	2.0	2.6	0.0	26.9
Belgium	13.6	4.1	1.4	9.6	0.0	2.4	1.9	0.1	33.0
Canada	0.0	0.0	0.0	26.3	0.1	6.4	4.9	0.3	38.1
Denmark	7.4	2.6	1.3	7.2	0.0	1.7	2.1	0.5	22.7
Finland	5.4	2.8	0.7	4.1	0.0	1.6	0.6	0.1	15.3
France	64.8	22.5	6.2	43.6	0.0	15.6	4.7	0.2	157.6
Germany	73.9	25.9	7.7	72.7	0.0	19.8	9.6	0.2	209.9
Greece	6.4	0.0	0.0	2.4	0.0	1.7	0.1	0.0	10.6
Ireland	4.3	0.0	0.6	7.2	0.0	0.8	0.9	0.1	13.9
Italy	45.7	0.8	0.0	2.8	0.0	23.1	15.0	0.2	87.7
Japan	0.0	14.6	39.7	96.9	0.0	51.3	6.6	0.4	209.4
Luxembourg	0.9	0.0	1.3	0.9	0.0	0.2	0.3	0.0	3.5
Netherlands	15.9	8.6	2.4	5.9	0.0	10.8	6.6	0.7	50.8
New Zealand	0.0	0.0	0.0	0.8	0.0	0.7	0.0	0.0	1.6
Norway	0.0	0.0	0.0	10.7	0.0	1.6	5.0	0.6	17.8
Portugal	4.4	2.2	1.3	1.4	0.0	1.1	0.3	0.0	10.6
Spain	29.8	0.0	0.0	19.1	0.0	17.8	8.9	0.3	76.0
Sweden	9.0	12.3	1.9	15.5	0.0	2.3	8.2	0.6	49.7
Switzerland	0.0	6.2	1.6	14.1	0.0	2.6	2.6	0.2	27.3
United Kingdom	59.8	24.9	12.9	84.7	0.0	13.5	9.0	0.7	205.4
United States	0.0	0.2	14.8	80.7	0.0	46.0	6.4	0.6	148.7
Total DAC countries	349.3	131.7	100.5	529.4	0.2	226.7	96.7	5.8	1440.3
EU institutions				4.6		0.4	12.5		17.5

Table 3 presents donor countries' total commitments to agriculture, including both bilateral and imputed multilateral contributions. For a number of countries, taking imputed contributions into account makes a significant difference in the total, which needs to be born in mind when consulting Part 2 of this report which presents donor profiles covering bilateral aid only. In particular, for Austria, Greece, Italy, Portugal and the United Kingdom, imputed multilateral contributions represent at least two-thirds of their total support for the agricultural sector. For most of these countries, the level of their imputed multilateral contributions is more than double the level of their bilateral contributions.

Table 3. DAC members' total aid to agriculture

2002-07, bilateral commitments, imputed multilateral contributions, bilateral disbursements,
annual averages, USD million, constant 2007 prices

	Bilateral commitments			Imputed multilateral contributions			Total commitments to agriculture			Bilateral disbursements		
	2002-03	2004-05	2006-07	2002-03	2004-05	2006-07	2002-03	2004-05	2006-07	2002-03	2004-05	2006-07
Australia	114	84	82	17	14	24	132	98	106	86	83	92
Austria	7	9	12	16	21	27	23	30	39	6	6	9
Belgium	95	93	91	25	34	33	120	127	124	73	57	66
Canada	115	213	114	27	27	38	142	240	152	84	91	122
Denmark	67	204	83	28	26	23	95	230	106	32	73	79
Finland	22	48	34	14	14	15	35	61	49	37	..	23
France	255	184	451	142	147	158	397	331	608	257	218	399
Germany	239	229	317	151	167	210	391	396	527	230	236	238
Greece	2	1	5	8	7	11	10	9	15	2	1	5
Ireland	23	22	33	5	7	14	28	29	47	20	22	32
Italy	48	27	45	71	147	88	118	173	132	15	26	42
Japan	820	702	821	149	95	209	968	797	1030	457	498	687
Luxembourg	4	11	9	2	3	4	6	14	13	0	11	9
Netherlands	226	148	121	58	63	51	284	211	172	142	121	116
New Zealand	5	9	11	1	1	2	7	10	13	3	5	6
Norway	83	115	117	33	23	18	116	138	135	80	105	100
Portugal	6	3	2	13	8	11	19	12	12	6	3	2
Spain	136	115	163	44	59	76	180	174	239	77	105	111
Sweden	44	79	144	51	34	50	95	113	193	52	64	117
Switzerland	85	70	69	20	24	27	104	94	96	77	79	64
United Kingdom	221	143	113	115	126	205	336	269	319	161	150	153
United States	337	669	932	119	181	149	455	850	1081	216	514	472
Total DAC countries	2953	3179	3768	1109	1229	1440	4062	4408	5209	2114	2468	2944
EU institutions	471	288	444	24	6	17	494	294	462	111	238	359
Total DAC members	3423	3467	4213	1133	1235	1458	4557	4702	5670	2225	2705	3302

Note that the imputed multilateral commitments shown above do not include earmarked contributions to specific projects and programmes administered by multilateral agencies. These earmarked contributions are reported as bilateral aid.

DAC members' disbursements

DAC members' disbursements to agriculture increased steadily over the period 2002-07, and amounted to **USD 3.3 billion** in 2006-07. The ranking of donors is slightly modified when based on disbursements instead of commitments: Japan ranks first instead of the United States.

Chart 3 illustrates the relation between commitments and disbursements at the DAC total level. Commitment data reflect donors' programming and changes in their policies; they thus give an indication about future flows. Disbursements show actual payments in each year. Disbursements and commitments cannot be compared for a same year, as commitments are multi-year and subsequent disbursements span several years.

Disbursements follow the slightly increasing trend of commitments over the period 2002-07. A gap between disbursement and commitment figures is noticeable in years 2005 and 2007 where large increases in aid allocations (commitments) occurred and would be visible in disbursement data only with a few years' time lag. Out of DAC members' total bilateral commitments in 2007, approximately one-third was disbursed the same year; the remaining two-thirds will only be disbursed in subsequent years[2].

---

[2] This figure relates to countries for which it is possible to track project implementation i.e. link disbursements to the original commitments.

Chart 3. DAC countries' bilateral and imputed multilateral aid for agriculture,

2002-2007, constant 2007 prices

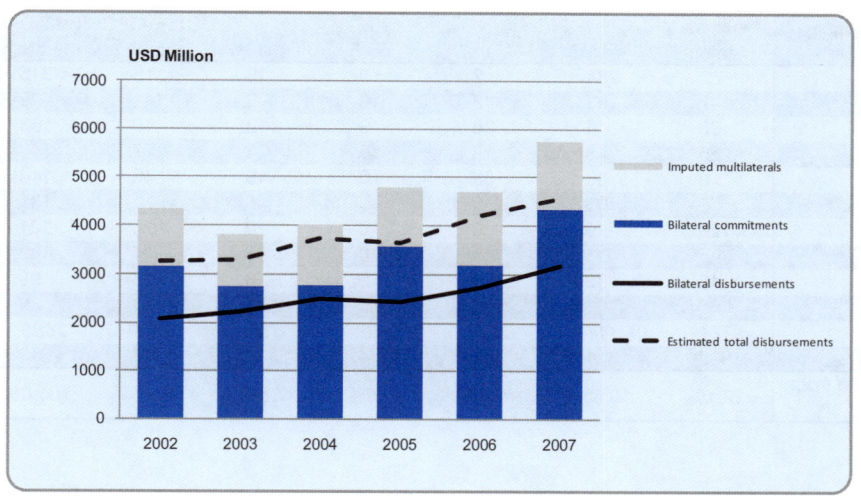

I.3 Main characteristics of DAC members' bilateral aid to the agricultural sector

Geographical focus and concentration of aid

Table 4 below summarizes the geographical focus of DAC members' aid for agriculture. Aid is concentrated in a relatively small number of very large commitments: the 100 largest activities reported by members for 2006-07 accounted for a half of the total commitments during these two years. Activities of more than USD 50 million include the following:

- The United States financed agricultural alternative development projects to reduce illicit drug cultivation in **Afghanistan** (USD 216 million) and **Colombia** (USD 75 million). The United States also contributed through the Millennium Challenge Account[3] to various countries and agricultural sub-sectors: **Armenia** (USD 150 million for irrigation), **Ghana** (USD 67 million for agricultural education, and USD 58 million for financial services) and **Mali** (USD 150 million for policy and administrative management); another large contribution from the United States was its support to agricultural productivity in **Iraq** (USD 63 million).

- Japan extended new loans to **China**: afforestation projects in Jilin (USD 81 million) and Henan (USD 63 million) and a forestry development project in Qinghai (USD 53 million); **India**: construction of irrigation facilities in Andhra Pradesh (USD 204 million), restoration of degraded forests in Gujarat (USD 149 million), Orissa (USD 117 million), and Tripura (USD 66 million); **Peru**: irrigation (USD 50 million); and **Philippines**: agrarian reform (USD 84 million)

- France contributed to agricultural research in **Brazil** (USD 71 million) and **Madagascar** (USD 53 million), and extended a loan to **Viet Nam** for the development of its rural central provinces (USD 55 million).

- Other activities above USD 50 million included an agricultural commodities programme for all **ACP countries** (African Caribbean Pacific) by the EU (USD 51 million); grants from Germany to **Egypt** for irrigation (USD 125 million); and a contribution to agricultural research by the **United Kingdom** (USD 77 million).

---

[3] Presidential initiative to reduce poverty through economic growth in poor countries.

Table 4. Main donors and recipients of DAC members' bilateral aid to agriculture
2006-07 average commitments in million USD, constant 2007 prices

USD million, average 2006-07	United States	Japan	France	EC	Germany	Other DAC members	Total DAC members	% of agriculture to all recipients
India	2	284	2	0	18	10	316	8%
Afghanistan	147	3	0	16	2	41	209	5%
Ghana	96	6	28	3	12	18	163	4%
Mali	80	3	18	10	8	27	148	4%
Viet Nam	0	30	59	0	10	48	146	3%
China	0	108	0	0	8	21	137	3%
Iraq	120	0	0	0	0	0	121	3%
Egypt	6	11	1	0	87	4	108	3%
Peru	45	32	1	0	8	13	99	2%
Colombia	60	1	4	5	11	4	85	2%
Other recipients	374	344	338	410	153	1061	2680	64%
Total amount	932	821	451	444	317	1248	4213	100%
% of agriculture aid from all DAC members	22.1%	19.5%	10.7%	10.5%	7.5%	29.6%	100.0%	

Financial instruments

At DAC total level, aid activities in the agricultural sector are financed primarily through **grants which represented 82% of total aid over the period 2002-07**. This share has not changed in recent years. However, Germany and Japan make use of concessional loans in higher proportions to support the agricultural sector: 61% of Japanese aid to agriculture is in the form of loans, and the share is 39% for Germany (period 2006-07). The other members that make use of loans are: EU institutions (1%), Finland (2%), France (18%), Italy (17%), and Spain (15%).

A few donors have also reported as ODA bilateral development finance institutions' investments in agricultural enterprises (agriculture and fisheries). The annual average of ODA equities reported for this sector in years 2006-07 was USD 20 million. These include investments by CDC (United Kingdom), DEG (Germany), EIB (EU institutions), FINNFUND (Finland) and NORFUND (Norway).

The figures presented above deal only with aid, i.e. grants and soft loans. But developing countries also receive unsubsidised loans from major bilateral and multilateral lenders. The major providers of these loans to the agricultural sector in 2007 were IBRD (USD 777 million), and the Asian (USD 260 million) and African Development Banks (USD 97 million).

Aid modalities

Aid modalities cannot currently be properly identified in CRS reporting, but donors will be requested to report on them starting with 2010 flows. Based on existing reporting, a few facts and figures can already be gathered.

Donors' use of programmatic approaches in the field of agriculture is particularly noticeable for Nordic countries and the EU institutions. Denmark extends 86% of its contributions to agriculture in the framework of sector programmes; the share is 59%, 47%, 27%, and 20% for Norway, Finland, Sweden and the EU institutions respectively.

- Over 2006-07, Denmark contributed to sector programmes in Bangladesh (fisheries and livestock development, agriculture), Mali (agriculture), Tanzania (forestry) and Vietnam (agriculture and rural development).
- The EU institutions provide sector budget support in the context of accompanying measures for sugar protocol countries affected by reform of EU sugar regime (Barbados, Guyana, Jamaica, Mauritius, Mozambique, Trinidad and Tobago).
- Contributions to sector wide approaches can be identified in CRS reporting for the Mozambican national programme for agricultural development – ProAgri (contributions can be identified for Austria, EU institutions, Ireland, Sweden) and the Nicaraguan rural productive development programme – ProRural (Norway).
- Other common funding arrangements for agricultural development included contributions to multi-donor trust funds: the TerrAfrica leveraging fund for Africa (Norway); and the forestry trust fund for Nicaragua (Finland, Netherlands).

Donors' general budget support, once integrated in developing countries' domestic budgets, will contribute to the development of the agricultural sector, but the amounts cannot be precisely specified since the support is not earmarked for any specific use but may be accompanied by various exclusions or understandings on the government's development strategy. These donors do not control the spending but monitor the implementation of the recipient's strategy as a whole on the basis of an agreed set of indicators.

A number of donors claim to increasingly channel their aid to developing countries through general budget support instead of undertaking specific projects in identified sectors. Some make estimates of their contribution to various sectors through general support based on the proportion of recipient governments' domestic spending on these sectors. The resulting figures are not used in standard DAC statistical presentations.

Sub-sectors

Chart 4. DAC members' aid to agriculture by sub-sectors

average 2006-07, constant 2007 prices

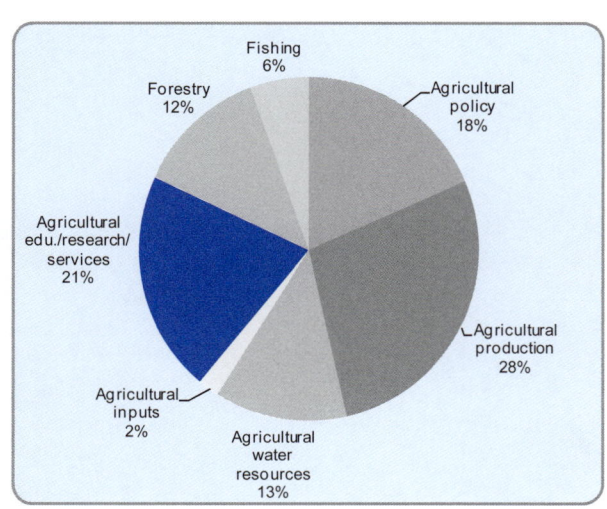

Chart 4 highlights the shares of forestry and fishing in total aid to agriculture: 12% and 6% respectively. Members have different approaches, some placing emphasis more on forestry (Finland: 67%; Japan and Netherlands: 36%; Australia: 24%), and others on fishing (Greece: 38%; Spain: 31%; Denmark, New Zealand and Norway: 21%).

See Annex for description of the sub-sectors.

Agricultural production accounted for 28% of DAC members' contributions to agriculture in 2006-07. Agricultural development is included under this heading, and covers integrated projects and farm development; it captures 14% of members' aid to agriculture; agricultural alternative development is another significant component representing 7% of total aid to agriculture (27% of total aid from the United States, 8% from Canada, 7% from the Netherlands); food crop production and livestock attract only 3% of total aid, the same share as for export crops[4].

Agricultural policy represents 18% of total aid, and is a significant of component of a number of members' aid to agriculture (Ireland, Switzerland, United Kingdom and the United States). Specific support to land resources is mainly undertaken by Ireland (13%) and Italy (17%), and to agrarian reform by Belgium and Sweden (5%) and Japan (6%).

As regards the remaining sub-sectors, members seem to concentrate their support on different areas:

- agricultural water resources account for 13% on average, but for more than one-third of aid from Germany, and more than 20% of aid from Luxembourg and Japan;
- the supply of agricultural inputs is relatively small at total level (2%), but more significant for Finland (10%), Japan (6%), Luxembourg and Norway (5%);
- the 16% average share allocated to education and research rises to much higher level in the case of France (69%); Sweden (42%); United kingdom (37%); Australia and Portugal (30%) [see also Box 1 on research];
- the small 5% average share allocated to the various agricultural services is nevertheless exceptionally large for a number of members New Zealand (28% of total aid dedicated to services such as marketing policies and organisations); Portugal (9% to plant and post-harvest protection and pest control); Norway (8% to agricultural co-operatives); Greece and New Zealand (22% and 12% respectively for veterinary services).

---

[4] Percentages for these specific sub-sectors are likely to be underestimated, as integrated projects or programmes that do include related components will be recorded under larger defined sub-sectors, e.g. agricultural development.

Box 1. Aid to agricultural research

Aid to research in agriculture 2006-2007 average

	USD million	Share in total aid to agriculture
Australia	21	25%
Austria	0	1%
Belgium	5	5%
Canada	2	2%
Denmark	4	4%
Finland	1	3%
France	305	68%
Germany	23	7%
Greece	0	1%
Ireland	1	3%
Italy	2	5%
Japan	0	0%
Luxembourg	0	0%
Netherlands	6	5%
New Zealand	0	3%
Norway	9	7%
Portugal	1	29%
Spain	2	1%
Sweden	34	23%
Switzerland	9	13%
United Kingdom	1	1%
United States	26	3%
EU institutions	17	4%
Total DAC members	**468**	**11%**

Activities of the **Australian** Centre for International Agricultural Research (ACIAR) target specific issues and recipient countries, e.g. developing molecular markers to enable selection against chalk in rice in the Philippines or using pathogen-tested planting materials to improve sustainable sweet potato production in the Solomon Islands and Papua New Guinea.

France reports ODA-eligible research undertaken by the CIRAD ("Centre de coopération internationale en recherche agronomique pour le développement ").

Germany supports International Agricultural Research Centres (IARCs), especially those backed by the Consultative Group for International Agricultural Research (CGIAR) by providing targeted funding to strengthen cooperation between German and International research institutions.

Sweden committed in 2006-07 ODA grant research activities principally through the CGIAR and an "Upland Research and Capacity Development Programme) in Laos.

The **United States** principally support the CGIAR, but also work directly on specific projects such as the "Former Soviet Union Cooperative Research programme" that aims at redirecting efforts of former Soviet biological weapons scientists to peaceful agricultural research.

Gender equality focus

Women's role in agriculture is critical, as in many developing countries, women provide most of the agricultural labour and produce most food. Information on the extent to which members have taken particular measures in their aid activities for agriculture to advance gender equality and women's empowerment is captured through the "gender equality policy marker". On the basis of available data[5], almost half of aid for agriculture targeted the objective of gender equality. Chart 5 shows the individual members with the highest shares of gender-equality focused aid, along with the DAC total.

---

[5] Only members with a marker coverage above 80% were included.

Chart 5 Gender equality focus of aid activities in the agricultural sector

average 2006-07, constant 2007 prices

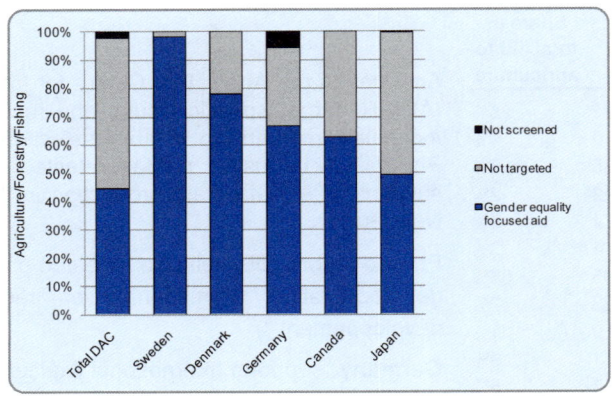

Aid to agriculture and the objectives of the Rio Conventions

DAC members pay attention to environmental concerns when designing aid activities in the field of agriculture. For those reporting on the "Rio markers", data shown in Chart 6 indicate that 20% of aid to agriculture also targeted the objectives of the Convention on desertification in recent years, the share was also 20% for biodiversity, and 17% for climate change[6].

Chart 6. Focus on the objectives of the Rio Conventions within the agricultural sector

average 2006-07, constant 2007 prices

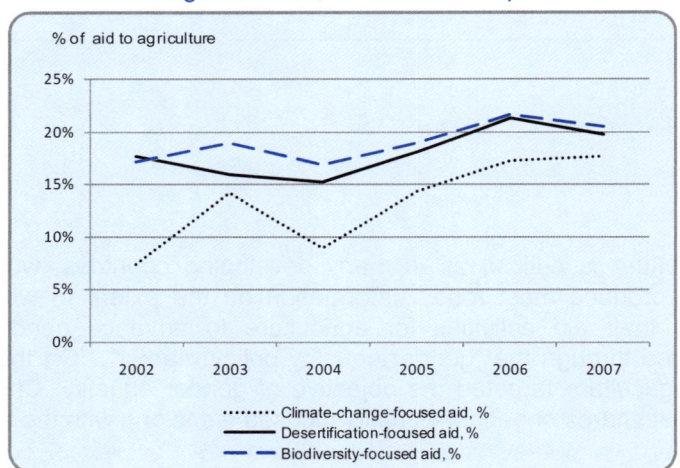

Not all donors report on Rio markers and the charts shown exclude Luxembourg, Norway, United States (all markers), and Switzerland (desertification). Germany and the Netherlands had not yet reported on Rio markers for 2007 data at the time of this publication, and 2006 figures are shown for 2007 data.

---

[6] Activities can target the objectives of more than one of the Conventions at the same time, so these percentages should not be added up.

II. Degree of targeting to countries most in need

Tables 5.a. and 5.b. show the top ten recipients of aid for agriculture in volume and per capita terms. The figures represent total ODA flows including both bilateral donors and multilateral agencies presented in Table 1, except FAO which does not report total flows by recipient (hence the difference in totals between Table 1 and Tables 5).

In 2006-07, India was the top recipient in terms of volume, benefiting especially from grants and loans extended by IDA, Japan and Germany. In per capita terms, Armenia ranks first with a very large contribution by the Millennium Challenge Account of the United States.

Tables 5.a. and 5.b. Top ten recipients of aid to agriculture
2006-07 average

Table 5.a. Aid in volume

	Total USD amount USD million	per capita USD
1 India	563	0.5
2 Afghanistan	248	7.6
3 Viet Nam	236	2.8
4 Mali	186	13.8
5 Kenya	176	5.1
6 Ethiopia	166	2.1
7 Ghana	163	7.0
8 China	155	0.1
9 Indonesia	140	0.6
10 Pakistan	133	0.9
Other	4004	
Total	6169	

Table 5.b. Aid per capita[7]

	per capita USD	Total USD amount USD million
1 Armenia	29.4	88
2 Mauritius	15.5	19
3 Mali	13.8	186
4 Swaziland	12.4	14
5 Bolivia	10.7	98
6 Gabon	9.4	13
7 Laos	7.8	46
8 Afghanistan	7.6	248
9 Timor-Leste	7.2	8
10 Burkina Faso	7.1	104
Other		5345
Total		6169

Note: The difference between totals in Table 1 and Tables 5.a./5.b. is due to FAO totals that cannot be distributed by country (only estimated totals are available).

Charts 7.a. and 7.b. illustrate the distribution of aid commitments by region and income group over the period 2002-07. Aid flows to agriculture primarily targeted Sub-Saharan Africa (31%) and South and Central Asia (23%). For both these regions, the share has increased over the last decade, from 27% in 1998-99 to 33% in 2006-07 for Sub-Saharan Africa, and from 19% to 23% for South and Central Asia. The region that was progressively less targeted over the same period was Far East Asia: its share decreased from 22% to 12%.

---

[7] Amongst countries of more than 1 million inhabitants.

Charts 7.a. and 7.b. Aid to agriculture, distribution by region and income group, 2002-07

Chart 7.a. Distribution by region **Chart 7.b. Distribution by income group**

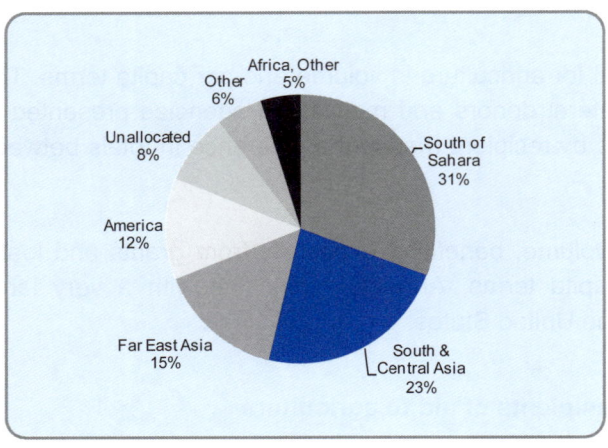 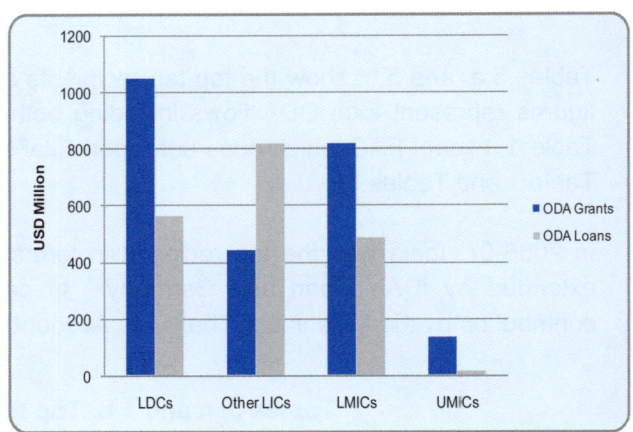

Least Developed Countries (LDCs) and Other Low Income Countries (Other LICs) received two thirds of total aid to agriculture (excluding regional/multi-country aid that cannot be allocated to income groups), indicating a clear focus on the poorest countries. LDCs alone received 37% of total agricultural aid.

Concessional loans represented 35% of agricultural aid to LDCs, and were mostly extended by IDA, the African and Asian Development Funds, and IFAD. For Other LICs, the high proportion of loans (65%) originates from these same agencies, but also from two large bilateral donors: 83% of aid from Japan to Other LICs was in the form of loans, and the share was 62% for France.

Comparing levels of aid to agriculture with indicators of malnourishment gives another perspective on the needs focus of this aid. A useful benchmark is Millennium Development Target 1.C: *Halve, by 2015, the proportion of people who suffer from hunger.* Progress against Target 1.C is measured using the indicator of prevalence of underweight children (under five years of age). This declined from 33% in 1990 to 26% in 2007, but provisional estimates indicate that it rose by a percentage point in 2008. The 2009 UN MDG report states that "the rate of progress is insufficient to meet the goal of reducing underweight prevalence by half - even without taking into account higher food prices and the economic crisis that developed in the meantime." The report recommends implementing measures to increase the availability of food, including raising production.

As illustrated in Map 1, the most affected regions are Southern Asia with almost 50% of children who are underweight, and Sub-Saharan Africa with 28%. South Asia alone accounts for more than half of the world's undernourished children. Maps 1 and 2 allow a comparison of the degree of malnourishment in developing countries and the levels of donors' support to agriculture. Sub-Saharan African countries, appear to receive among the highest levels of aid to agriculture per capita, whereas South Asian countries also severely affected receive less in per capita terms. Aid to these countries may be large in volume terms, but high population figures leave little per inhabitant: India is for example the top recipient in volume terms, but only receives USD 0.3 (30 cents) per capita.

Map 1. Proportion of population under five years old suffering from underweight 2007, source: World Bank – UNICEF

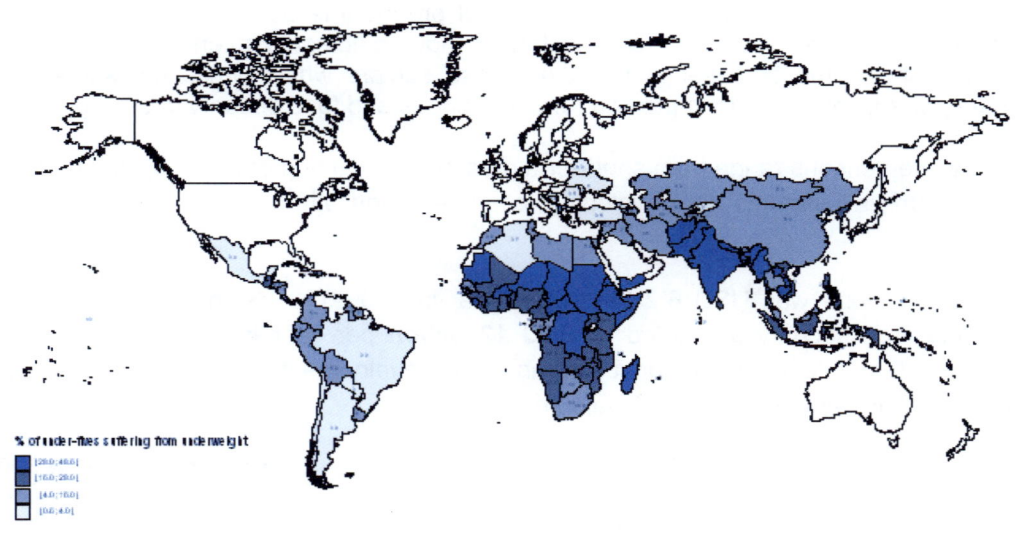

Map 2. Aid to agriculture per capita, average 2006-07, USD

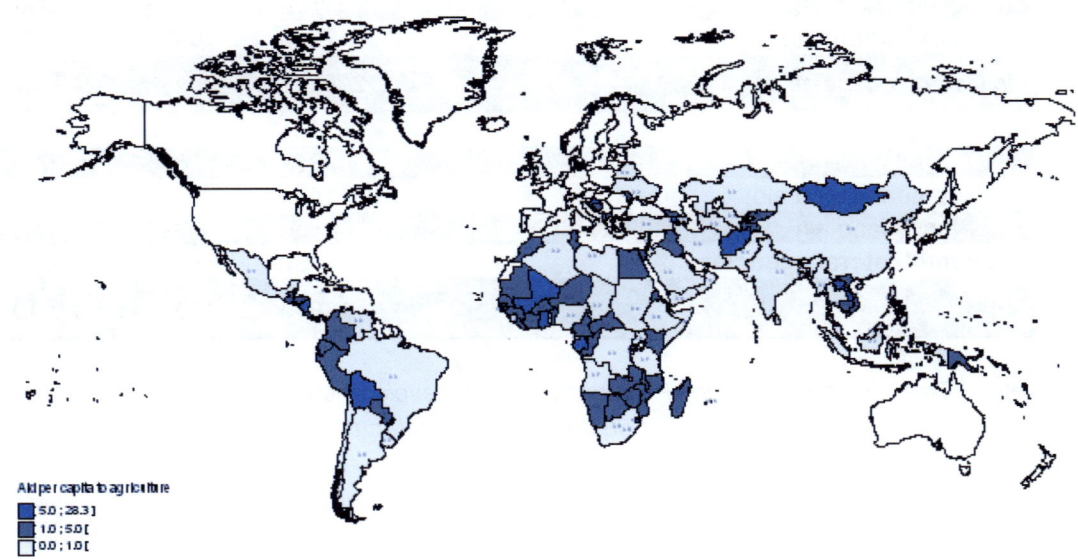

III. Aid flows to agriculture and other food-security-related sectors

As explained in section I.1, the DAC definition of aid to agriculture excludes rural development (classified as multi-sector aid), developmental food aid (general programme assistance) and emergency food aid (humanitarian assistance). Aid to these sectors is not reflected in main statistics presented above. Table 6 below shows the data series including these sectors which, while not supporting agricultural development *per se*, contribute to improving rural livelihoods and food security in developing countries.

Food security interventions range from short-term measures, including emergency and developmental food aid, to long-term investments in developing countries' agriculture and rural development to raise productivity.

Aid to agriculture alone was USD 6.2 billion per year in 2006-07. Using the broader measure, aid to agriculture and food security amounted to **USD 12 billion** per year in 2006-07, with **USD 7.6 billion** allocated to long-term interventions (agriculture and rural development), and the remaining **USD 4.4 billion** to short term food aid (emergency and developmental) .

Table 6. Aid to agriculture and other food-security-related sectors in 2002-07
Annual average commitments, constant 2007 prices

DAC countries	2002-03	2004-05	2006-07
Agriculture/Forestry/Fishing	2953	3179	3768
Rural development	825	599	884
Developmental food aid	1644	1198	1060
Emergency food aid	1798	2015	1811
Total DAC countries	**7219**	**6990**	**7523**
Multilateral agencies	2002-03	2004-05	2006-07
Agriculture/Forestry/Fishing	2161	1875	2400
Rural development	872	655	589
Developmental food aid	1401	1190	1013
Emergency food aid	356	322	393
Total multilateral agencies	**4790**	**4042**	**4395**
Total	**12009**	**11033**	**11919**

Note: The distribution of WFP flows between developmental (84%) and emergency (16%) food aid was obtained for 2008 flows and pro-rated to total figures for earlier years.

Chart 8.a. shows that, over the last decade, bilateral aid to agriculture and other food-security-related sectors actually rose from USD 5.9 billion in 1996-97 to USD 7.5 billion in 2006-07, with the decline in aid to agriculture itself being compensated by increased aid to other food- security-related sectors, especially short-term food aid.

Chart 8.b. illustrates a different trend for multilateral agencies which slightly increased their aid to agriculture alone over the period, and also aid to other food-security-related sectors. The particular roles played by the three UN agencies based in Rome that address food-security are briefly described in Box 2.

Charts 8a. and 8.b. Aid flows to agriculture and other food-security-related sectors

Annual average commitments in millions of USD, constant 2007 prices

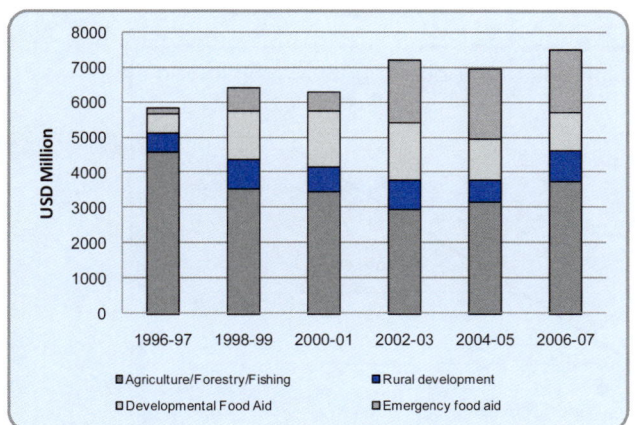

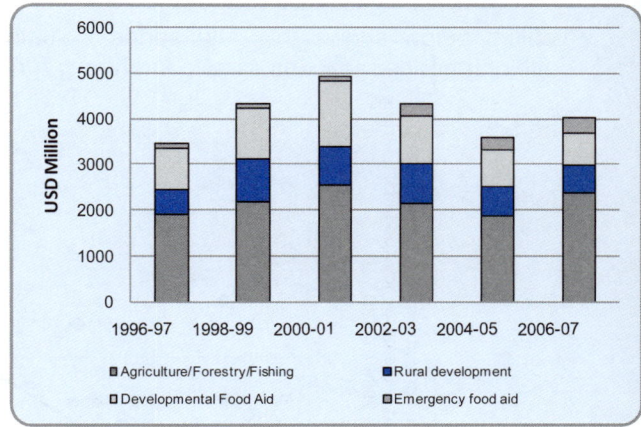

Box 2. "The three Rome-based agencies"

The **World Food Programme (WFP)** classifies its activities into four categories: i) *Emergency Operations* (emergency food aid); ii) *Protracted Relief and Recovery Operations* (food aid also in support of longer-term rehabilitation and post-conflict efforts), iii) *Development* (including school feeding programmes); and iv) *Special operations* (logistical support to the delivery of food aid). Most of WFP budget consists of DAC countries' earmarked funding recorded in DAC statistics as part of their bilateral aid. WFP activities financed out of its core budget fall under multilateral outflows (USD 372 million per year in 2006-07), and are classified under emergency food aid [category i), 14% of the total], developmental food aid [categories ii) and iii), 86%]. Category iv) represents a small share. WFP flows were taken into account in Table 6.

The **Food and Agriculture Organisation (FAO)** runs both normative and operational activities. Expenditures in headquarters are mainly targeted to normative work including through research and pilot projects. Such normative work benefits both developed and developing countries, and is therefore not counted as ODA in full (an ODA coefficient of 51% was last reviewed by the DAC in 2003, and will be re-examined in 2009). The regional and country offices' main tasks are to liaise with the recipient governments, do advocacy for global work and run technical assistance activities in countries. In this study, the 51% ODA-eligible share of FAO budget (USD 203 million per year in 2006-07) was taken into account in Tables 1 and 6 (aid to agriculture).

The **International Fund for Agricultural Development (IFAD)** is an international financial institution that provides both concessional and non-concessional loans, as well as grants for agricultural and rural development in developing countries. Funds are mainly recorded under agriculture (44% for years 2006-07) and rural development (40%). There are also activities that support an enabling environment for agriculture, through improved access to financial institutions (12%). IFAD recently reported to the CRS/DAC a more detailed sectoral breakdown for its historical data that provides more precise indications on the distribution of its aid flows: the share for agriculture is broadly confirmed (40%) but the share for rural development is only 7%, while financial institutions get 27%. Other important sectors are governance and civil society (10%) and industry (7%).

Chart 9. Aid to agriculture and other food-security-related sectors in relation to malnourishment

Aid to agriculture and other food-security-related sectors per capita and prevalence of underweight children under five years of age

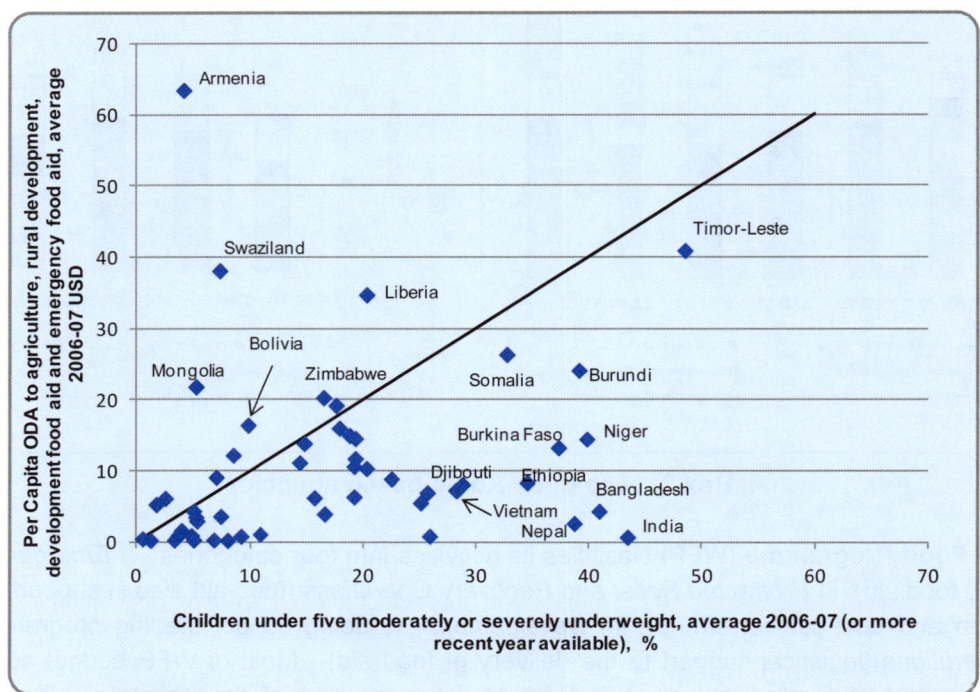

Chart 9 highlights that in per capita terms, aid to agriculture and other food-security-related sectors is generally not concentrated on malnourished populations (see the line that separates "over-aided" from "under-aided" countries). Among the most affected countries, only Timor-Leste, Somalia and Burundi received more than USD 20 per capita in the period 2006-07. Numerous other severely affected countries received very little during the same period (in particular India, Bangladesh, Nepal and Togo) while countries in a better situation (left side of the chart) received more (e.g. Armenia: USD 63).

Annex. List of CRS purpose codes for aid to agriculture and agriculture-related sectors Groupings used in Chart 4

AID TO AGRICULTURE

AGRICULTURE

Agricultural policy (grouping 1)

Agricultural policy and administrative management (31110)
Sector policy, planning and programmes; aid to agricultural ministries; institution capacity building and advice; unspecified agriculture.
Agricultural land resources (31130)
Including soil degradation control; soil improvement; drainage of water logged areas; soil desalination; agricultural land surveys; land reclamation; erosion control, desertification control.
Agrarian reform (31164)
Including agricultural sector adjustment.

Agricultural production (grouping 2)

Agricultural development (31120)
Integrated projects; farm development.
Food crop production (31161)
Including grains (wheat, rice, barley, maize, rye, oats, millet, sorghum); horticulture; vegetables; fruit and berries; other annual and perennial crops. [Use code 32161 for agro-industries.]
Industrial crops/export crops (31162)
Including sugar; coffee, cocoa, tea; oil seeds, nuts, kernels; fibre crops; tobacco; rubber. [Use code 32161 for agro-industries.]
Livestock (31163)
Animal husbandry; animal feed aid.
Agricultural alternative development (31165)
Projects to reduce illicit drug cultivation through other agricultural marketing and production opportunities (see code 43050 for non-agricultural alternative development).

Agricultural water resources (grouping 3) (31140)

Irrigation, reservoirs, hydraulic structures, ground water exploitation for agricultural use.

Agricultural inputs (grouping 4) (31150)

Supply of seeds, fertilizers, agricultural machinery/equipment.

Agricultural education/research/services (grouping 5)

Agricultural extension (31166)
Non-formal training in agriculture.
Agricultural education/training (31181)
Agricultural research (31182)
Plant breeding, physiology, genetic resources, ecology, taxonomy, disease control, agricultural bio-technology; including livestock research (animal health, breeding and genetics, nutrition, physiology).
Agricultural services (31191)
Marketing policies & organisation; storage and transportation, creation of strategic reserves.
Plant and post-harvest protection and pest control (31192)

Including integrated plant protection, biological plant protection activities, supply and management of agrochemicals, supply of pesticides, plant protection policy and legislation.
Agricultural financial services (31193)
Financial intermediaries for the agricultural sector including credit schemes; crop insurance.
Agricultural co-operatives (31194)
Including farmers' organisations.
Livestock/veterinary services (31195)
Animal health and management, genetic resources, feed resources.

FORESTRY (grouping 6)

Forestry policy and administrative management (31210)
Forestry sector policy, planning and programmes; institution capacity building and advice; forest surveys; unspecified forestry and agro-forestry activities.
Forestry development (31220)
Afforestation for industrial and rural consumption; exploitation and utilisation; erosion control, desertification control; integrated forestry projects.
Fuelwood/charcoal (31261)
Forestry development whose primary purpose is production of fuelwood and charcoal.
Forestry education/training (31281)
Forestry research (31282)
Including artificial regeneration, genetic improvement, production methods, fertilizer, harvesting.
Forestry services (31291)

FISHING (grouping 8)

Fishing policy and administrative management (31310)
Fishing sector policy, planning and programmes; institution capacity building and advice; ocean and coastal fishing; marine and freshwater fish surveys and prospecting; fishing boats/equipment; unspecified fishing activities.
Fishery development (31320)
Exploitation and utilisation of fisheries; fish stock protection; aquaculture; integrated fishery projects.
Fishery education/training (31381)
Fishery research (31382)
Pilot fish culture; marine/freshwater biological research.
Fishery services (31391)
Fishing harbours; fish markets; fishery transport and cold storage.

AID TO OTHER FOOD-SECURITY-RELATED SECTORS

Agro-industries (32161)
Staple food processing, dairy products, slaughter houses and equipment, meat and fish processing and preserving, oils/fats, sugar refineries, beverages/tobacco, animal feeds production.

Forest industries (32162)
Wood production, pulp/paper production.

Rural development (43040)
Integrated rural development projects; e.g. regional development planning; promotion of decentralised and multi-sectoral competence for planning, co-ordination and management; implementation of regional development and measures (including natural reserve management); land management; land use planning; land settlement and resettlement activities [excluding resettlement of refugees and internally displaced person (72010)]; functional integration of rural and urban areas; geographical information systems.

Developmental food aid (52010)
Supply of edible human food under national or international programmes including transport costs; cash payments made for food supplies; project food aid and food aid for market sales when benefiting sector not specified; excluding emergency food aid.

Emergency food aid (72040)
Food aid normally for general free distribution or special supplementary feeding programmes; short-term relief to targeted population groups affected by emergency situations. Excludes non-emergency food security assistance programmes/food aid (52010).

PARTIE 1

Analyse de l'aide à l'agriculture

Période 2002-07

www.oecd.org/dac/stats/agriculture

Principales observations

Depuis le milieu des années 80, l'aide à l'agriculture a été divisée par deux. En 2006-07, les engagements d'aide bilatérale des pays membres du CAD pour l'agriculture s'élevaient annuellement à **3,8 milliards USD**. En prenant en compte les apports des agences multilatérales, le montant total de l'aide à l'agriculture atteignait **6,2 milliards USD en moyenne annuelle.**

La part de l'aide à l'agriculture dans les programmes d'aide des membres du CAD diminue de façon régulière : de 17% à la fin des années 80 à 6% ces dernières années, révélant un **abandon relatif clair de ce secteur.**

Les tendances récentes indiquent un ralentissement de ce déclin et même la perspective d'un redressement : sur la période 2002-07, l'aide bilatérale à l'agriculture a augmenté à un taux de variation annuel moyen de 5 % en termes réels.

Parmi les membres du CAD, les principaux donneurs en 2006-07 ont été **les Etats-Unis (en moyenne 932 millions USD par an), le Japon (821 millions USD) et la France (451 millions USD).**

Au cours de la période 2002-07, les apports d'aide à l'agriculture ont surtout ciblé **l'Afrique subsaharienne (31 %) et l'Asie centrale et du Sud (23 %).** Les Pays les moins avancés (PMA) et les autres Pays à faible revenu (autres PFR) reçoivent les deux tiers de l'aide totale à l'agriculture, indiquant un ciblage clair des pays les plus pauvres.

Le total de l'aide à l'agriculture et à la sécurité alimentaire, y compris le développement rural, l'aide alimentaire à des fins développementales et l'aide alimentaire d'urgence, **s'est élevée à 12 milliards USD** par an en 2006-07, dont **7,6 milliards USD** alloués à des interventions à long terme (agriculture et développement rural), et les **4,4 milliards USD** restants à l'aide alimentaire de court terme (d'urgence et de développement).

Presque **la moitié du total de l'aide à l'agriculture est sous forme de prêts**, principalement par les grands organismes internationaux, dont la Banque Mondiale, les banques régionales de développement et le FIDA.

Même si **deux tiers de l'aide à l'agriculture sont dirigés vers les pays les moins avancés et les autres pays à faible revenu**, elle n'est pas concentrée sur les pays ayant les taux de malnutrition les plus élevés.

www.oecd.org/dac/stats/agriculture

Introduction

Contexte

Le présent rapport passe au crible les statistiques sur l'aide à l'agriculture au cours des six années de la période 2002-07. Il a été élaboré par le Secrétariat du Comité d'aide au développement (CAD) en collaboration avec le Groupe de travail du CAD sur les Statistiques.

La sécurité alimentaire est un problème constant, et la récente crise des prix des denrées alimentaires est venue rappeler l'importance de cet enjeu. Les sommets atteints par les cours des denrées mi-2008 ont replongé dans la pauvreté quelque 100 millions de personnes, portant à environ 1 milliard le nombre des personnes qui souffrent de la faim. L'agriculture fournit les ressources et les revenus nécessaires pour garantir un approvisionnement en denrées alimentaires à des conditions de prix abordables. Sa contribution à la réduction de la pauvreté est mise en valeur dans les Objectifs du millénaire pour le développement (OMD) par l'OMD 1. *Réduction de l'extrême pauvreté et de la faim*, l'*Objectif 1.C* ciblant plus particulièrement la faim : *Réduire de moitié, entre 1990 et 2015, la proportion de la population qui souffre de la faim.*

Structure du présent rapport

Ce rapport comprend trois parties. La Partie 1 analyse **l'aide à l'agriculture** en présentant les tendances de l'aide des donneurs, le ciblage géographique des apports et un panorama des interventions à court et long terme des donneurs face aux enjeux de la sécurité alimentaire (y compris l'aide alimentaire et le développement rural). La Partie 2 met en évidence les **profils individuels des donneurs** à partir des statistiques synthétiques sur leur aide bilatérale à l'agriculture, sous la forme de graphiques et de tableaux. La Partie 3 contient **la liste des activités d'aide** au secteur agricole, précisant leurs montants et descriptifs, notifiées par les donneurs bilatéraux et les agences multilatérales en 2007.

Les statistiques présentées incluent des données jusqu'en 2007. Les engagements pris par les donneurs en réponse à la crise du prix des denrées alimentaires en 2008 et 2009 n'apparaissent donc pas dans ce rapport. Les données relatives aux apports de 2008 seront disponibles à partir de la fin de 2009 et seront publiées début 2010 sur notre page Internet dédiée à l'agriculture sous la forme d'une brochure statistique.

Source des informations

Les statistiques se fondent sur les rapports d'engagements et de versements d'Aide publique au développement (APD) des donneurs compilés dans la base de données sur les activités d'aide du Système de notification des pays créanciers (SNPC). Les données ont été confirmées par les membres, sur requête spéciale, au printemps 2009.

Les données sous-tendant le présent rapport peuvent être consultées sur le site Internet du CAD.

www.oecd.org/dac/stats/agriculture
www.oecd.org/dac/stats/data

I. Tendances de l'aide à l'agriculture

I.1 Définitions et terminologie

Champ de l' « aide à l'agriculture »

La définition statistique du CAD de l'aide à l'agriculture couvre la politique, la planification et les programmes agricoles, les ressources en terres cultivables, les ressources en eau à usage agricole, le développement agricole et l'approvisionnement en produits à usage agricole, la production agricole, l'élevage, les services agricoles, l'éducation et la formation ainsi que la recherche dans le domaine agricole, le renforcement des capacités institutionnelles et le conseil. La sylviculture et la pêche sont identifiées comme des secteurs à part depuis 1996, mais sont incluses dans l'agriculture dans le présent rapport.

La définition exclut le développement rural (classé dans l'aide plurisectorielle), l'aide au secteur agroalimentaire et à l'industrie forestière (qui font partie de l'aide à l'industrie), l'aide alimentaire à des fins de développement (une sous catégorie de l'aide-programme générale) et l'aide alimentaire d'urgence (incluse dans l'aide humanitaire). Ces secteurs ne sont pas pris en compte dans les statistiques principales qui figurent dans ce rapport, mais, du fait de leur pertinence pour la sécurité alimentaire, ils sont aussi comptabilisés dans la Section III et apparaissent séparément dans les profils des donneurs de la Partie 2.

On trouvera en Annexe la définition de l'ensemble des secteurs et sous-secteurs.

Aide ventilable par secteur

Afin de mieux refléter l'orientation sectorielle des programmes des donneurs, dans le calcul de la part de l'aide bilatérale totale consacrée à l'agriculture, les contributions non ventilables par secteur sont exclues du dénominateur (soutien budgétaire général, actions relatives à la dette, aide humanitaire, coûts administratifs et autres transactions internes dans le pays donneur).

Contributions multilatérales imputées des membres du CAD

En sus de leurs activités d'aide bilatérale dans le secteur de l'agriculture, les membres du CAD contribuent aux agences multilatérales actives dans le domaine de l'agriculture. Afin de fournir une image complète de l'effort total d'APD consenti par le donneur au titre de l'aide au secteur agricole, les données relatives à l'aide multilatérale imputée des membres du CAD affectée à l'agriculture ont été comptabilisées.

Le calcul des contributions multilatérales imputées s'effectue en deux étapes[8] : 1/ on calcule la moyenne sur trois ans de la part de l'aide totale ventilable de chaque agence multilatérale allouée à l'agriculture (en considérant uniquement leurs ressources régulières), 2/ en multipliant ce résultat par la contribution du membre concerné aux ressources régulières de l'agence en question pour les années considérées, on obtient le montant de l'apport imputé de ce donneur au secteur agricole par le biais de l'agence multilatérale visée. Ce montant reste toujours approximatif.

---

[8] Voir aussi « *The OECD methodology for calculating imputed multilateral ODA* » à l'adresse www.oecd.org/dac/stats/methodology.

I.2 Aide bilatérale et multilatérale à l'agriculture

Cette section présente les statistiques sur l'aide à l'agriculture des donneurs bilatéraux (pays membres du CAD) et des institutions multilatérales.

Le Graphique 1 illustre la tendance à long terme de l'aide à l'agriculture tandis que le Graphique 2 porte plus précisément sur les dix dernières années. L'aide à l'agriculture a été divisée par deux depuis le milieu des années 80, passant de 12 milliards USD en 1984-85 à 6,2 milliards USD en 2006-07 (sur la base des engagements bilatéraux et multilatéraux).

Graphique 1. Tendances de l'APD à l'agriculture

1971-2007, moyennes mobiles sur 5 ans et données annuelles, à prix constants de 2007

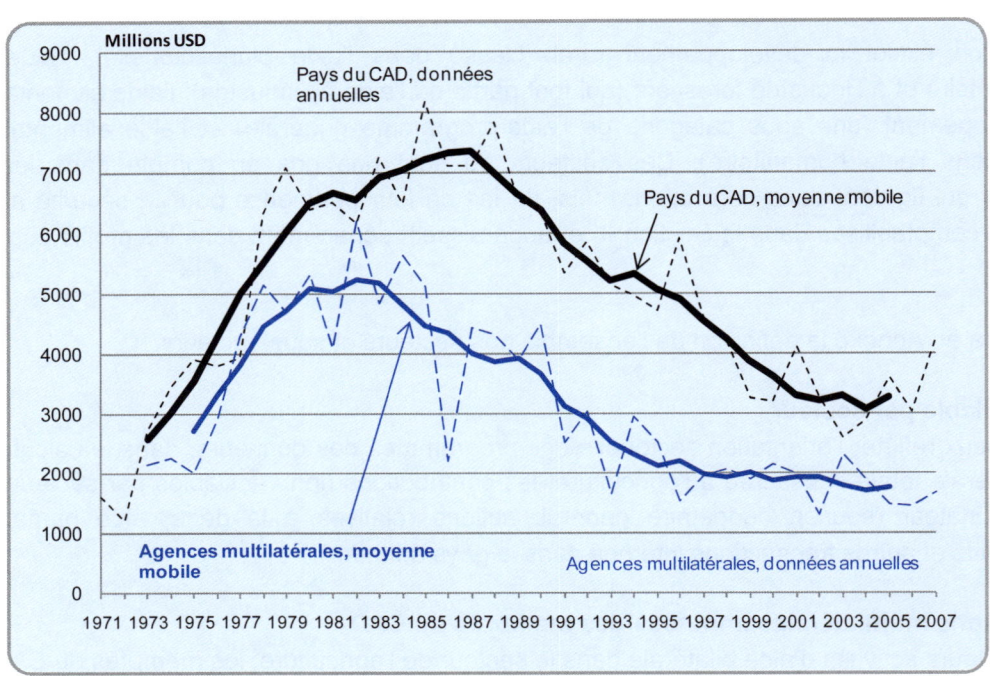

Le Graphique 2 montre que le déclin ralentit sur la dernière décennie et que s'amorce même un redressement : sur la période 2002-07, l'aide bilatérale à l'agriculture a augmenté à un taux de variation moyen annuel de 5 % en termes réels. En 2006-07, l'aide bilatérale annuelle à l'agriculture s'est élevée à **3,8 milliards USD.**

La part de l'aide bilatérale totale ventilable consacrée à l'agriculture fournit une indication de l'importance des programmes d'aide des donneurs ciblant l'agriculture et de la priorité donnée à ce secteur. Parmi les pays membres du CAD, cette proportion a chuté de 13 % à 6 % au cours de la dernière décennie (1995-2007). Ce déclin semble maintenant enrayé, la part dévolue à l'agriculture s'étant stabilisée aux alentours de 6 % depuis le début des années 2000.

Graphique 2. Tendances de l'APD à l'agriculture des pays membres du CAD

1995-2007, moyennes mobiles sur 3 ans, à prix constants de 2007

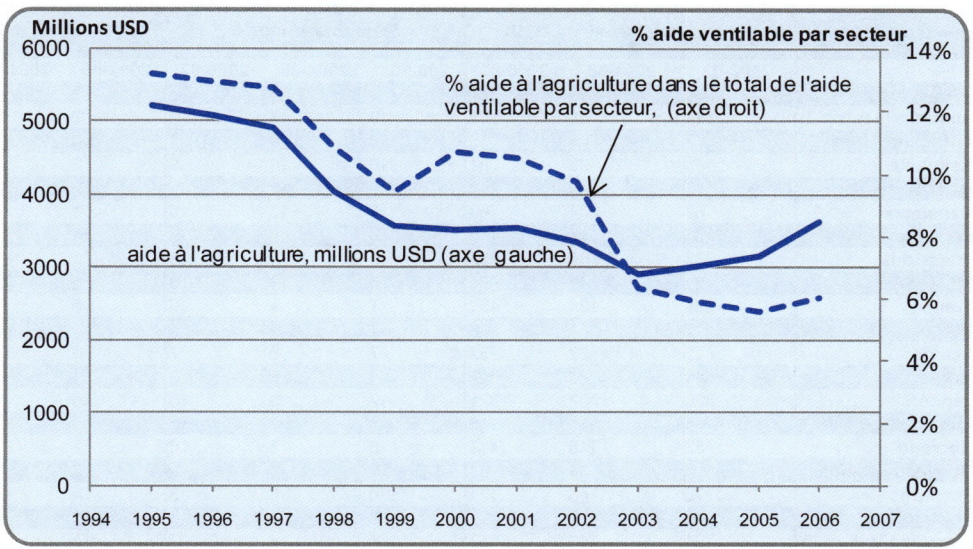

Le Tableau 1 ci-dessous montre la ventilation de l'aide à l'agriculture par donneur et institution multilatérale. Sur un total de 3,8 milliards USD d'aide à l'agriculture consentis par les pays membres du CAD, 25 % proviennent des Etats-Unis, qui apparaissent comme le principal donneur, devant le Japon (22 %) et la France (12 %). L'IDA est le premier donneur multilatéral pour le secteur de l'agriculture, avec 45 % de l'aide multilatérale, devant les institutions de l'UE (19 %).

Les États-Unis, la Grèce, le Luxembourg, la Nouvelle-Zélande et la Suède ont plus que doublé leurs engagements d'aide en faveur du secteur de l'agriculture entre 2002 et 2007. L'Autriche, la Finlande, la France, l'Irlande et la Norvège ont aussi notablement renforcé leur soutien à l'agriculture, tandis que l'Allemagne, le Danemark et l'Espagne l'ont augmenté, mais dans des proportions bien moindres. L'Australie, le Royaume-Uni, les Pays-Bas, le Portugal et la Suisse ont diminué leur soutien à l'agriculture, tandis que l'on ne note aucune évolution significative du schéma d'aide de la Belgique, du Canada, de l'Italie, du Japon et des institutions de l'UE.

Certains donneurs consacrent une part relativement importante de leur aide à des projets dans le domaine de l'agriculture : le Japon (10 %), la Belgique, le Danemark, la France et la Suisse (9 % chacun), la Finlande et la Suède (8 % chacun) et l'Espagne et l'Irlande (7 % chacun) se situent au-dessus de la moyenne de 6 % observée pour les membres du CAD pour les années 2006-07.

Tableau 1. Aide des donneurs bilatéraux et multilatéraux en faveur de l'agriculture
2002-07, engagements annuels moyens et part du total de l'aide ventilable par secteur, prix constants de 2007

Donneurs	Engagements, million USD			% du total donneur			% tous donneurs		
	2002-03	2004-05	2006-07	2002-03	2004-05	2006-07	2002-03	2004-05	2006-07
Allemagne	239	229	317	5	5	5	5	5	5
Australie	114	84	82	10	7	5	2	2	1
Autriche	7	9	12	2	4	4	0	0	0
Belgique	95	93	91	10	10	9	2	2	1
Canada	115	213	114	7	11	6	2	4	2
Danemark	67	204	83	8	13	9	1	4	1
Espagne	136	115	163	9	9	7	3	2	3
Etats-Unis	337	669	932	3	4	5	7	13	15
Finlande	22	48	34	6	11	8	0	1	1
France	255	184	451	7	4	9	5	4	7
Grèce	2	1	5	1	1	3	0	0	0
Irlande	23	22	33	7	6	7	0	0	1
Italie	48	27	45	7	4	6	1	1	1
Japon	820	702	821	12	9	10	16	14	13
Luxembourg	4	11	9	3	8	6	0	0	0
Norvège	83	115	117	6	8	6	2	2	2
Nouvelle Zélande	5	9	11	5	6	5	0	0	0
Pays-Bas	226	148	121	10	6	3	4	3	2
Portugal	6	3	2	2	1	1	0	0	0
Royaume Uni	221	143	113	6	4	2	4	3	2
Suède	44	79	144	3	5	8	1	2	2
Suisse	85	70	69	13	9	9	2	1	1
Total pays du CAD	2953	3179	3768	7	6	6	58	63	61
FAfD	306	207	245	23	16	15	6	4	4
FAsD	286	226	223	16	15	14	6	4	4
FAO	219	213	203	100	100	100	4	4	3
IDA	725	718	1075	10	10	12	14	14	17
BID F. Sp.	48	42	0	8	10	0	1	1	0
FIDA	107	175	209	24	40	44	2	3	3
Institutions UE	471	288	444	7	3	4	9	6	7
PNUD	0	7	1	-	2	0	0	0	0
Total Multilatéral	2161	1875	2400	12	9	10	42	37	39
Total	5114	5054	6169	8	7	7	100	100	100

Notes :

Le **soutien budgétaire général,** une fois intégré aux budgets nationaux des pays en développement, contribue au développement du secteur agricole, mais cette contribution n'est pas précisée, ni prise en compte dans les chiffres mentionnés ci-avant (se reporter également à la Section I.3 intitulée « Modalités de l'aide »).

Les données de la **FAO** ont été estimées par le Secrétariat à partir du coefficient d'APD de l'agence et de son budget régulier.

La **Belgique** utilise en interne un système de codes-objet multiples qui aboutit à des chiffres supérieurs : 122 millions USD pour 2002-03, 121 millions USD pour 2004-05 et 124 millions USD pour 2006-07.

Les données montrent que le **FIDA** ne consacre pas la totalité de son aide à l'agriculture (44 % en 2006-07). La plupart des apports restants concernent le développement rural selon la nomenclature sectorielle du CAD et sont pris en compte dans le calcul plus large de l'aide à l'agriculture et à la sécurité alimentaire (cf. Section III).

Contributions multilatérales imputées des membres du CAD en faveur de l'agriculture

Comme indiqué à la Section I.1, il est possible d'évaluer les contributions multilatérales imputées des membres du CAD (cf. Tableau 2) et de les ajouter à leur contribution bilatérale (cf. Tableau 3).

Le Tableau 2 montre que sur le montant total des contributions des membres du CAD au budget régulier des agences multilatérales, **1,4 milliard USD** peut être imputé, en moyenne annuelle, à l'agriculture en 2006-07. L'IDA est le premier canal multilatéral, représentant 37 % de ce total, devant les institutions de l'UE (24 %).

Le total moyen annuel des contributions multilatérales imputées pour 2006-07, 1,4 milliard USD, ne correspond pas au total des apports multilatéraux indiqué au Tableau 1 (2,4 milliards USD sur 2006-07), parce que les apports multilatéraux d'une année donnée ne proviennent pas directement des contributions des donneurs pour cette année-là. Dans le cas des institutions financières internationales (IFI), le montant des prêts émis annuellement est bien supérieur aux contributions des donneurs (dons), car il est aussi alimenté par les remboursements de principal, les intérêts acquittés et les transferts de capitaux entre les IFI. Une autre raison est le décalage temporel entre la réception des fonds par les agences multilatérales et leur versement aux pays en développement.

Tableau 2. Contributions multilatérales imputées des membres du CAD en faveur de l'agriculture

moyennes annuelles 2006-07, en millions USD et à prix constants de 2007

	à travers IDA	à travers les institutions de l'UE	à travers FAfD	à travers FIDA	à travers FAsD	à travers FAO	à travers PNUD	à travers BID	Total des contributions multilatérales imputées
Allemagne	72.7	73.9	25.9	9.6	7.7	19.8	0.2	0.0	209.9
Australie	13.9	0.0	0.0	0.7	5.6	3.7	0.1	0.0	24.0
Autriche	9.1	8.0	4.1	2.6	1.2	2.0	0.0	0.0	26.9
Belgique	9.6	13.6	4.1	1.9	1.4	2.4	0.1	0.0	33.0
Canada	26.3	0.0	0.0	4.9	0.0	6.4	0.3	0.1	38.1
Danemark	7.2	7.4	2.6	2.1	1.3	1.7	0.5	0.0	22.7
Espagne	19.1	29.8	0.0	8.9	0.0	17.8	0.3	0.0	76.0
Etats-Unis	80.7	0.0	0.2	6.4	14.8	46.0	0.6	0.0	148.7
Finlande	4.1	5.4	2.8	0.6	0.7	1.6	0.1	0.0	15.3
France	43.6	64.8	22.5	4.7	6.2	15.6	0.2	0.0	157.6
Grèce	2.4	6.4	0.0	0.1	0.0	1.7	0.0	0.0	10.6
Irlande	7.2	4.3	0.0	0.9	0.6	0.8	0.1	0.0	13.9
Italie	2.8	45.7	0.8	15.0	0.0	23.1	0.2	0.0	87.7
Japon	96.9	0.0	14.6	6.6	39.7	51.3	0.4	0.0	209.4
Luxembourg	0.9	0.9	0.0	0.3	1.3	0.2	0.0	0.0	3.5
Norvège	10.7	0.0	0.0	5.0	0.0	1.6	0.6	0.0	17.8
Nouvelle Zélande	0.8	0.0	0.0	0.0	0.0	0.7	0.0	0.0	1.6
Pays-Bas	5.9	15.9	8.6	6.6	2.4	10.8	0.7	0.0	50.8
Portugal	1.4	4.4	2.2	0.3	1.3	1.1	0.0	0.0	10.6
Royaume Uni	84.7	59.8	24.9	9.0	12.9	13.5	0.7	0.0	205.4
Suède	15.5	9.0	12.3	8.2	1.9	2.3	0.6	0.0	49.7
Suisse	14.1	0.0	6.2	2.6	1.6	2.6	0.2	0.0	27.3
Total pays du CAD	**529.4**	**349.3**	**131.7**	**96.7**	**100.5**	**226.7**	**5.8**	**0.2**	**1440.3**
Institutions de l'UE	4.6			12.5		0.4			17.5

Le Tableau 3 fait apparaître le total des engagements en faveur de l'agriculture, y compris les contributions bilatérales et multilatérales imputées. La prise en compte des contributions imputées modifie sensiblement l'ampleur du soutien apporté par certains pays ; il convient d'en tenir compte en consultant le profil des donneurs décrit dans la Partie 2 du présent rapport, qui ne comptabilise que l'aide bilatérale. De façon plus spécifique, les contributions multilatérales imputées de l'Autriche, de la Grèce, de l'Italie, du Portugal et du Royaume-Uni représentent au moins deux tiers de leur appui total au secteur agricole. Pour la plupart de ces pays, le niveau des contributions multilatérales imputées est plus de deux fois supérieur à celui de leur contribution bilatérale.

Tableau 3. Aide totale des membres du CAD à l'agriculture

2002-07, engagements bilatéraux, contributions multilatérales imputées, versements bilatéraux
moyennes annuelles, en millions USD et à prix constants de 2007

	Engagements bilatéraux			Contributions multilatérales imputées			Engagements totaux			Versements bilatéraux		
	2002-03	2004-05	2006-07	2002-03	2004-05	2006-07	2002-03	2004-05	2006-07	2002-03	2004-05	2006-07
Allemagne	239	229	317	151	167	210	391	396	527	230	236	238
Australie	114	84	82	17	14	24	132	98	106	86	83	92
Autriche	7	9	12	16	21	27	23	30	39	6	6	9
Belgique	95	93	91	25	34	33	120	127	124	73	57	66
Canada	115	213	114	27	27	38	142	240	152	84	91	122
Danemark	67	204	83	28	26	23	95	230	106	32	73	79
Espagne	136	115	163	44	59	76	180	174	239	77	105	111
Etats-Unis	337	669	932	119	181	149	455	850	1081	216	514	472
Finlande	22	48	34	14	14	15	35	61	49	37	0	23
France	255	184	451	142	147	158	397	331	608	257	218	399
Grèce	2	1	5	8	7	11	10	9	15	2	1	5
Irlande	23	22	33	5	7	14	28	29	47	20	22	32
Italie	48	27	45	71	147	88	118	173	132	15	26	42
Japon	820	702	821	149	95	209	968	797	1030	457	498	687
Luxembourg	4	11	9	2	3	4	6	14	13	0	11	9
Norvège	83	115	117	33	23	18	116	138	135	80	105	100
Nouvelle Zélande	5	9	11	1	1	2	7	10	13	3	5	6
Pays-Bas	226	148	121	58	63	51	284	211	172	142	121	116
Portugal	6	3	2	13	8	11	19	12	12	6	3	2
Royaume Uni	221	143	113	115	126	205	336	269	319	161	150	153
Suède	44	79	144	51	34	50	95	113	193	52	64	117
Suisse	85	70	69	20	24	27	104	94	96	77	79	64
Total pays du CAD	2953	3179	3768	1109	1229	1440	4062	4408	5209	2225	2468	2944
Institutions de l'UE	471	288	444	24	6	17	494	294	462	111	238	359
Total membres du CAD	3423	3467	4213	1133	1235	1458	4557	4702	5670	2114	2705	3302

A noter : le recours des membres du CAD au canal multilatéral ne se limite pas à leurs contributions au budget régulier des agences multilatérales. Ils effectuent aussi des versements spécifiquement affectés à des fonds gérés par des institutions multilatérales et dédiés au secteur de l'agriculture, ainsi que des versements affectés au financement d'activités d'aide particulières. Ces contributions préaffectées sont considérées comme de l'aide bilatérale.

Versements des membres du CAD

Les versements des membres du CAD à l'appui de l'agriculture ont régulièrement progressé sur la période 2002-07 et se sont élevés à **3,3 milliards USD** en moyenne annuelle en 2006-07. Le classement des donneurs en fonction des versements diffère légèrement de celui fondé sur les engagements : le Japon arrive en tête, à la place des Etats-Unis.

Le Graphique 3 illustre la relation entre les engagements et versements à l'échelle globale du CAD. Les données relatives aux engagements reflètent les programmations des donneurs et l'évolution de leurs politiques : elles constituent par conséquent une indication des apports futurs. Celles portant sur les versements indiquent les montants réellement versés chaque année. Les versements et les engagements d'une même année ne sont pas comparables, car les engagements sont pluriannuels et les versements correspondants s'étalent sur plusieurs années.

Les versements suivent l'orientation légèrement haussière des engagements sur la période 2002-07. Un écart apparaît en 2005 et en 2007 entre le montant des versements et celui des engagements, reflétant un accroissement important des montants alloués à l'aide (engagements) ne trouvant sa contrepartie au niveau des versements que dans les années suivantes. Sur le total des engagements bilatéraux des membres du CAD en 2007, un tiers a été versé la même année et les deux tiers restants ne seront versés que les années suivantes[9].

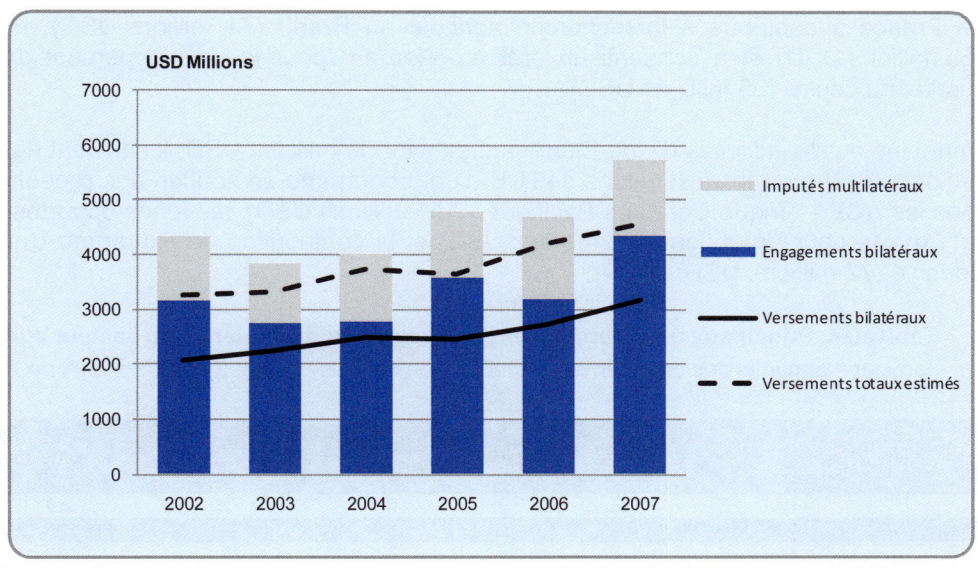

Graphique 3. Aide bilatérale et multilatérale imputée des pays membres du CAD pour l'agriculture

2002-2007, prix constants de 2007

I.3 Principales caractéristiques de l'aide bilatérale des membres du CAD au secteur agricole

Ventilation et ciblage géographique de l'aide

Le Tableau 4 ci-après synthétise le ciblage géographique de l'aide des membres du CAD pour l'agriculture. L'aide se concentre sur un nombre relativement réduit d'engagements de très grande portée : les 100 premières activités d'aide des membres pour 2006-07 totalisent la moitié du total des engagements de ces deux années. Les activités suivantes correspondent à des engagements de plus de 50 millions USD :

- Les Etats-Unis ont financé des projets de développement agricole alternatif afin de réduire les cultures illicites (drogue) en **Afghanistan** (216 millions USD) et en **Colombie** (75 millions USD). Ils ont également procuré un soutien à divers pays et sous-secteurs agricoles au travers du *Millennium Challenge Account*[10] et notamment à l'**Arménie** (150 millions USD pour l'irrigation), au **Ghana** (67 millions USD pour l'enseignement agricole et 58 millions USD pour les services financiers) et au **Mali** (150 millions USD pour la gestion politique et administrative) ; les Etats-Unis ont également apporté une contribution importante en soutenant la productivité agricole en **Irak** (63 millions USD).

---

[9] Cette analyse ne porte que sur les pays pour lesquels il est possible de relier les versements aux engagements d'origine.

[10] Initiative présidentielle pour réduire la pauvreté en stimulant la croissance économique des pays pauvres.

- Le Japon a consenti de nouveaux prêts à la **Chine** pour des projets de reboisement dans le Jilin (81 millions USD) et le Henan (63 millions USD) et un autre projet de développement forestier dans le Qinghai (53 millions USD); à l'**Inde**, pour la construction d'infrastructures d'irrigation en Andhra Pradesh (204 millions USD), la restauration des forêts dégradées dans le Gujarat (149 millions USD), l'Orissa (117 millions USD), et le Tripura (66 millions USD) ; au **Pérou**, pour l'irrigation (50 millions USD) et aux **Philippines**, pour la réforme agraire (84 millions USD).

- La France a contribué à la recherche agricole au **Brésil** (71 millions USD) et à **Madagascar** (53 millions USD) et a consenti un prêt au **Vietnam** pour le développement de ses provinces rurales du centre (55 millions USD).

- Parmi les autres activités d'aide mobilisant plus de 50 millions USD, il convient également de citer le financement par les institutions de l'UE d'un programme de soutien aux produits agricoles pour tous les ACP - Afrique Caraïbes Pacifique - (51 millions USD), les fonds apportés par l'Allemagne à l'**Egypte** pour l'irrigation (125 millions USD) et la contribution du **Royaume-Uni** à la recherche agricole (77 millions USD).

Tableau 4. Principaux donneurs et bénéficiaires de l'aide bilatérale au secteur agricole
moyenne annuelle des engagements 2006-07, en millions USD et à prix constants de 2007

USD Million, moyenne 2006/07	Etats-Unis	Japon	France	Institutions de l'UE	Allemagne	Autres membres du CAD	Total membres du CAD	% dans l'aide totale à l'agriculture
Inde	2	284	2	0	18	10	316	8%
Afghanistan	147	3	0	16	2	41	209	5%
Ghana	96	6	28	3	12	18	163	4%
Mali	80	3	18	10	8	27	148	4%
Viet Nam	0	30	59	0	10	48	146	3%
Chine	0	108	0	0	8	21	137	3%
Irak	120	0	0	0	0	0	121	3%
Egypte	6	11	1	0	87	4	108	3%
Pérou	45	32	1	0	8	13	99	2%
Colombie	60	1	4	5	11	4	85	2%
Autres receveurs	374	344	338	410	153	1061	2680	64%
Total	932	821	451	444	317	1248	4213	100%
% dans l'aide à l'agriculture par l'ensemble des membres du CAD	22.1%	19.5%	10.7%	10.5%	7.5%	29.6%	100.0%	

Instruments financiers

A l'échelle de l'ensemble du CAD, les activités d'aide dans le secteur de l'agriculture sont principalement financées par des **dons qui ont représenté 82 % de l'aide totale sur la période 2002-07.** Cette part n'a pas changé ces dernières années. Toutefois, l'Allemagne et le Japon ont eu davantage recours aux prêts concessionnels pour apporter leur soutien à l'agriculture : 61 % de l'aide japonaise à l'agriculture est consentie sous la forme de prêts et cette proportion est de 39 % pour l'Allemagne (sur la période 2006-07). Les autres membres qui ont eu recours aux prêts concessionnels sont l'Espagne (15 %), la Finlande (2 %), la France (18 %), les institutions de l'UE (1 %) et l'Italie (17 %).

Quelques donneurs ont également notifié comme APD les investissements d'institutions de financement du développement au sein d'entreprises agricoles (agriculture et pêche). Le montant moyen de l'APD au secteur agricole sous la forme de prises de participation s'élève à 20 millions USD au titre des années 2006-07. Elles comprennent des investissements de la BEI (institutions UE), de CDC (Royaume-Uni), DEG (Allemagne), FINNFUND (Finlande) et NORFUND (Norvège).

Les données présentées jusqu'ici concernent seulement l'aide, c'est-à-dire, des dons et des prêts concessionnels. Cependant, les pays en développement reçoivent aussi des prêts non subventionnés de la part de bailleurs de fonds bilatéraux et multilatéraux. Les plus grands pourvoyeurs de ce genre de prêts sont la BIRD (777 millions USD), et les banques asiatique et africaine de développement (respectivement 206 et 97 millions USD).

Modalités de l'aide

Actuellement, les modalités de l'aide ne sont pas correctement identifiées dans le SNPC, mais il sera demandé aux donneurs de les communiquer à partir des données 2010. Les notifications existantes permettent cependant déjà de calculer certains chiffres et de réaliser certaines observations.

L'utilisation d'approches-programmes par les donneurs est particulièrement notable dans le domaine de l'agriculture parmi les pays scandinaves et les institutions de l'UE. Le Danemark apporte 86 % de ses contributions à l'agriculture dans le cadre de programmes sectoriels ; ce pourcentage est de 59 % pour la Norvège, 47 % pour la Finlande, 27 % pour la Suède et 20 % pour les institutions de l'UE.

- Sur 2006-07, le Danemark a contribué à des programmes sectoriels au Bangladesh (développement de la pêche et de l'élevage, agriculture), au Mali (agriculture), en Tanzanie (forêts) et au Vietnam (agriculture et développement rural).
- Les institutions de l'UE apportent un soutien budgétaire sectoriel dans le contexte des mesures d'accompagnement pour les pays du Protocole Sucre affectés par la réforme du régime communautaire du sucre (Barbade, Guyane, Jamaïque, Ile Maurice, Mozambique, Trinidad et Tobago).
- Les contributions aux approches sectorielles peuvent être identifiées dans le SNPC pour le programme national du Mozambique pour le développement agricole – ProAgri (les contributions peuvent être identifiés pour l'Autriche, les institutions de l'UE, l'Irlande et la Suède) et le programme de développement du secteur productif rural du Nicaragua – ProRural (pour la Norvège).
- Les contributions des donneurs à des fonds multi-donneurs sont aussi des arrangements de financements groupés employés pour soutenir le développement agricole : fonds de mobilisation pour l'Afrique TerrAfrica (Norvège), fonds forestier pour le Nicaragua (Finlande, Pays-Bas).

Le soutien budgétaire général, une fois intégré aux budgets nationaux des pays en développement, contribue au développement du secteur agricole, mais cette contribution n'est pas précisée. Le soutien budgétaire général n'est pas préaffecté à des usages spécifiques, mais s'accompagne de considérations et accords sur la stratégie de développement des pays bénéficiaires. Les donneurs ne maîtrisent pas les dépenses, mais suivent la mise en oeuvre de la stratégie du pays bénéficiaire dans son ensemble sur la base d'un jeu d'indicateurs prédéterminé.

Plusieurs donneurs indiquent qu'ils ont tendance à utiliser davantage le canal du soutien budgétaire général pour apporter leur soutien aux pays en développement, plutôt que de mettre en oeuvre des projets spécifiques dans des secteurs identifiés. Ils calculent des estimations de leur contribution aux divers secteurs par le biais du soutien général sur la base de la ventilation de dépenses publiques des pays bénéficiaires entre ces secteurs. Ces estimations ne sont pas utilisées dans les présentations courantes de statistiques du CAD.

Sous-secteurs

Graphique 4. Aide des membres du CAD à l'agriculture par sous-secteur

moyenne 2006-07, prix constants de 2007

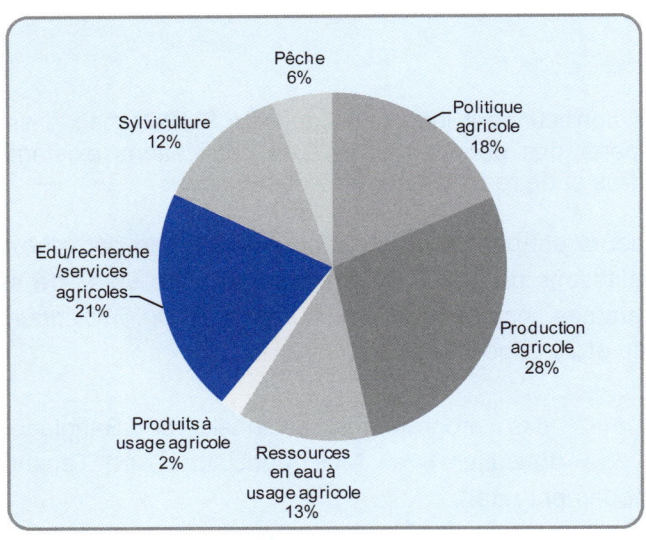

Le Graphique 4 met en évidence la part de la sylviculture et de la pêche dans l'aide totale à l'agriculture : respectivement 12 % et 6 %. Les membres ont différentes approches, certains plaçant davantage l'accent sur la sylviculture (Finlande : 67 % ; Japon et Pays-Bas : 36 % ; Australie : 24 %), et d'autres sur la pêche (Grèce : 38 % ; Espagne : 31 % ; Danemark, Nouvelle-Zélande et Norvège : 21 %).

Voir en Annexe la description des sous-secteurs.

La production agricole représente 28 % des contributions des membres du CAD à l'agriculture en 2006-07. Le développement agricole entre dans cette catégorie et recouvre les projets intégrés et le développement des exploitations agricoles ; il reçoit 14 % de l'aide des membres à l'agriculture ; le développement agricole alternatif est une autre composante significative, représentant 7 % de l'aide totale à l'agriculture (27 % de l'aide totale des Etats-Unis, 8 % de celle du Canada et 7 % de celle des Pays-Bas) ; la production de cultures alimentaires et l'élevage n'atteignent que 3 % de l'aide totale, à égalité avec les cultures d'exportation.[11]

La politique agricole représente 18 % de l'aide totale et est une composante significative de l'aide apportée par plusieurs membres à l'agriculture (Etats-Unis, Irlande, Royaume-Uni et Suisse). Les ressources agraires bénéficient d'un soutien spécifique principalement de l'Irlande (13 %) et de l'Italie (17 %) et la réforme agraire, de celui de la Belgique et de la Suède (5 %) et du Japon (6 %).

Concernant les autres sous-secteurs, les membres semblent concentrer leur aide dans différents domaines :

- les ressources en eau agricole reçoivent en moyenne 13 % de l'aide prodiguée, mais plus d'un tiers de l'aide de l'Allemagne et plus de 20 % de celle du Luxembourg et du Japon ;
- la part des produits à usage agricole est relativement réduite au niveau global (2 %), mais plus significative pour la Finlande (10 %), le Japon (6 %), le Luxembourg et la Norvège (5 %).
- les 16 % alloués en moyenne à l'éducation et la recherche atteignent des niveaux bien supérieurs dans le cas de la France (69 %), de la Suède (42 %) du Royaume-Uni (37 %) de l'Australie et du Portugal (30 %) [voir aussi l'Encadré 1 sur la recherche] ;

---

[11] Les pourcentages alloués à ces segments spécifiques sont probablement sous-évalués dans la mesure où les programmes intégrés qui comprennent des composantes liées à ces segments spécifiques sont comptabilisés à un niveau sectoriel plus général, par exemple, dans le développement agricole.

- la part réduite de 5 % en moyenne affectée aux divers services agricoles prend des proportions beaucoup plus importantes pour plusieurs membres et notamment pour la Nouvelle-Zélande (28 % de l'aide totale est dédiée à des services comme les organisations et politiques de marketing), le Portugal (9 % à la protection des cultures et des récoltes et aux traitements antiparasitaires), la Norvège (8 % aux coopératives agricoles), la Grèce et la Nouvelle-Zélande (respectivement 22 % et 12 % aux services vétérinaires).

Encadré 1. Aide à l'agriculture sous la forme de recherche

Aide à la recherche dans le secteur agricole
moyenne 2006-07

	millions USD	Part dans le total de l'aide à l'agriculture
Allemagne	23	7%
Australie	21	25%
Autriche	0	1%
Belgique	5	5%
Canada	2	2%
Danemark	4	4%
Espagne	2	1%
Etats-Unis	26	3%
Finlande	1	3%
France	305	68%
Grèce	0	1%
Institutions de l'UE	17	4%
Irlande	1	3%
Italie	2	5%
Japon	0	0%
Norvège	9	7%
Nouvelle Zélande	0	3%
Pays-Bas	6	5%
Portugal	1	29%
Royaume Uni	1	1%
Suède	34	23%
Suisse	9	13%
Total membres du CAD	**468**	**11%**

Les activités d'aide du Centre australien pour la recherche internationale de l'agriculture (ACIAR) visent de façon spécifique quelques enjeux et pays receveurs d'aide, par exemple, le développement de marqueurs moléculaires pour permettre la sélection contre la maladie du riz aux Philippines, ou l'emploi d'équipement de plantation testé contre les pathogènes pour améliorer la production durable de patate douce dans les Iles Salomon et en Papouasie Nouvelle-Guinée.

La France notifie la recherche éligible à l'APD conduite *par le* Centre de coopération internationale en recherche agronomique pour le développement (CIRAD).

L'Allemagne contribue au financement des Centres internationaux de recherche agricole (CIRA) et plus particulièrement à ceux soutenus par le Groupe consultatif pour la recherche agricole internationale (GCRAI) en apportant des financements ciblés pour renforcer la coopération entre les instituts de recherche allemands et internationaux.

La Suède a fourni des subventions APD de recherche en 2006-07 principalement par le canal du GCRAI et dans le cadre d'un programme de développement des capacités et de la recherche dans les zones de montagne au Laos.

Les Etats-Unis soutiennent principalement le GCRAI, mais travaillent aussi directement à des projets spécifiques comme le programme de recherche coopérative de l'ex-Union soviétique qui entend créer divers projets coopératifs avec les scientifiques anciennement employés à l'armement biologique soviétique afin de redéployer leurs efforts vers une recherche agricole à applications pacifiques.

Attention portée à l'égalité homme-femme

Les femmes jouent un rôle crucial dans l'agriculture : dans de nombreux pays en développement, elles constituent la majorité de la main-d'oeuvre agricole et elles produisent la plupart des aliments. Le « marqueur de l'égalité homme-femme » permet d'évaluer le degré d'intérêt et les mesures particulières prises par les membres du CAD pour faire progresser l'égalité homme-femme dans le cadre de leurs activités d'aide à l'agriculture. Sur la base des données disponibles[12] Parmi les membres qui ont fourni des informations sur cet aspect de leur programme d'aide, presque la moitié de l'aide allouée visait à améliorer l'égalité. Le Graphique 5 met en évidence les pays membres qui accordent le plus d'attention à l'égalité homme-femme dans leurs programmes d'aide, ainsi que le total pour le CAD.

Graphique 5. Attention portée à l'égalité homme-femme dans les activités d'aide au secteur agricole
moyenne 2006-07, prix constants de 2007

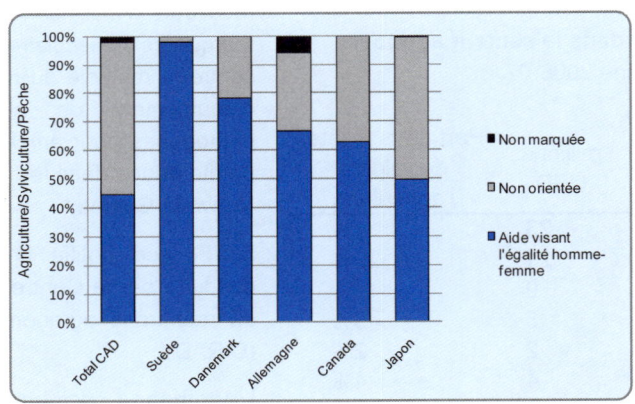

L'aide à l'agriculture et les objectifs des Conventions de Rio

Les membres du CAD sont attentifs aux enjeux environnementaux dans la conception de leurs activités d'aide dans le domaine de l'agriculture. Les données des pays ayant notifié des informations sur les « marqueurs Rio », présentées au Graphique 6, montrent que 20 % de l'aide à l'agriculture a également ciblé les objectifs de la Convention sur la désertification ces dernières années ; un pourcentage égal a aussi ciblé la biodiversité et 17 % le changement climatique.[13]

---

[12] Seuls les membres avec une couverture supérieure à 80% pour le marqueur ont été inclus.

[13] Les activités pouvant cibler simultanément les objectifs de plus d'une des Conventions, ces pourcentages ne s'additionnent pas.

Graphique 6. Attention portée aux objectifs des Conventions de Rio au sein du secteur de l'agriculture
%, moyenne 2006-07

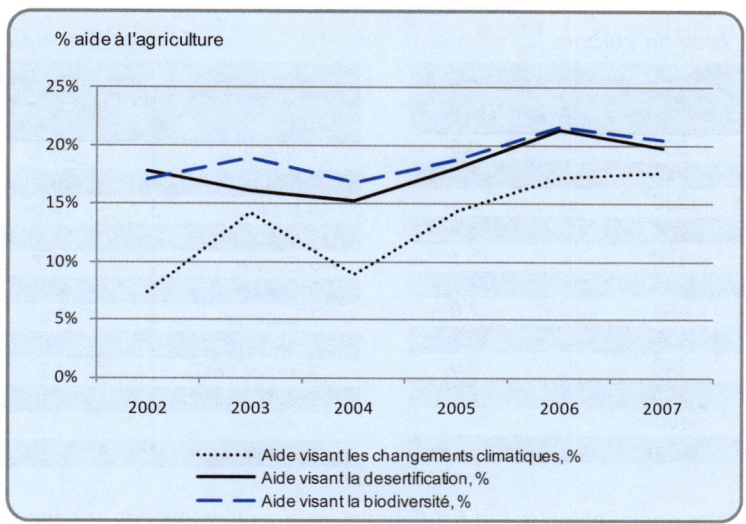

Tous les donneurs ne notifient pas d'informations sur les marqueurs de Rio et les graphiques présentés excluent les Etats-Unis, le Luxembourg et la Norvège (pour tous les marqueurs) et la Suisse (pour la désertification.

II. Ciblage des pays où la situation est la plus critique

Les Tableaux 5.a. et 5.b. présentent les dix premiers bénéficiaires de l'aide à l'agriculture en volume et par habitant. Dans cette section, la perspective adoptée est celle du pays bénéficiaire ; les données portent sur l'ensemble des apports d'APD, qu'elle soit bilatérale ou accordée par les institutions multilatérales présentées au Tableau 1, à l'exception de la FAO qui ne divulgue pas le total des apports par bénéficiaire (ce qui explique l'écart entre les totaux des Tableaux 1 et 5).

En 2006-07, l'Inde a été le premier bénéficiaire de l'aide en volume, ayant reçu des dons et des prêts de l'IDA, du Japon et de l'Allemagne. Par habitant, l'Arménie arrive au premier rang, avec une contribution très importante du *Millennium Challenge Account* des États-Unis.

Tableaux 5.a et 5.b. Les dix premiers bénéficiaires de l'aide à l'agriculture

moyenne 2006-07

Tableau 5.a Aide en volume

	Total millions USD	USD par habitant
1 Inde	563	0.5
2 Afghanistan	248	7.6
3 Viet Nam	236	2.8
4 Mali	186	13.8
5 Kenya	176	5.1
6 Ethiopie	166	2.1
7 Ghana	163	7.0
8 Chine	155	0.1
9 Indonésie	140	0.6
10 Pakistan	133	0.9
Autre	4004	
Total	6169	

Tableau 5.b Aide par habitant[14]

	USD par habitant	Total million USD
1 Arménie	29.4	88
2 Maurice	15.5	19
3 Mali	13.8	186
4 Swaziland	12.4	14
5 Bolivie	10.7	98
6 Gabon	9.4	13
7 Laos	7.8	46
8 Afghanistan	7.6	248
9 Timor-Leste	7.2	8
10 Burkina Faso	7.1	104
Autre		5345
Total		6169

Note : L'écart entre les totaux des Tableaux 1 et 5.a./5.b. est dû aux montants des apports de la FAO qui ne peuvent être ventilés par pays (seules des estimations sont disponibles).

Les Graphiques 7.a et 7.b. illustrent la ventilation des engagements d'aide par région et catégorie de revenu sur la période 2002-07. Les apports d'aide à l'agriculture ont surtout ciblé l'Afrique subsaharienne (31 %) et l'Asie centrale et du Sud (23 %). La part de ces deux régions a augmenté au cours de la dernière décennie, passant de 27 % en 1998-99 à 33 % en 2006-07 pour l'Afrique subsaharienne et de 19 % à 23 % pour l'Asie centrale et du Sud. Une région a été progressivement moins ciblée sur cette période : l'Asie d'Extrême-Orient, dont la part chute de 22 % à 12 %.

Graphiques 7.a et 7.b. Aide à l'agriculture ventilée par région et catégorie de revenu, 2002-07
Graphique 7.a. Ventilation par région — **Graphique 7.b. Ventilation par catégorie de revenu**

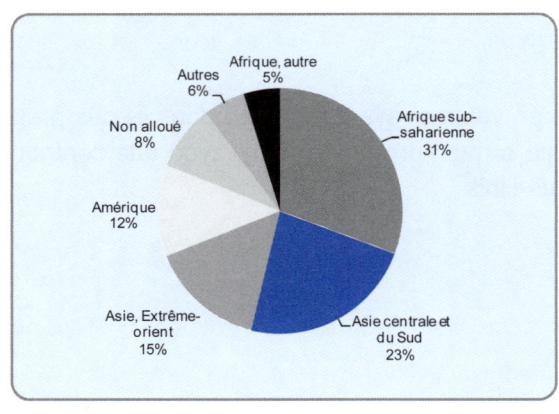
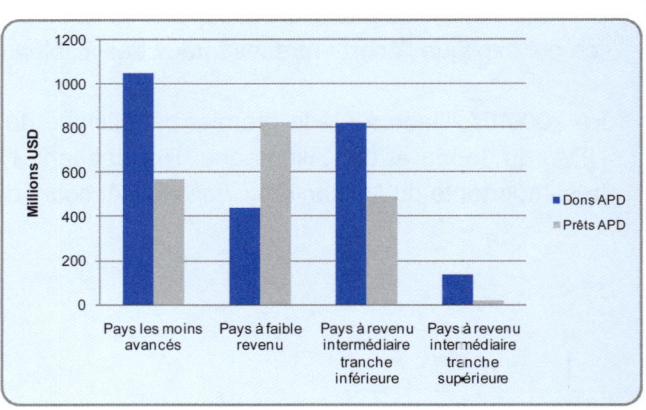

Les Pays les moins avancés (PMA) et les autres Pays à faible revenu (autres PFR) reçoivent les deux tiers de l'aide totale à l'agriculture (en excluant l'aide régionale/multi-pays qui ne peut être ventilée entre les catégories de revenu), indiquant un ciblage clair des pays les plus pauvres. A eux seuls, les PMA reçoivent 37 % de l'aide totale.

---

[14] Parmi les pays comptant plus d'1 million d'habitants.

Les prêts concessionnels représentent 35 % de l'aide aux PMA ; ils proviennent principalement de l'IDA, des Fonds africain et asiatique de développement et du FIDA. La proportion de prêts est de 65% pour les les autres PFR qui les reçoivent de ces mêmes agences, mais aussi de deux grands donneurs bilatéraux : 83 % de l'aide du Japon aux autres PFR prend la forme de prêts et ce pourcentage est de 62 % pour la France.

La comparaison du niveau de l'aide à l'agriculture et des indicateurs de malnutrition des pays permet d'analyser sous un angle différent le ciblage des pays dont la situation est la plus critique. Un point de référence utile est l'objectif 1.C des Objectifs du Millenium : *Réduire de moitié, entre 1990 et 2015, la proportion de la population qui souffre de la faim.* Les progrès concernant l'OMD 1.C sont mesurés à l'aide de l'indicateur de la prévalence de l'insuffisance pondérale parmi les enfants (de moins de 5 ans). Cet indicateur a décliné de 33% dans les années 90 à 26% en 2007, cependant des estimations prévisionnelles indiquent qu'il a à nouveau augmenté d'un point en 2008. Le rapport 2009 des Nations unies sur les OMD affirme que « le taux de progrès observé est insuffisant pour atteindre l'objectif de réduire de moitié en 2015 la proportion de la population qui souffre de la faim – même sans tenir compte de la hausse des denrées alimentaires et de la crise économique qui se sont développées entretemps». Le rapport détaille un ensemble de mesures pouvant augmenter la disponibilité de nourriture, y compris une augmentation de la production.

Comme le montre la Carte 1, les régions les plus affectées sont l'Asie du Sud, où près de 50 % des enfants souffrent d'insuffisance pondérale, et l'Afrique subsaharienne, où ils sont 28 %. A elle seule, l'Asie du Sud abrite plus de la moitié des enfants de la planète victimes de malnutrition. Les Cartes 1 et 2 permettent de comparer le degré de malnutrition dans les pays en développement et le niveau des apports des donneurs pour l'agriculture. Il apparaît que les pays d'Afrique subsaharienne reçoivent le niveau le plus élevé d'aide à l'agriculture par habitant, alors que les pays de l'Asie du Sud, qui sont aussi gravement touchés, ne bénéficient pas d'une aide aussi importante par habitant. L'aide accordée à ces pays peut être considérable, en volume, mais la population est nombreuse, ce qui réduit les montants par habitant : l'Inde, par exemple, est le premier bénéficiaire de l'aide en volume, mais ne perçoit que 0,3 USD (30 cents) par habitant.

**Carte 1. Proportion des enfants de moins de cinq ans souffrant d'insuffisance pondérale
2007, source : Banque mondiale - UNICEF**

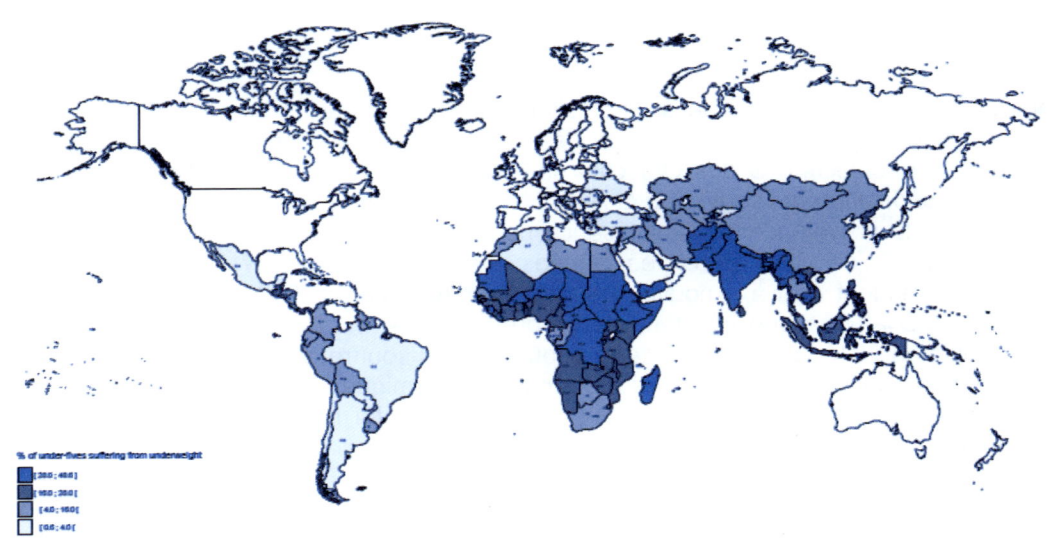

Carte 2. Aide à l'agriculture par habitant, moyenne 2006-07, en USD

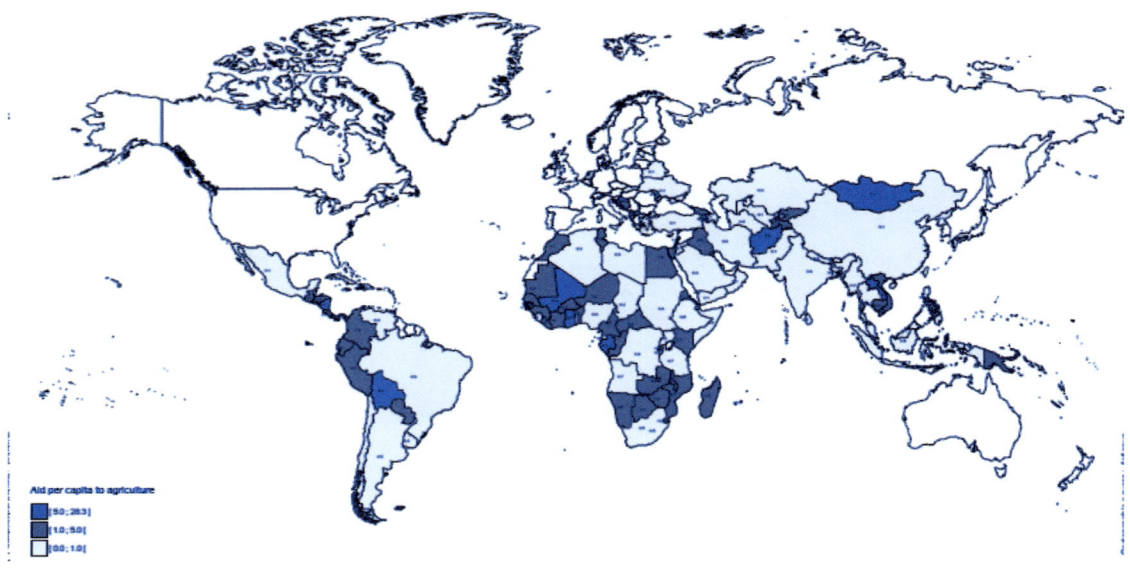

III. Apports d'aide à l'agriculture et aux autres secteurs liés à la sécurité alimentaire

Comme indiqué à la section I.1, le CAD exclut dans sa définition de l'aide à l'agriculture le développement rural (classé dans l'aide plurisectorielle), l'aide au secteur agroalimentaire et à l'industrie forestière (qui font partie de l'aide à l'industrie), l'aide alimentaire à des fins de développement (une sous catégorie de l'aide-programme générale) et l'aide alimentaire d'urgence (incluse dans l'aide humanitaire). L'aide à ces secteurs n'est pas reflétée dans les statistiques générales présentées plus haut. Le Tableau 6 ci-après montre les séries de données incluant ces secteurs qui, sans soutenir directement le développement agricole, contribuent à l'amélioration de la vie rurale et de la sécurité alimentaire des pays en développement.

Les interventions des donneurs bilatéraux et des agences multilatérales pour améliorer la sécurité alimentaire se répartissent entre les mesures de court terme, comprenant l'aide alimentaire d'urgence et de développement, et l'investissement dans l'agriculture et le développement rural des pays pour assurer des provisions alimentaires adéquates sur le long terme.

A elle seule, l'aide à l'agriculture a représenté 6,2 milliards USD en 2006-07. Sur la base de l'indicateur élargi, l'aide à l'agriculture et à la sécurité alimentaire s'est élevée à **12 milliards USD** par an en 2006-07, dont **7,6 milliards USD** alloués à des interventions à long terme (pour l'agriculture et le développement rural) et les **4,4 milliards USD** restant à l'aide alimentaire de court terme (d'urgence et de développement).

Tableau 6. Aide à l'agriculture et aux autres secteurs liés à la sécurité alimentaire sur 2002-07

Engagements annuels moyens, en millions USD et à prix constants de 2007

Pays du CAD	2002-03	2004-05	2006-07
Agriculture/Sylviculture/Pêche	2953	3179	3768
Développement rural	825	599	884
Aide alimentaire à des fins de développement	1644	1198	1060
Aide alimentaire d'urgence	1798	2015	1811
Total pays du CAD	**7219**	**6990**	**7523**
Agences multilatérales	2002-03	2004-05	2006-07
Agriculture/Sylviculture/Pêche	2161	1875	2400
Développement rural	872	655	589
Aide alimentaire à des fins de développement	1401	1190	1013
Aide alimentaire d'urgence	356	322	393
Total agences multilatérales	**4790**	**4042**	**4395**
Total	**12009**	**11033**	**11919**

Note : La ventilation des apports du PAM entre aide alimentaire à des fins de développement (84 %) et aide alimentaire d'urgence (16 %) est celle observée en 2008 ; ces pourcentages ont ensuite été appliqués aux totaux des années antérieures.

Le Graphique 8.a. révèle que, sur la dernière décennie, l'aide bilatérale à l'agriculture et aux autres secteurs liés à la sécurité alimentaire a augmenté, passant de 5,9 milliards USD en 1996-97 à 7,5 milliards USD en 2006-07, la diminution de l'aide à l'agriculture ayant été plus que compensée par l'accroissement de l'aide aux autres secteurs liés à la sécurité alimentaire, en particulier l'aide alimentaire à court terme.

Le Graphique 8.b. illustre une tendance différente parmi les agences multilatérales, qui ont légèrement accru leur aide à l'agriculture sur la période, mais aussi leur aide aux autres secteurs liés à la sécurité alimentaire. Le rôle particulier joué par les trois agences des Nations Unies basées à Rome dans le domaine de la sécurité alimentaire est présenté succinctement dans l'Encadré 2.

Graphiques 8.a. et 8.b. Apports d'aide à l'agriculture et aux autres secteurs liés à la sécurité alimentaire

Engagements annuels moyens, en millions USD et à prix constants de 2007

Graphique 8.a. Donneurs bilatéraux

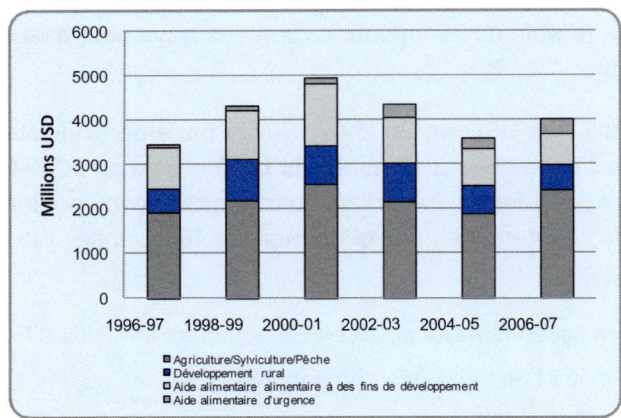

Graphique 8.b. Agences multilatérales

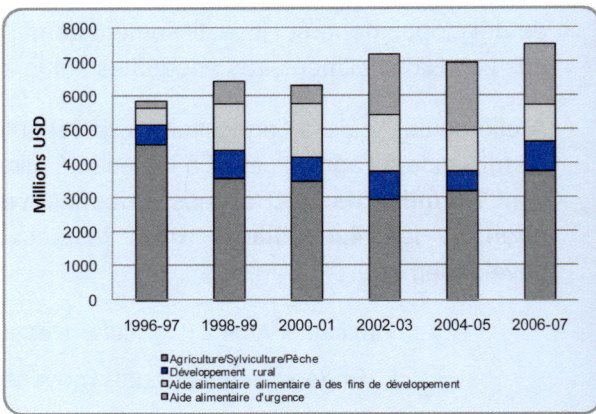

Encadré 2. « Les trois agences basées à Rome »

Le **Programme alimentaire mondial (PAM)** divise ses activités en quatre catégories : i) *Les opérations d'urgence* (aide alimentaire d'urgence), ii) *l'aide prolongée et les opérations de redressement* (aide alimentaire en appui des efforts de réhabilitation à long terme et après les conflits), iii) *le développement* (y compris les programmes alimentaires dans les écoles) et iv) *les opérations spéciales* (soutien logistique à l'apport d'aide alimentaire). Le budget du PAM se compose essentiellement de fonds préaffectés des pays membres du CAD comptabilisés dans les statistiques du CAD comme aide bilatérale. Les activités du PAM financées à partir du budget régulier sont comptabilisées dans les apports multilatéraux (372 millions USD en 2006-07) et se répartissent entre l'aide alimentaire d'urgence [catégorie i), 14 % du total] et l'aide alimentaire de développement [catégories ii) et iii), 86 %]. La Catégorie iv) représente une part réduite des activités du PAM. Les apports du PAM sont pris en compte dans le Tableau 6.

L'**Organisation des Nations Unies pour l'alimentation et l'agriculture (FAO)** réalise un travail normatif et opérationnel. Les dépenses du siège sont principalement axées sur les activités normatives, y compris la recherche et les projets pilotes. Ce travail normatif bénéficie à la fois aux pays développés et aux pays en développement et n'est donc pas comptabilisé intégralement en APD (un coefficient d'APD de 51 % est appliqué depuis la dernière estimation du CAD en 2003 et fera l'objet d'une révision en 2009). Les principales missions des bureaux régionaux et nationaux sont la liaison avec les autorités des pays bénéficiaires, la promotion des efforts mondiaux et la mise en oeuvre de l'assistance technique dans les pays. Dans cette étude, les 51 % du budget de la FAO éligibles à l'APD (203 millions USD par an en 2006-07) ont été pris en compte dans les Tableaux 1 et 6 (aide à l'agriculture).

Le **Fonds international pour le développement agricole (FIDA)** est une institution financière internationale qui fournit des prêts, concessionnels et non concessionnels, ainsi que des dons pour le développement agricole et rural des pays en développement. Les fonds sont principalement affectés à l'agriculture (44 % au cours des années 2006-07) et au développement rural (40 %). Certaines activités visent également à promouvoir un environnement favorable à l'agriculture, grâce à un accès facilité aux institutions de financement (12 %). Le FIDA a récemment notifié au SNPC/CAD une ventilation sectorielle plus détaillée de ses données historiques qui fournit des indications plus précises sur la répartition des apports d'aide : la part de l'agriculture reste globalement inchangée (40 %), mais celle du développement rural est réduite à 7 %, tandis que les institutions de financement reçoivent 27 %. D'autres secteurs apparaissent importants : la gouvernance et la société civile (10 %) et l'industrie (7 %).

Graphique 9. L'aide à l'agriculture et aux autres secteurs liés à la sécurité alimentaire en relation avec la malnutrition

Aide à l'agriculture et aux autres secteurs liés à la sécurité alimentaire par habitant et prévalence de l'insuffisance pondérale parmi les enfants de moins de cinq ans

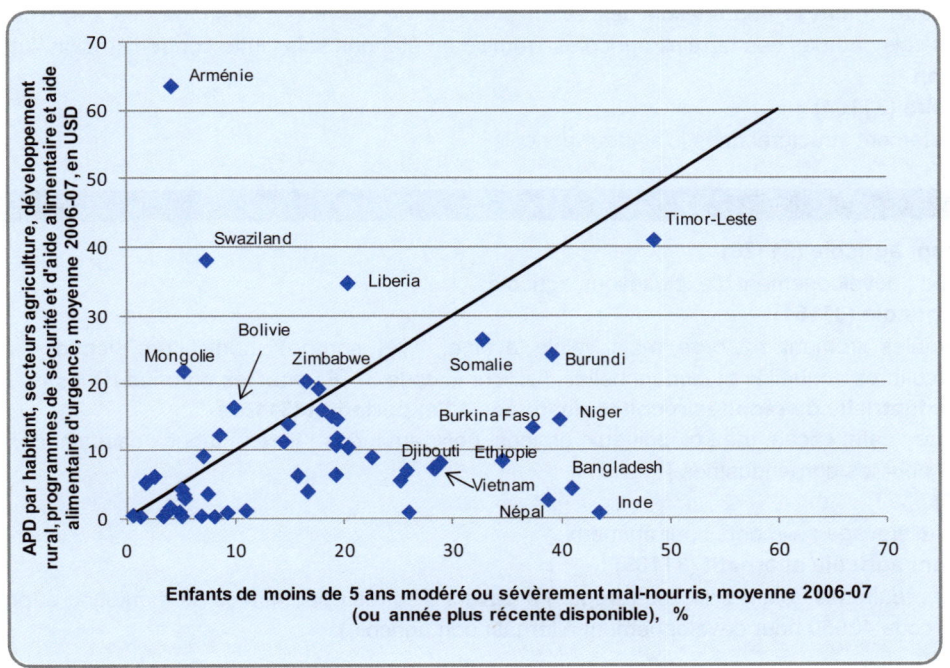

Le Graphique 9 souligne qu'en données par habitant, l'aide à l'agriculture et aux autres secteurs liés à la sécurité alimentaire n'est généralement pas allouée en priorité aux populations souffrant de malnutrition (voir la ligne départageant les pays « sous-aidés » des pays « sur-aidés »). Parmi les pays les plus affectés, seuls le Timor-Leste, la Somalie et le Burundi ont reçu plus de 20 USD par habitant durant la période 2006-07. De nombreux autres pays gravement touchés ont reçu très peu d'aide durant la même période (en particulier l'Inde, le Bangladesh, le Népal, le Togo), tandis que des pays dans des situations plus favorables en ont reçu davantage. C'est par exemple le cas de l'Arménie avec 63 USD par habitant.

Annexe. Liste des codes SNPC pour l'aide à l'agriculture et autres secteurs liés à la sécurité alimentaire
Groupements utilisés dans le Graphique 4

AIDE A L'AGRICULTURE

AGRICULTURE

Politique agricole (groupe 1)

Politique agricole et gestion administrative (31110)
Politique agricole, planification et programmes ; aide aux ministères de l'agriculture ; renforcement des capacités institutionnelles et conseils ; activités d'agriculture non spécifiées.

Ressources en terres cultivables (31130)
Y compris la lutte contre la dégradation des sols ; amélioration des sols ; drainage des zones inondées ; dessalage des sols ; études des terrains agricoles ; remise en état des sols ; lutte contre l'érosion, lutte contre la désertification.

Réforme agraire (31164)
Y compris ajustement structurel dans le secteur agricole.

Production agricole (groupe 2)

Développement agricole (31120)
Projets intégrés ; développement d'exploitations agricoles.

Production agricole (31161)
Y compris céréales (froment, riz, orge, mais, seigle, avoine, millet, sorgho) ; horticulture ; légumes ; fruits et baies ; autres cultures annuelles et pluriannuelles. [Utiliser le code 32161 pour les agro-industries.]

Production industrielle de récoltes/récoltes destinées à l'exportation (31162)
Y compris sucre ; café, cacao, thé ; oléagineux, graines, noix, amandes ; fibres ; tabac ; caoutchouc. [Utiliser le code 32161 pour les agro-industries.]

Bétail (31163)
Toutes formes d'élevage ; aliments pour animaux.

Développement agricole alternatif (31165)
Projets afin de réduire les cultures illicites (drogue) à travers d'autres opportunités de marketing et production agricoles (voir code 43050 pour développement alternatif non agricole).

Ressources en eau à usage agricole (groupe 3) (31140)
Irrigation, réservoirs, structures hydrauliques, exploitation de nappes phréatiques.

Produits à usage agricole (groupe 4) (31150)
Approvisionnement en semences, engrais, matériel et outillage agricoles.

Éducation/recherche/services agricoles (groupe 5)

Vulgarisation agricole (31166)
Formation agricole non formelle.

Éducation et formation dans le domaine agricole (31181)

Recherche agronomique (31182)
Étude des espèces végétales, physiologie, ressources génétiques, écologie, taxonomie, lutte contre les maladies, biotechnologie agricole ; y compris recherche vétérinaire (dans les domaines génétiques et sanitaires, nutrition, physiologie).

Services agricoles (31191)
Organisation et politiques des marchés ; transport et stockage ; établissements de réserves stratégiques.

Protection des plantes et des récoltes, lutte antiacridienne (31192)

Y compris protection intégrée des plantes, les activités de protection biologique des plantes, la fourniture et la gestion de substances agrochimiques, l'approvisionnement en pesticides ; politique et législation de la protection des plantes.

Services financiers agricoles (31193)
Intermédiaires financiers du secteur agricole, y compris les plans de crédit ; assurance récoltes.

Coopératives agricoles (31194)
Y compris les organisations d'agriculteurs.

Services vétérinaires (bétail) (31195)
Santé des animaux, ressources génétiques et nutritives.

SYLVICULTURE (groupe 6)

Politique de la sylviculture et gestion administrative (31210)
Politique de la sylviculture, planification et programmes ; renforcement des capacités institutionnelles et conseils ; études des forêts ; activités sylvicoles et agricoles liées à la sylviculture non spécifiées.

Développement sylvicole (31220)
Boisement pour consommation rurale et industrielle ; exploitation et utilisation ; lutte contre l'érosion, lutte contre la désertification ; projets intégrés.

Reboisement (bois de chauffage et charbon de bois) (31261)
Développement sylvicole visant à la production de bois de chauffage et de charbon de bois.

Éducation et formation en sylviculture (31281)

Recherche en sylviculture (31282)
Y compris reproduction artificielle et amélioration des espèces, méthodes de production, engrais, coupe et ramassage du bois.

Services sylvicoles (31291)

PECHE (groupe 7)

Politique de la pêche et gestion administrative (31310)
Politique de la pêche, planification et programmes ; renforcement des capacités institutionnelles et conseils ; pêche hauturière et côtière ; évaluation, études et prospection du poisson en milieu marin et fluvial ; bateaux et équipements de pêche ; activités de pêche non spécifiées.

Développement de la pêche (31320)
Exploitation et utilisation des pêcheries ; sauvegarde des bancs de poisson ; aquaculture ; projets aquatique.

Éducation et formation dans le domaine de la pêche (31381)

Recherche dans le domaine de la pêche (31382)
Pisciculture pilote ; recherche biologique aquatique.

Services dans le domaine de la pêche (31391)
Ports de pêche ; vente des produits de la pêche ; transport et entreposage frigorifique du poisson.

PART 2

Donor profiles on aid to agriculture

Statistics for each DAC member

www.oecd.org/dac/stats/agriculture

PARTIE 2

Profils d'aide des donneurs dans le secteur de l'agriculture

Statistiques pour chacun des donneurs du CAD

Donor profiles on aid to agriculture

The following profiles describe statistical aspects of DAC members' aid to agriculture through a few summary statistics, charts and tables.

Statistics are based on members' reporting to the CRS and DAC, and comply with the definition of aid to agriculture given in section I.1 of this report. Information shown includes the total volume of aid to agriculture and trends over time, the share of aid to agriculture in total bilateral sector-allocable aid, the top ten recipients, the geographical distribution and the sub-sector distribution. One table also shows statistics based on a wider definition of aid to agriculture that includes agro-industries, forest-industries, rural development, food aid and emergency food aid.

Profils d'aide des donneurs dans le secteur de l'agriculture

Les profils suivants décrivent les aspects statistiques de l'aide à l'agriculture par les membres du CAD à travers quelques graphiques et tableaux.

Les Statistiques sont basées sur la notification des membres au SNPC et CAD, et utilisent la définition d'aide à l'agriculture tel qu'elle est présentée dans la section I.1 de ce document. Les statistiques comprennent le total du volume d'aide à l'agriculture, les tendances ces dernières années, l'importance de l'aide à l'agriculture dans le total de l'aide allouée par secteur, les dix principaux pays bénéficiaires, la répartition géographique ainsi que la répartition par sous-secteurs. L'un des tableaux montre également les statistiques basées sur une définition plus large incluant les agro-industries, les industries forestières, le développement rural, l'aide alimentaire ainsi que l'aide alimentaire d'urgence.

AUSTRALIA
Aid at a glance - Agriculture

Summary statistics

If not otherwise stated, figures refer to 2006-2007 annual average commitments expressed in 2007 constant prices.

	Total aid to agriculture		
	USD million	Aid to agriculture by Australia as a share of total aid by Australia*	Aid to agriculture by Australia as a share of total DAC members' aid to agriculture
Australia	81.8	5%	2%
For reference, total DAC	*4213*	*6%*	*100%*

Top ten recipients of aid to agriculture			
		Aid to agriculture by Australia to that recipient as a share of	
	USD million	total aid by Australia to that recipient*	total DAC members' aid to agriculture to that recipient
Papua New Guinea	17.8	7%	75%
Indonesia	15.7	7%	24%
Cambodia	3.2	16%	11%
Philippines	3.0	3%	4%
Viet Nam	2.7	9%	2%
China	2.6	11%	2%
Myanmar	2.4	56%	31%
India	2.0	25%	1%
Tonga	1.1	9%	94%
Laos	1.0	5%	3%

Aid to all agriculture-related sectors	
	USD million
Agriculture/Forestry/Fishing	81.8
Agro-industries	0.0
Forest-industries	0.1
Rural development	6.9
Food aid	13.4
Emergency food aid	2.8
Total	*105.0*

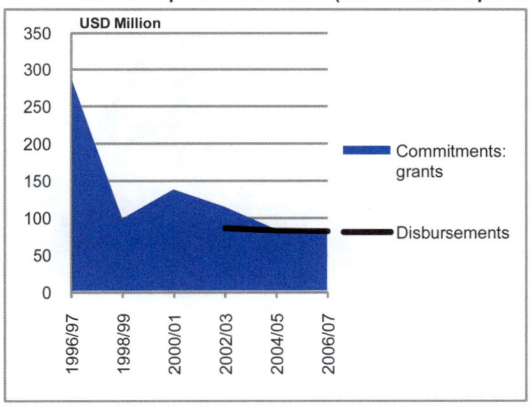
Trend of aid to agriculture over the period 1995-2007 (2007 constant prices)

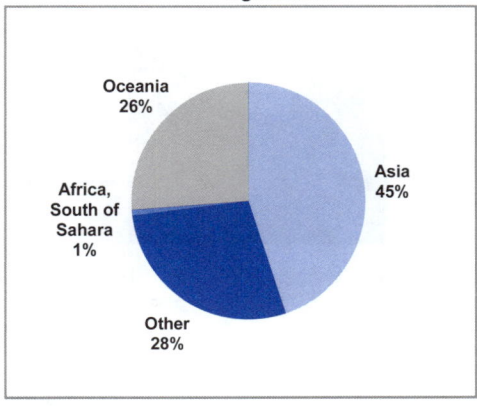
Regional distribution of aid to agriculture

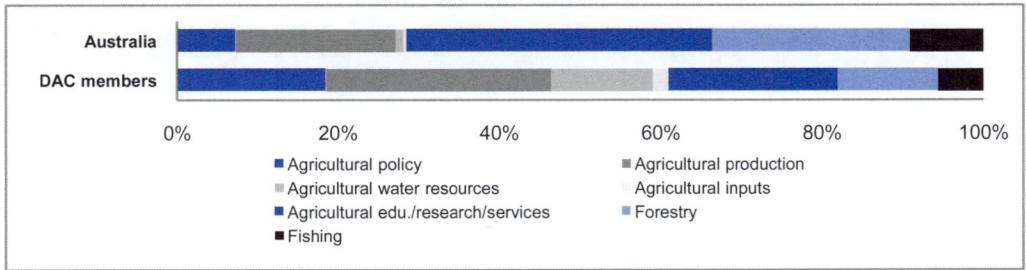
Agriculture aid by subsector

* % of sector allocable aid

AUSTRIA
Aid at a glance - Agriculture

Summary statistics
If not otherwise stated, figures refer to 2006-2007 annual average commitments expressed in 2007 constant prices.

	Total aid to agriculture		
	USD million	Aid to agriculture by Austria as a share of total aid by Austria*	Aid to agriculture by Austria as a share of total DAC members' aid to agriculture
Austria	12.1	4%	0%
For reference, total DAC	4213	6%	100%

Top ten recipients of aid to agriculture			
		Aid to agriculture by Austria to that recipient as a share of	
	USD million	total aid by Austria to that recipient*	total DAC members' aid to agriculture to that recipient
Nicaragua	1.9	31%	7%
China	1.5	4%	1%
Ethiopia	1.4	17%	3%
Mozambique	1.0	27%	1%
Tanzania	0.6	44%	2%
Senegal	0.6	12%	1%
Moldova	0.4	11%	4%
El Salvador	0.4	29%	5%
Brazil	0.3	11%	1%
Ukraine	0.3	6%	5%

Aid to all agriculture-related sectors	
	USD million
Agriculture/Forestry/Fishing	12.1
Agro-industries	0.2
Forest-industries	0.2
Rural development	10.6
Food aid	1.3
Emergency food aid	1.1
Total	25.5

Trend of aid to agriculture over the period 1995-2007 (2007 constant prices)

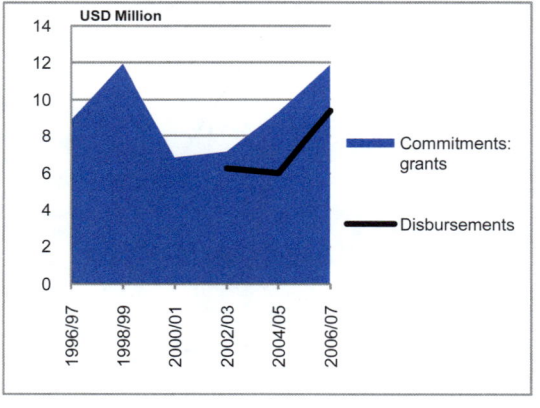

Regional distribution of aid to agriculture

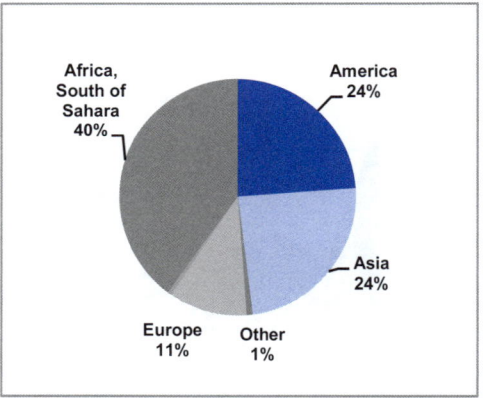

Agriculture aid by subsector

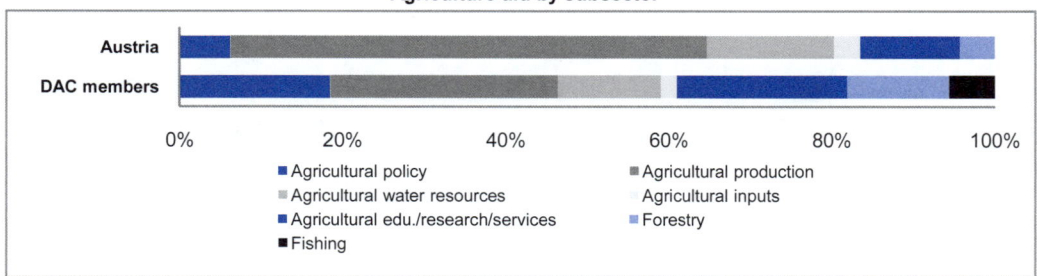

* % of sector allocable aid

BELGIUM
Aid at a glance - Agriculture

Summary statistics
If not otherwise stated, figures refer to 2006-2007 annual average commitments expressed in 2007 constant prices.

	Total aid to agriculture		
	USD million	Aid to agriculture by Belgium as a share of total aid by Belgium*	Aid to agriculture by Belgium as a share of total DAC members' aid to agriculture
Belgium	91.3	9%	2%
For reference, total DAC	4213	6%	100%

Top ten recipients of aid to agriculture			
		Aid to agriculture by Belgium to that recipient as a share of	
	USD million	total aid by Belgium to that recipient*	total DAC members' aid to agriculture to that recipient
Ecuador	13.9	44%	43%
Congo, Dem. Rep.	11.2	8%	57%
South Africa	8.5	34%	37%
Niger	6.1	29%	30%
Rwanda	5.8	8%	34%
Tanzania	5.0	27%	18%
Benin	4.5	28%	15%
Burkina Faso	3.4	21%	6%
Peru	3.1	18%	3%
Malawi	2.9	64%	8%

Aid to all agriculture-related sectors	
	USD million
Agriculture/Forestry/Fishing	91.3
Agro-industries	0.4
Forest-industries	0.0
Rural development	50.8
Food aid	17.4
Emergency food aid	26.1
Total	*185.9*

Trend of aid to agriculture over the period 1995-2007 (2007 constant prices)

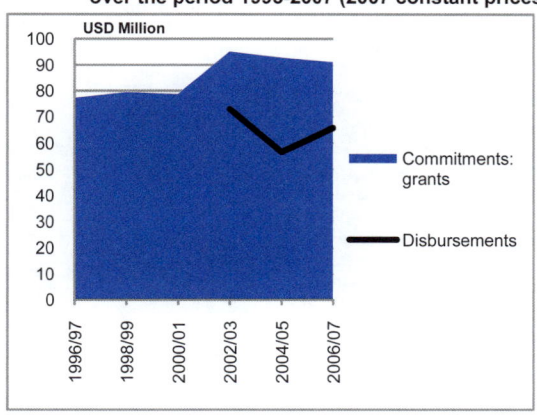

Regional distribution of aid to agriculture

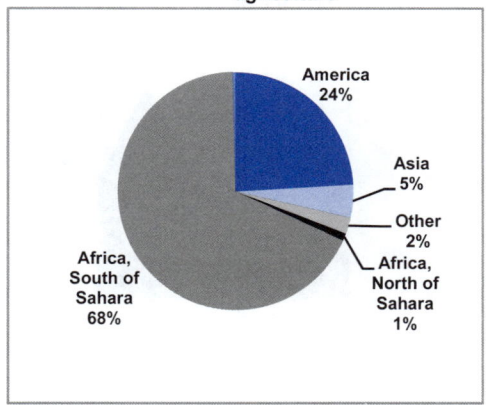

Agriculture aid by subsector

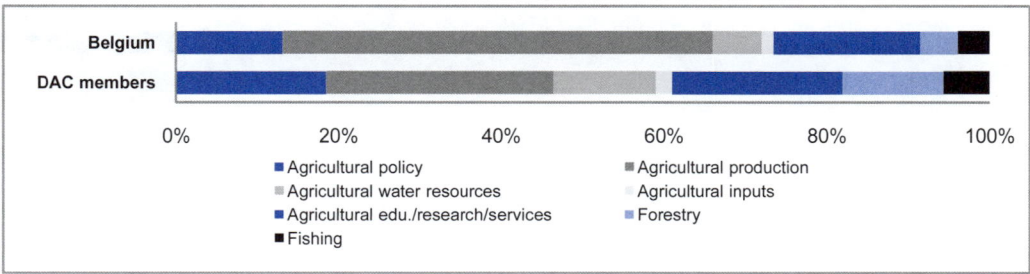

* % of sector allocable aid

CANADA
Aid at a glance - Agriculture

Summary statistics
If not otherwise stated, figures refer to 2006-2007 annual average commitments expressed in 2007 constant prices.

	Total aid to agriculture		
	USD million	Aid to agriculture by Canada as a share of total aid by Canada*	Aid to agriculture by Canada as a share of total DAC members' aid to agriculture
Canada	114.4	6%	3%
For reference, total DAC	*4213*	*6%*	*100%*

Top ten recipients of aid to agriculture			
		Aid to agriculture by Canada to that recipient as a share of	
	USD million	total aid by Canada to that recipient*	total DAC members' aid to agriculture to that recipient
Afghanistan	29.1	11%	14%
Ethiopia	10.8	18%	24%
Haiti	10.3	11%	50%
Senegal	8.8	26%	19%
Pakistan	6.9	14%	32%
Bangladesh	6.3	8%	13%
Ukraine	4.1	19%	58%
Tajikistan	3.1	65%	26%
Ghana	1.4	17%	1%
Sri Lanka	1.2	4%	11%

Aid to all agriculture-related sectors	
	USD million
Agriculture/Forestry/Fishing	114.4
Agro-industries	0.7
Forest-industries	0.1
Rural development	18.3
Food aid	22.4
Emergency food aid	170.0
Total	325.8

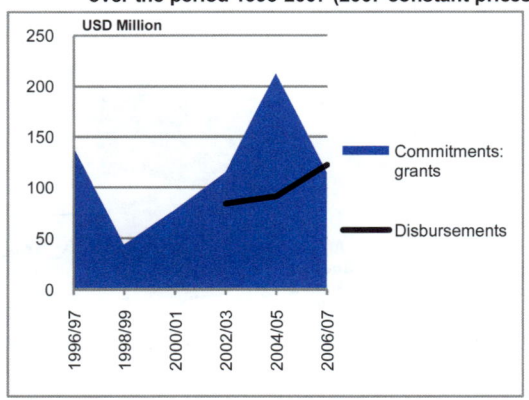
Trend of aid to agriculture over the period 1995-2007 (2007 constant prices)

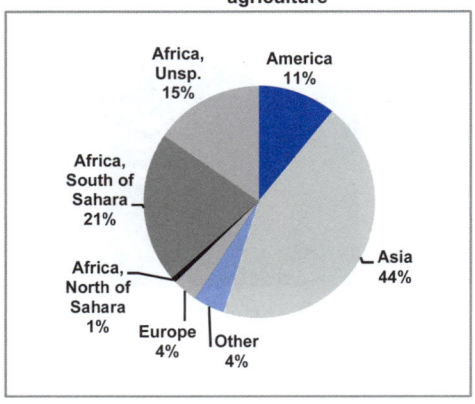
Regional distribution of aid to agriculture

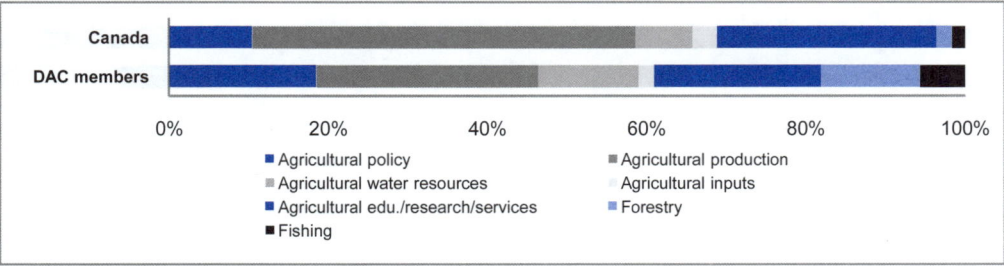
Agriculture aid by subsector

* % of sector allocable aid

DENMARK
Aid at a glance - Agriculture

Summary statistics
If not otherwise stated, figures refer to 2006-2007 annual average commitments expressed in 2007 constant prices.

Total aid to agriculture			
	USD million	Aid to agriculture by Denmark as a share of total aid by Denmark*	Aid to agriculture by Denmark as a share of total DAC members' aid to agriculture
Denmark	82.8	9%	2%
For reference, total DAC	*4213*	*6%*	*100%*

Top ten recipients of aid to agriculture			
		Aid to agriculture by Denmark to that recipient as a share of	
	USD million	total aid by Denmark to that recipient*	total DAC members' aid to agriculture to that recipient
Bangladesh	34.8	61%	72%
Viet Nam	23.2	26%	16%
Mali	13.8	36%	9%
Tanzania	3.2	13%	11%
Bolivia	1.3	8%	2%
Uganda	0.9	2%	3%
Honduras	0.8	100%	8%
Nepal	0.7	3%	7%
Nicaragua	0.6	27%	2%
India	0.5	15%	0%

Aid to all agriculture-related sectors	
	USD million
Agriculture/Forestry/Fishing	82.8
Agro-industries	4.9
Forest-industries	0.0
Rural development	3.9
Food aid	0.2
Emergency food aid	11.3
Total	103.1

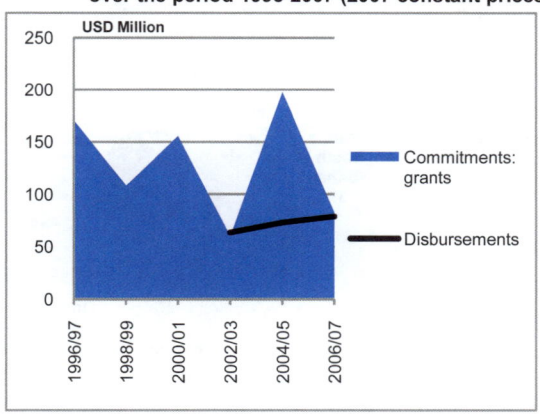

Trend of aid to agriculture over the period 1995-2007 (2007 constant prices)

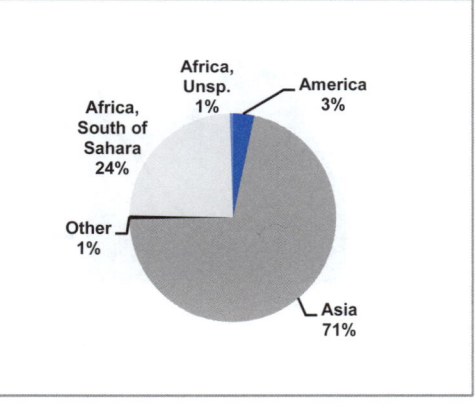

Regional distribution of aid to agriculture

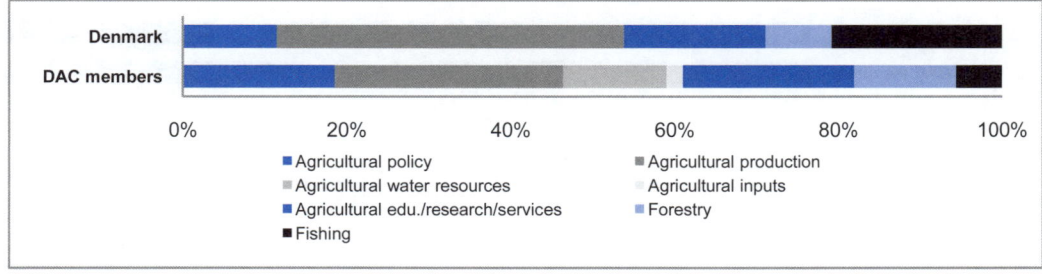

Agriculture aid by subsector

* % of sector allocable aid

EU institutions
Aid at a glance – Agriculture

Summary statistics
If not otherwise stated, figures refer to 2006-2007 annual average commitments expressed in 2007 constant prices.

	Total aid to agriculture		
	USD million	Aid to agriculture by EU inst. as a share of total aid by EU inst.*	Aid to agriculture by EU inst. as a share of total DAC members' aid to agriculture
EU institutions	444.3	4%	11%
For reference, total DAC	*4213*	*6%*	*100%*

Top ten recipients of aid to agriculture			
		Aid to agriculture by EU inst. to that recipient as a share of	
	USD million	total aid by EU inst. to that recipient*	total DAC members' aid to agriculture to that recipient
Mozambique	31.9	32%	45%
Guyana	22.5	63%	98%
Croatia	21.7	13%	89%
Cote d'Ivoire	20.9	76%	77%
Jamaica	16.7	45%	99%
Afghanistan	15.6	9%	7%
Swaziland	13.5	75%	95%
Bolivia	11.7	27%	14%
Mauritius	11.2	37%	60%
Mali	10.3	42%	7%

Aid to all agriculture-related sectors	
	USD million
Agriculture/Forestry/Fishing	444.3
Agro-industries	12.6
Forest-industries	0.0
Rural development	208.6
Food aid	369.8
Emergency food aid	274.3
Total	*1309.6*

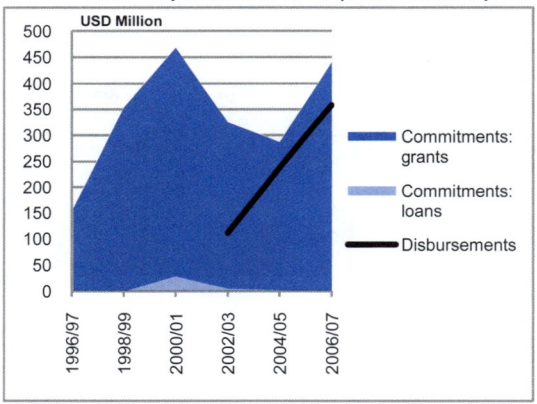

Trend of aid to agriculture over the period 1995-2007 (2007 constant prices)

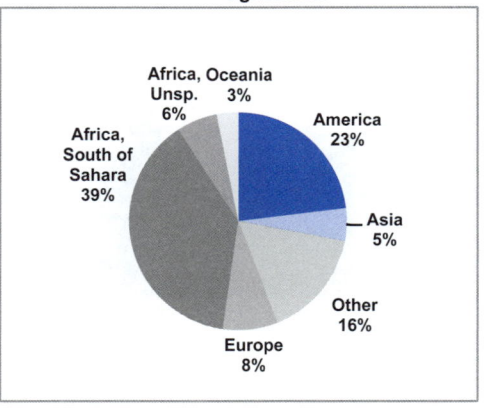

Regional distribution of aid to agriculture

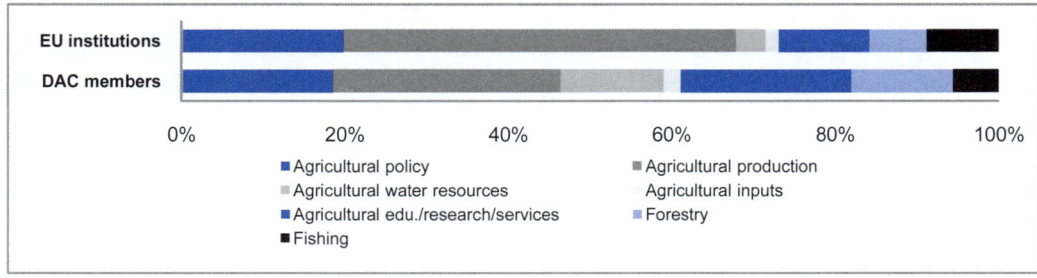

Agriculture aid by subsector

* % of sector allocable aid

FINLAND
Aid at a glance - Agriculture

Summary statistics
If not otherwise stated, figures refer to 2006-2007 annual average commitments expressed in 2007 constant prices.

	Total aid to agriculture		
	USD million	Aid to agriculture by Finland as a share of total aid by Finland*	Aid to agriculture by Finland as a share of total DAC members' aid to agriculture
Finland	34.1	8%	1%
For reference, total DAC	*4213*	*6%*	*100%*

Top ten recipients of aid to agriculture			
		Aid to agriculture by Finland to that recipient as a share of	
	USD million	total aid by Finland to that recipient*	total DAC members' aid to agriculture to that recipient
Viet Nam	8.7	24%	6%
Tanzania	8.1	27%	29%
China	5.7	58%	4%
Laos	1.6	57%	4%
Nicaragua	1.2	12%	4%
Indonesia	0.5	33%	1%
Nepal	0.3	2%	3%
Korea, Dem. Rep.	0.3	46%	9%
Tajikistan	0.3	22%	2%
Zimbabwe	0.2	51%	1%

Aid to all agriculture-related sectors	
	USD million
Agriculture/Forestry/Fishing	34.1
Agro-industries	0.1
Forest-industries	0.1
Rural development	18.3
Food aid	0.0
Emergency food aid	13.0
Total	65.7

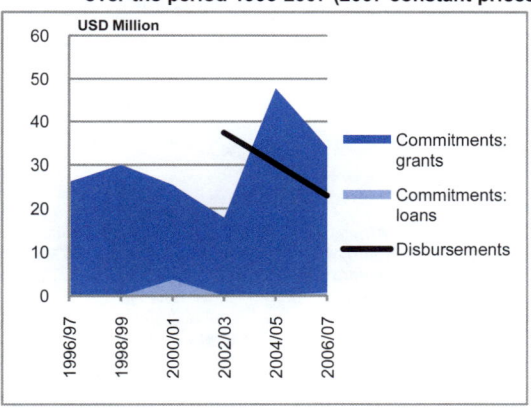

Trend of aid to agriculture over the period 1995-2007 (2007 constant prices)

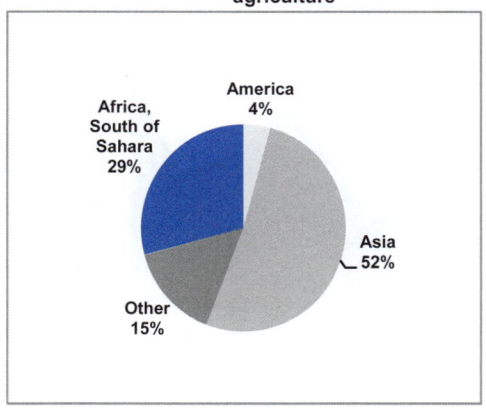

Regional distribution of aid to agriculture

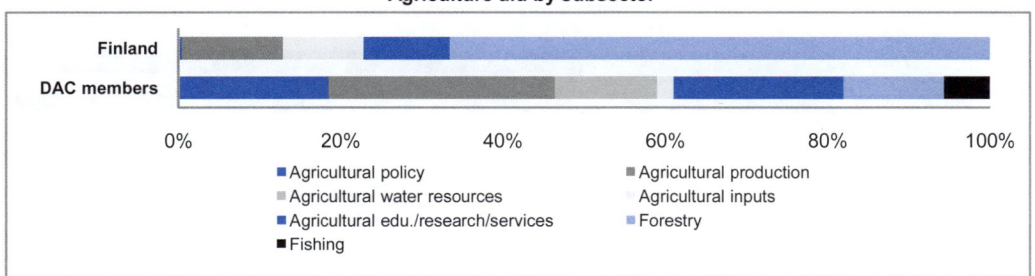

Agriculture aid by subsector

* % of sector allocable aid

FRANCE
Aid at a glance - Agriculture

Summary statistics
If not otherwise stated, figures refer to 2006-2007 annual average commitments expressed in 2007 constant prices.

	Total aid to agriculture		
	USD million	Aid to agriculture by France as a share of total aid by France*	Aid to agriculture by France as a share of total DAC members' aid to agriculture
France	**450.5**	**9%**	**11%**
For reference, total DAC	*4213*	*6%*	*100%*

Top ten recipients of aid to agriculture			
		Aid to agriculture by France to that recipient as a share of	
	USD million	total aid by France to that recipient*	total DAC members' aid to agriculture to that recipient
Viet Nam	58.7	13%	40%
Brazil	36.7	44%	65%
Burkina Faso	30.0	32%	56%
Madagascar	29.0	24%	58%
Ghana	27.7	52%	17%
Cameroon	26.1	24%	55%
Senegal	24.9	10%	52%
Mali	18.3	22%	12%
Thailand	14.9	23%	69%
Indonesia	13.5	52%	20%

Aid to all agriculture-related sectors	
	USD million
Agriculture/Forestry/Fishing	450.5
Agro-industries	0.0
Forest-industries	0.0
Rural development	29.0
Food aid	39.9
Emergency food aid	3.7
Total	523.1

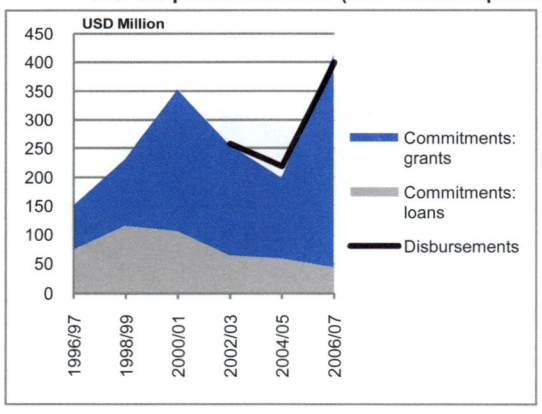
Trend of aid to agriculture over the period 1995-2007 (2007 constant prices)

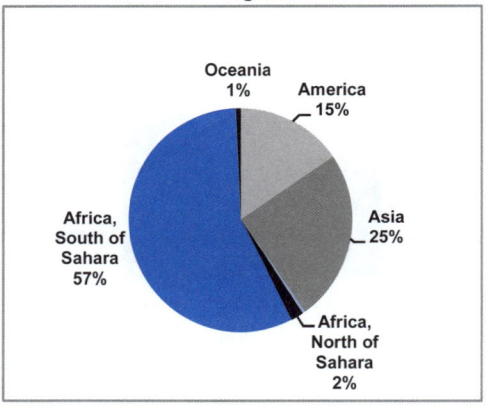
Regional distribution of aid to agriculture

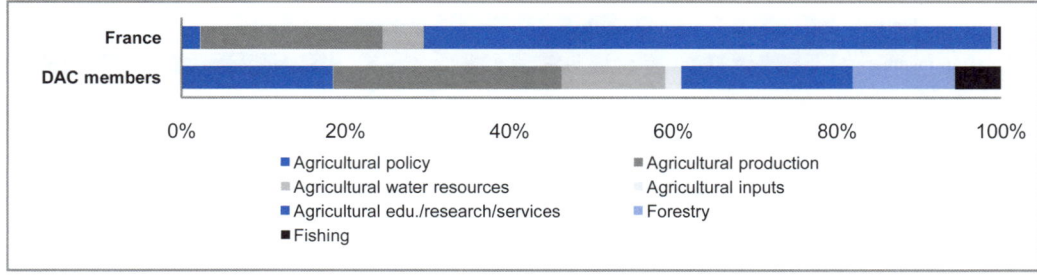
Agriculture aid by subsector

* % of sector allocable aid

GERMANY
Aid at a glance - Agriculture

Summary statistics
If not otherwise stated, figures refer to 2006-2007 annual average commitments expressed in 2007 constant prices.

	Total aid to agriculture		
	USD million	Aid to agriculture by Germany as a share of total aid by Germany*	Aid to agriculture by Germany as a share of total DAC members' aid to agriculture
Germany	317.2	5%	8%
For reference, total DAC	4213	6%	100%

Top ten recipients of aid to agriculture			
		Aid to agriculture by Germany to that recipient as a share of	
	USD million	total aid by Germany to that recipient*	total DAC members' aid to agriculture to that recipient
Egypt	86.8	51%	81%
India	18.4	5%	6%
Bolivia	12.1	24%	15%
Tunisia	11.7	25%	29%
Ghana	11.6	35%	7%
Colombia	11.0	27%	13%
Viet Nam	10.1	12%	7%
China	8.2	1%	6%
Mali	8.1	19%	6%
Peru	8.0	17%	8%

Aid to all agriculture-related sectors	
	USD million
Agriculture/Forestry/Fishing	317.2
Agro-industries	9.9
Forest-industries	0.0
Rural development	138.5
Food aid	31.8
Emergency food aid	70.6
Total	568.0

Trend of aid to agriculture over the period 1995-2007 (2007 constant prices)

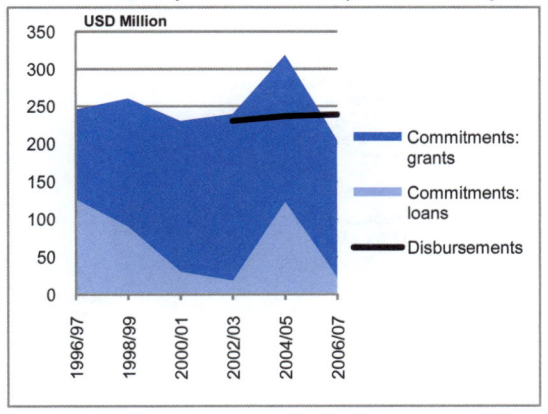

Regional distribution of aid to agriculture

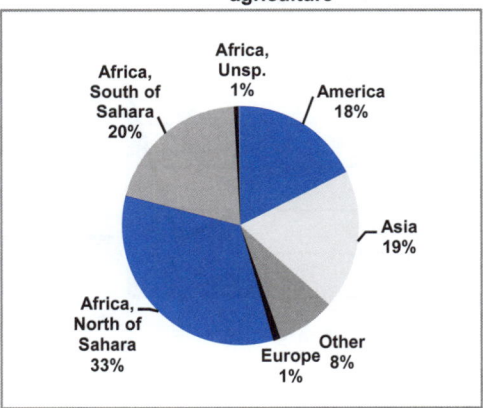

Agriculture aid by subsector

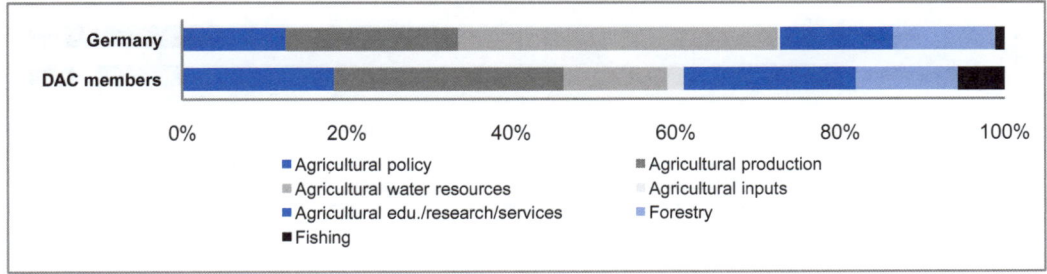

* % of sector allocable aid

GREECE
Aid at a glance - Agriculture

Summary statistics
If not otherwise stated, figures refer to 2006-2007 annual average commitments expressed in 2007 constant prices.

	Total aid to agriculture		
	USD million	Aid to agriculture by Greece as a share of total aid by Greece*	Aid to agriculture by Greece as a share of total DAC members' aid to agriculture
Greece	**4.9**	**3%**	**0%**
For reference, total DAC	4213	6%	100%

Top ten recipients of aid to agriculture			
		Aid to agriculture by Greece to that recipient as a share of	
	USD million	total aid by Greece to that recipient*	total DAC members' aid to agriculture to that recipient
Armenia	1.5	44%	2%
Sri Lanka	1.4	45%	12%
Albania	0.7	3%	14%
Mauritius	0.4	96%	2%
Cote d'Ivoire	0.2	72%	1%
Georgia	0.2	7%	4%
Lebanon	0.1	5%	3%
Congo, Dem. Rep.	0.1	8%	0%
Moldova	0.0	1%	0%
Syria	0.0	1%	1%

Aid to all agriculture-related sectors	
	USD million
Agriculture/Forestry/Fishing	4.9
Agro-industries	1.5
Forest-industries	0.5
Rural development	0.0
Food aid	0.4
Emergency food aid	3.8
Total	11.1

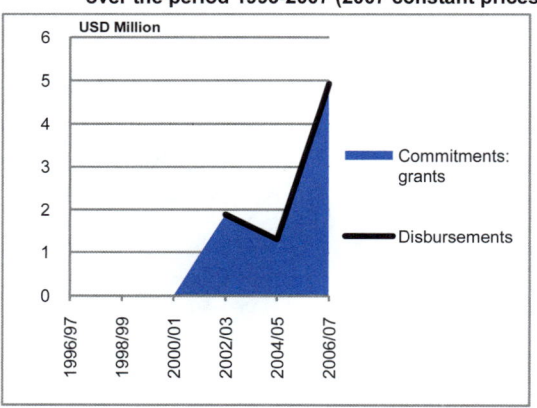

Trend of aid to agriculture over the period 1995-2007 (2007 constant prices)

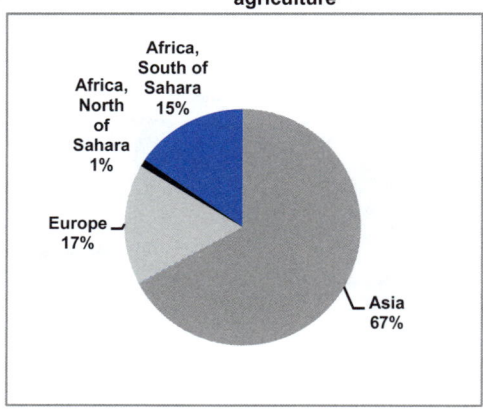

Regional distribution of aid to agriculture

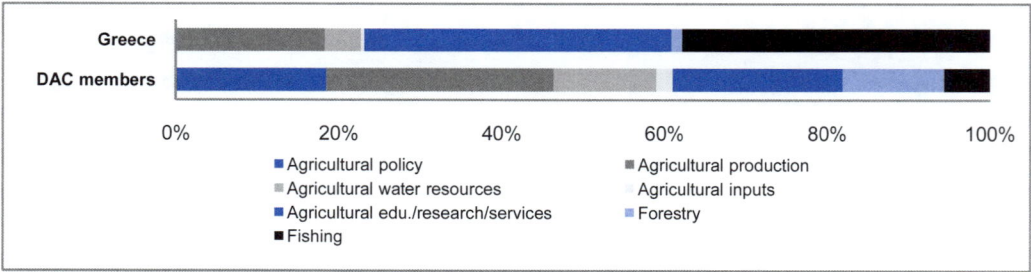

Agriculture aid by subsector

* % of sector allocable aid

IRELAND
Aid at a glance - Agriculture

Summary statistics
If not otherwise stated, figures refer to 2006-2007 annual average commitments expressed in 2007 constant prices.

	Total aid to agriculture		
	USD million	Aid to agriculture by Ireland as a share of total aid by Ireland*	Aid to agriculture by Ireland as a share of total DAC members' aid to agriculture
Ireland	32.8	7%	1%
For reference, total DAC	4213	6%	100%

Top ten recipients of aid to agriculture			
		Aid to agriculture by Ireland to that recipient as a share of	
	USD million	total aid by Ireland to that recipient*	total DAC members' aid to agriculture to that recipient
Mozambique	4.3	9%	6%
Ethiopia	4.0	10%	9%
Tanzania	3.6	12%	13%
Malawi	2.2	35%	6%
Uganda	1.6	3%	6%
Sudan	1.6	20%	12%
Cambodia	1.4	34%	5%
Kenya	1.2	12%	2%
Eritrea	1.2	70%	24%
Angola	1.1	54%	10%

Aid to all agriculture-related sectors	
	USD million
Agriculture/Forestry/Fishing	32.8
Agro-industries	0.0
Forest-industries	0.0
Rural development	9.0
Food aid	12.1
Emergency food aid	5.7
Total	59.6

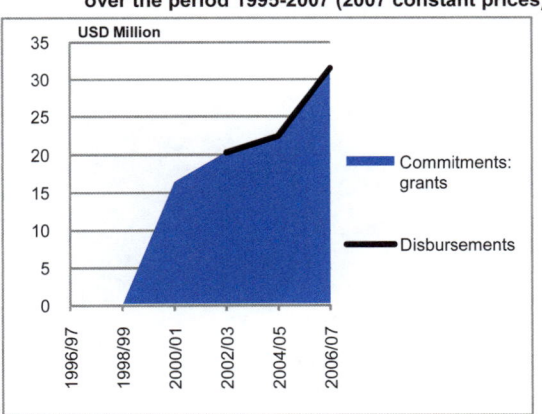
Trend of aid to agriculture over the period 1995-2007 (2007 constant prices)

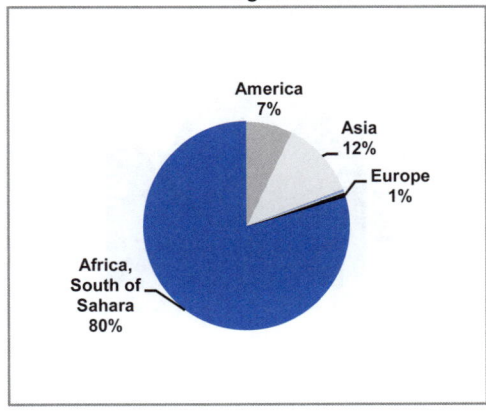
Regional distribution of aid to agriculture

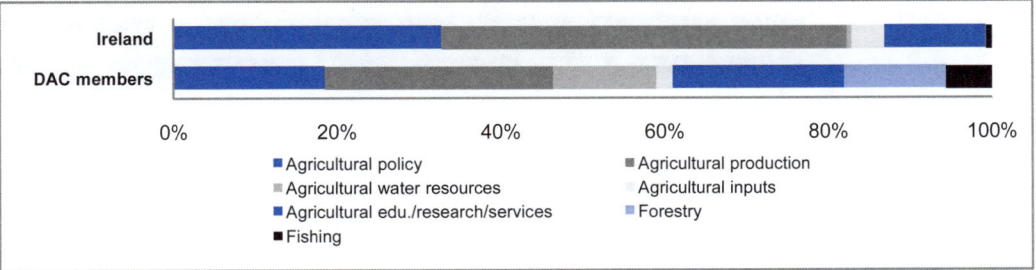
Agriculture aid by subsector

* % of sector allocable aid

ITALY
Aid at a glance - Agriculture

Summary statistics
If not otherwise stated, figures refer to 2006-2007 annual average commitments expressed in 2007 constant prices.

	Total aid to agriculture		
	USD million	Aid to agriculture by Italy as a share of total aid by Italy*	Aid to agriculture by Italy as a share of total DAC members' aid to agriculture
Italy	44.8	6%	1%
For reference, total DAC	4213	6%	100%

Top ten recipients of aid to agriculture			
		Aid to agriculture by Italy to that recipient as a share of	
	USD million	total aid by Italy to that recipient*	total DAC members' aid to agriculture to that recipient
China	7.8	6%	6%
Brazil	2.8	22%	5%
Niger	2.7	59%	13%
Mozambique	2.1	6%	3%
Tunisia	2.0	13%	5%
Libya	2.0	74%	98%
Rwanda	1.5	78%	9%
Lebanon	1.5	6%	34%
Bosnia-Herzegovina	1.4	28%	14%
Palestinian Adm. Areas	1.3	16%	7%

Aid to all agriculture-related sectors	
	USD million
Agriculture/Forestry/Fishing	44.8
Agro-industries	0.2
Forest-industries	0.0
Rural development	8.7
Food aid	8.4
Emergency food aid	8.1
Total	70.2

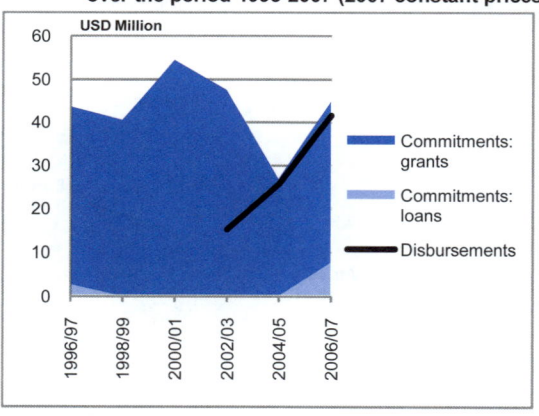
Trend of aid to agriculture over the period 1995-2007 (2007 constant prices)

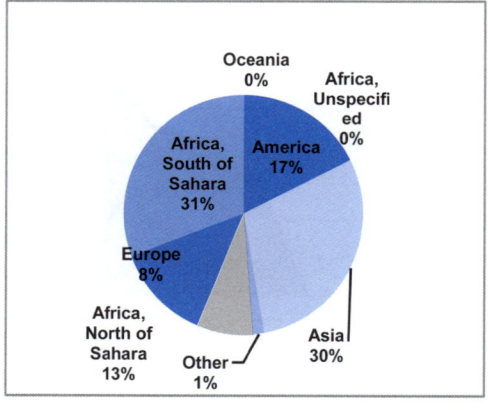
Regional distribution of aid to agriculture

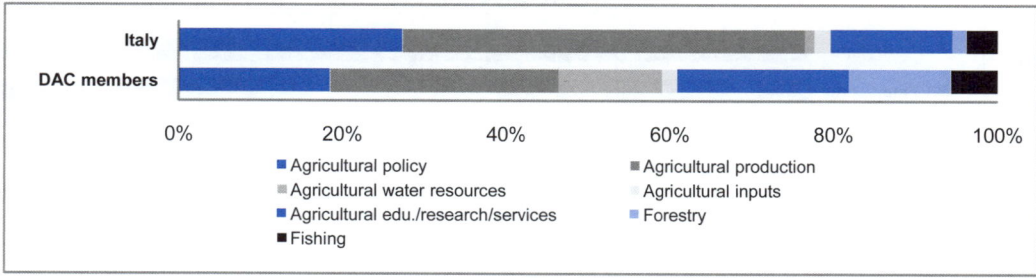
Agriculture aid by subsector

* % of sector allocable aid

JAPAN
Aid at a glance - Agriculture

Summary statistics
If not otherwise stated, figures refer to 2006-2007 annual average commitments expressed in 2007 constant prices.

Total aid to agriculture			
	USD million	Aid to agriculture by Japan as a share of total aid by Japan*	Aid to agriculture by Japan as a share of total DAC members' aid to agriculture
Japan	**821.0**	**10%**	**19%**
For reference, total DAC	4213	6%	100%

Top ten recipients of aid to agriculture			
		Aid to agriculture by Japan to that recipient as a share of	
	USD million	total aid by Japan to that recipient*	total DAC members' aid to agriculture to that recipient
India	284.0	17%	90%
China	107.7	10%	79%
Philippines	56.8	32%	83%
Peru	31.8	72%	32%
Viet Nam	29.5	3%	20%
Indonesia	23.6	3%	36%
Tunisia	23.6	24%	59%
Egypt	10.8	4%	10%
Malawi	7.9	29%	21%
Bolivia	7.6	26%	9%

Aid to all agriculture-related sectors	
	USD million
Agriculture/Forestry/Fishing	821.0
Agro-industries	4.6
Forest-industries	1.9
Rural development	121.7
Food aid	135.9
Emergency food aid	8.9
Total	1094.1

Trend of aid to agriculture over the period 1995-2007 (2007 constant prices)

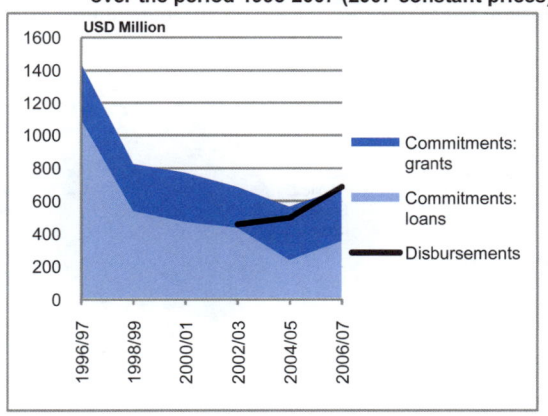

Regional distribution of aid to agriculture

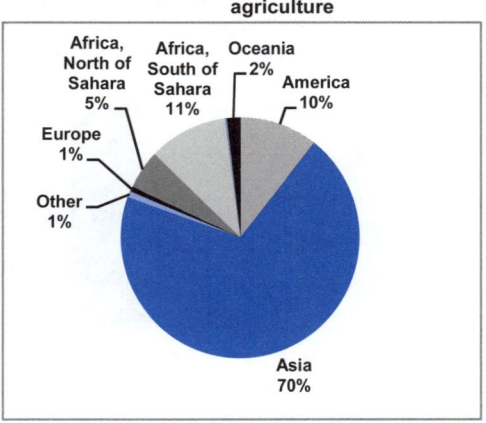

Agriculture aid by subsector

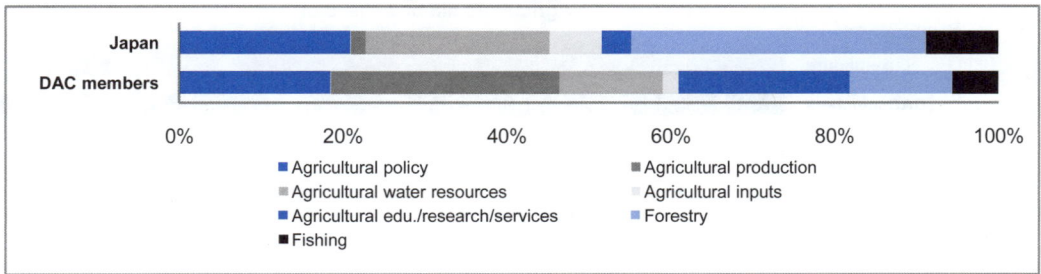

* % of sector allocable aid

LUXEMBOURG
Aid at a glance - Agriculture

Summary statistics
If not otherwise stated, figures refer to 2006-2007 annual average commitments expressed in 2007 constant prices.

	Total aid to agriculture		
	USD million	Aid to agriculture by Luxembourg as a share of total aid by Luxembourg*	Aid to agriculture by Luxembourg as a share of total DAC members' aid to agriculture
Luxembourg	9.1	6%	0%
For reference, total DAC	4213	6%	100%

Top ten recipients of aid to agriculture			
		Aid to agriculture by Luxembourg to that recipient as a share of	
	USD million	total aid by Luxembourg to that recipient*	total DAC members' aid to agriculture to that recipient
Rwanda	1.6	42%	9%
Mali	1.3	10%	1%
Serbia	1.0	17%	8%
Viet Nam	0.9	9%	1%
Burkina Faso	0.8	8%	2%
Peru	0.6	24%	1%
Montenegro	0.5	20%	29%
Laos	0.4	5%	1%
India	0.3	11%	0%
Namibia	0.2	3%	3%

Aid to all agriculture-related sectors	
	USD million
Agriculture/Forestry/Fishing	9.1
Agro-industries	1.1
Forest-industries	0.0
Rural development	11.3
Food aid	8.9
Emergency food aid	6.4
Total	36.8

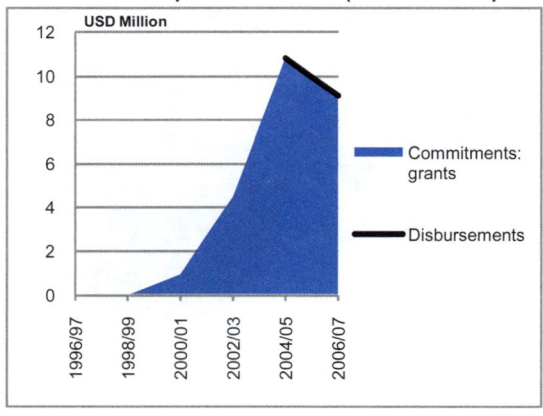

Trend of aid to agriculture over the period 1995-2007 (2007 constant prices)

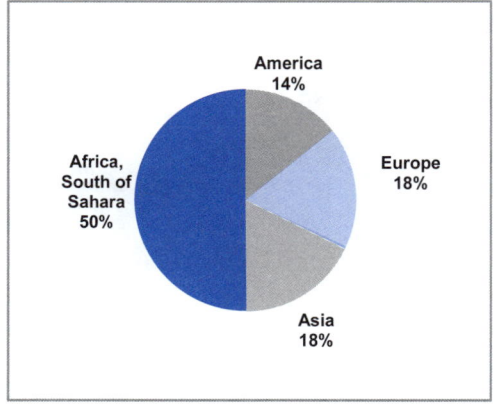

Regional distribution of aid to agriculture

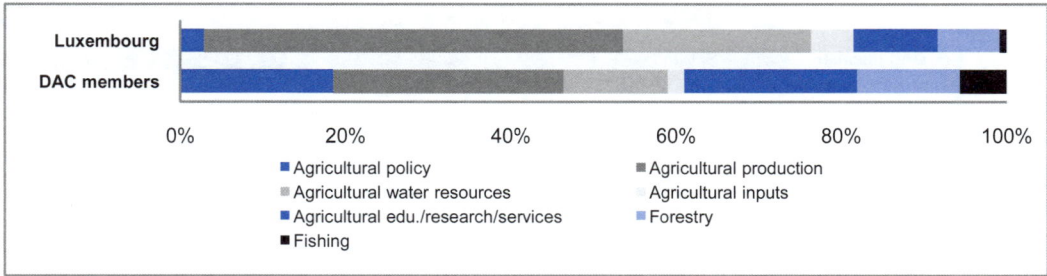

Agriculture aid by subsector

* % of sector allocable aid

NETHERLANDS
Aid at a glance - Agriculture

Summary statistics

If not otherwise stated, figures refer to 2006-2007 annual average commitments expressed in 2007 constant prices.

	Total aid to agriculture		
	USD million	Aid to agriculture by Netherlands as a share of total aid by Netherlands*	Aid to agriculture by Netherlands as a share of total DAC members' aid to agriculture
Netherlands	121.0	3%	3%
For reference, total DAC	4213	6%	100%

Top ten recipients of aid to agriculture			
		Aid to agriculture by Netherlands to that recipient as a share of	
	USD million	total aid by Netherlands to that recipient*	total DAC members' aid to agriculture to that recipient
Mali	8.5	11%	6%
Palestinian Adm. Areas	7.4	32%	40%
Bolivia	4.9	9%	6%
Benin	4.0	8%	13%
Afghanistan	3.6	3%	2%
Viet Nam	3.5	5%	2%
Mongolia	3.3	13%	22%
Egypt	3.2	10%	3%
Indonesia	3.1	2%	5%
Ghana	2.6	3%	2%

Aid to all agriculture-related sectors	
	USD million
Agriculture/Forestry/Fishing	121.0
Agro-industries	0.0
Forest-industries	0.0
Rural development	57.8
Food aid	1.2
Emergency food aid	35.4
Total	215.4

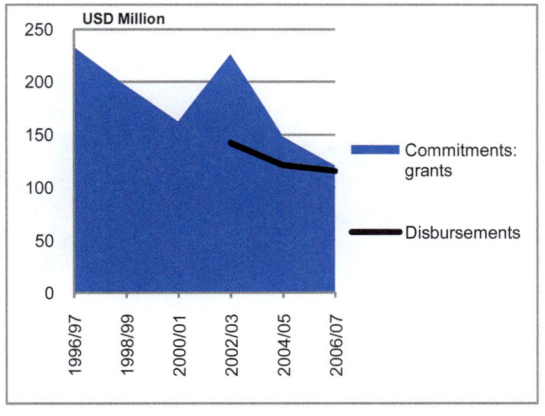

Trend of aid to agriculture over the period 1995-2007 (2007 constant prices)

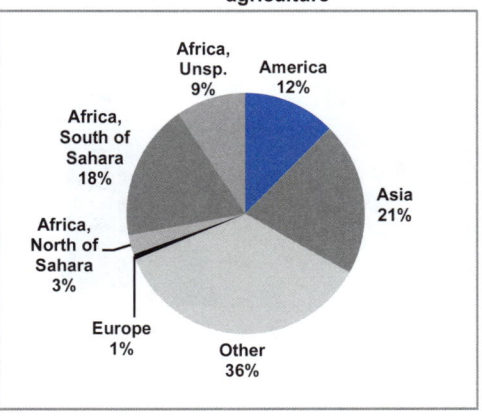

Regional distribution of aid to agriculture

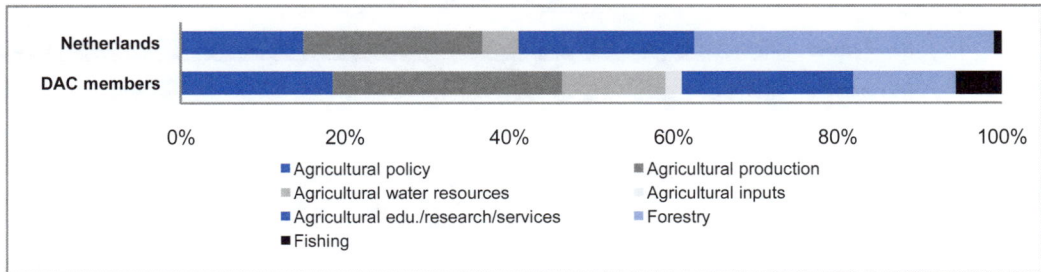

Agriculture aid by subsector

* % of sector allocable aid

NEW ZEALAND
Aid at a glance - Agriculture

Summary statistics
If not otherwise stated, figures refer to 2006-2007 annual average commitments expressed in 2007 constant prices.

	Total aid to agriculture		
	USD million	Aid to agriculture by New Zealand as a share of total aid by New Zealand*	Aid to agriculture by New Zealand as a share of total DAC members' aid to agriculture
New Zealand	11.0	5%	0%
For reference, total DAC	4213	6%	100%

Top ten recipients of aid to agriculture			
		Aid to agriculture by New Zealand to that recipient as a share of	
	USD million	total aid by New Zealand to that recipient*	total DAC members' aid to agriculture to that recipient
Papua New Guinea	3.1	13%	13%
Solomon Islands	1.9	5%	32%
Viet Nam	1.4	18%	1%
Indonesia	0.9	6%	1%
Philippines	0.3	10%	0%
China	0.3	38%	0%
Niue	0.3	4%	96%
Laos	0.2	11%	1%
Nicaragua	0.2	99%	1%
Nepal	0.2	26%	2%

Aid to all agriculture-related sectors	
	USD million
Agriculture/Forestry/Fishing	11.0
Agro-industries	0.1
Forest-industries	0.2
Rural development	5.6
Food aid	1.2
Emergency food aid	3.2
Total	21.3

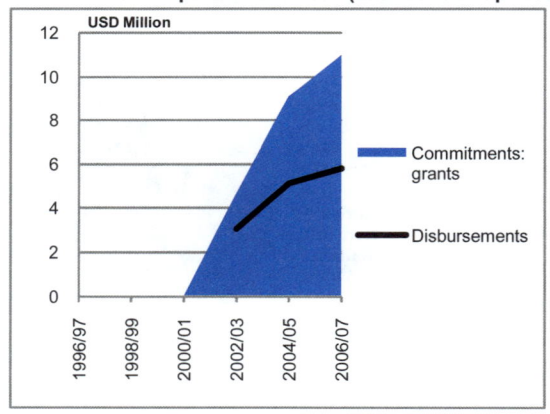
Trend of aid to agriculture over the period 1995-2007 (2007 constant prices)

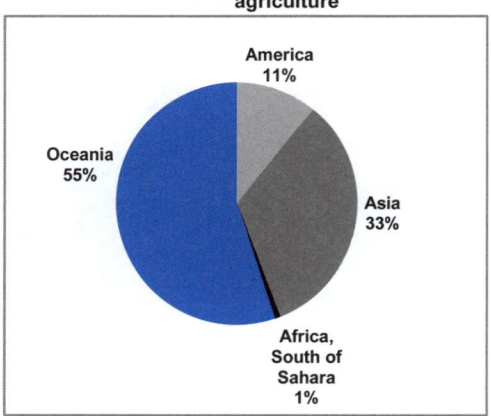
Regional distribution of aid to agriculture

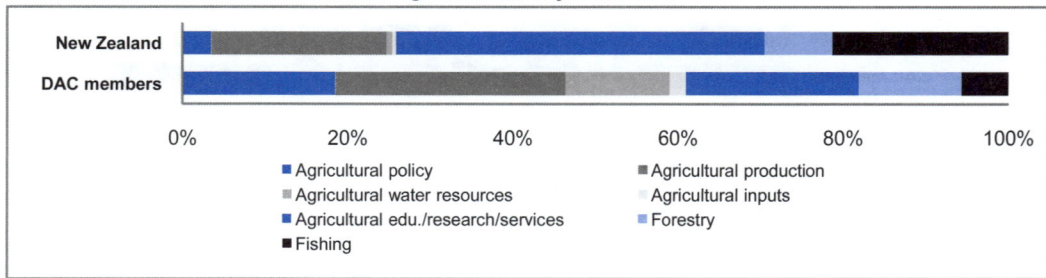
Agriculture aid by subsector

* % of sector allocable aid

NORWAY
Aid at a glance - Agriculture

Summary statistics
If not otherwise stated, figures refer to 2006-2007 annual average commitments expressed in 2007 constant prices.

	Total aid to agriculture		
	USD million	Aid to agriculture by Norway as a share of total aid by Norway*	Aid to agriculture by Norway as a share of total DAC members' aid to agriculture
Norway	**117.0**	**6%**	**3%**
For reference, total DAC	*4213*	*6%*	*100%*

Top ten recipients of aid to agriculture			
		Aid to agriculture by Norway to that recipient as a share of	
	USD million	total aid by Norway to that recipient*	total DAC members' aid to agriculture to that recipient
Zambia	17.4	47%	43%
Malawi	14.4	31%	38%
Sudan	4.9	13%	39%
Ethiopia	3.8	9%	8%
Nicaragua	3.6	28%	13%
Madagascar	3.4	20%	7%
Uganda	2.9	9%	11%
Serbia	2.5	7%	21%
Mozambique	2.2	4%	3%
Palestinian Adm. Areas	1.9	3%	10%

Aid to all agriculture-related sectors	
	USD million
Agriculture/Forestry/Fishing	117.0
Agro-industries	0.0
Forest-industries	0.0
Rural development	10.7
Food aid	4.1
Emergency food aid	44.0
Total	175.8

Trend of aid to agriculture over the period 1995-2007 (2007 constant prices)

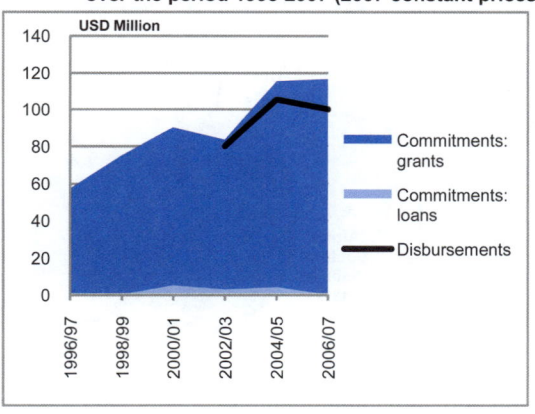

Regional distribution of aid to agriculture

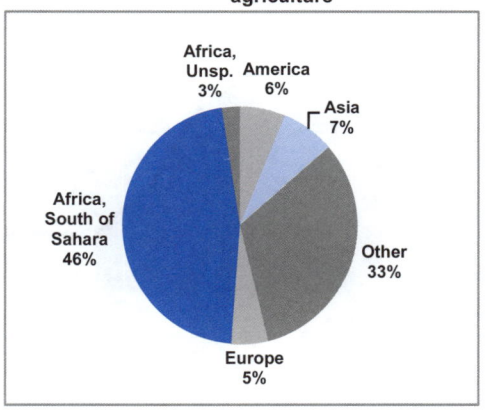

Agriculture aid by subsector

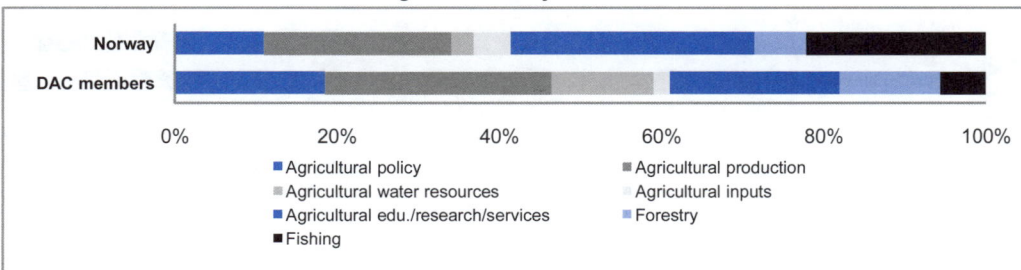

* % of sector allocable aid

PORTUGAL
Aid at a glance - Agriculture

Summary statistics
If not otherwise stated, figures refer to 2006-2007 annual average commitments expressed in 2007 constant prices.

	Total aid to agriculture		
	USD million	Aid to agriculture by Portugal as a share of total aid by Portugal*	Aid to agriculture by Portugal as a share of total DAC members' aid to agriculture
Portugal	1.9	1%	0%
For reference, total DAC	*4213*	*6%*	*100%*

Top recipients of aid to agriculture			
		Aid to agriculture by Portugal to that recipient as a share of	
	USD million	total aid by Portugal to that recipient*	total DAC members' aid to agriculture to that recipient
Timor-Leste	0.5	1%	7%
Angola	0.4	2%	4%
Cape Verde	0.1	0%	3%
Guinea-Bissau	0.1	1%	5%
Mozambique	0.1	0%	0%
Sao Tome & Principe	0.1	0%	25%

Aid to all agriculture-related sectors	
	USD million
Agriculture/Forestry/Fishing	1.9
Agro-industries	0.0
Forest-industries	0.0
Rural development	0.7
Food aid	0.0
Emergency food aid	0.0
Total	2.6

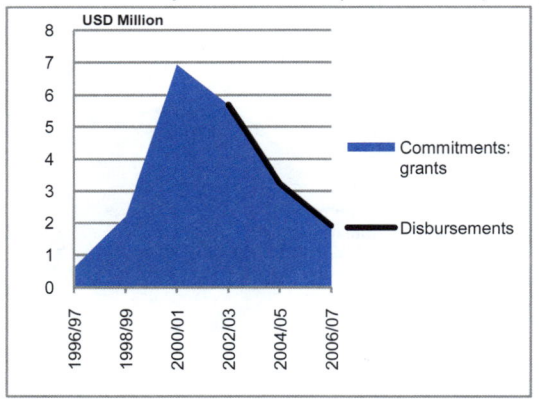

Trend of aid to agriculture over the period 1995-2007 (2007 constant prices)

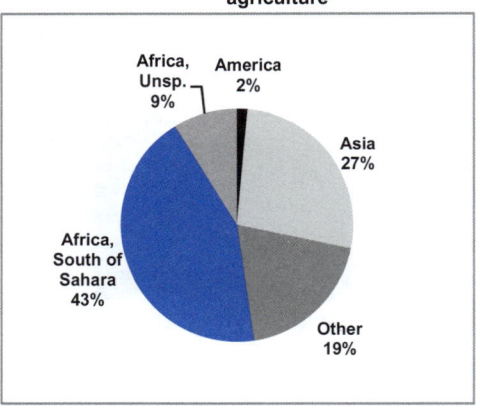

Regional distribution of aid to agriculture

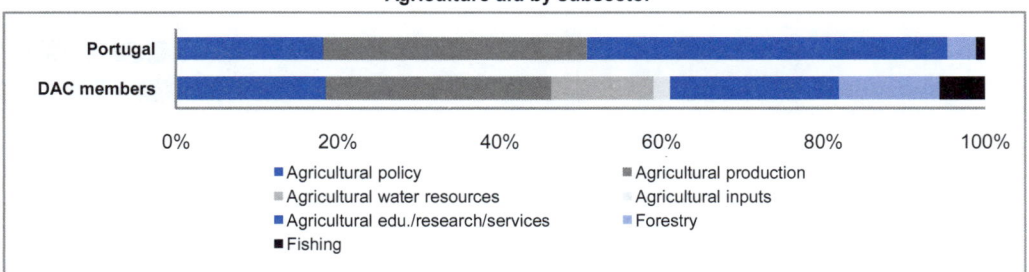

Agriculture aid by subsector

* % of sector allocable aid

SPAIN
Aid at a glance - Agriculture

Summary statistics

If not otherwise stated, figures refer to 2006-2007 annual average commitments expressed in 2007 constant prices.

	Total aid to agriculture		
	USD million	Aid to agriculture by Spain as a share of total aid by Spain*	Aid to agriculture by Spain as a share of total DAC members' aid to agriculture
Spain	163.3	7%	4%
For reference, total DAC	4213	6%	100%

Top ten recipients of aid to agriculture			
		Aid to agriculture by Spain to that recipient as a share of	
	USD million	total aid by Spain to that recipient*	total DAC members' aid to agriculture to that recipient
Nicaragua	11.3	14%	41%
Peru	7.8	10%	8%
Bolivia	5.7	10%	7%
Morocco	5.5	4%	17%
Ecuador	5.1	8%	16%
Ghana	4.8	35%	3%
Cameroon	4.5	26%	9%
Cuba	3.6	19%	78%
Senegal	3.5	13%	7%
Dominican Republic	3.3	14%	37%

Aid to all agriculture-related sectors	
	USD million
Agriculture/Forestry/Fishing	163.3
Agro-industries	3.7
Forest-industries	0.1
Rural development	44.1
Food aid	41.5
Emergency food aid	17.7
Total	270.3

Trend of aid to agriculture over the period 1995-2007 (2007 constant prices)

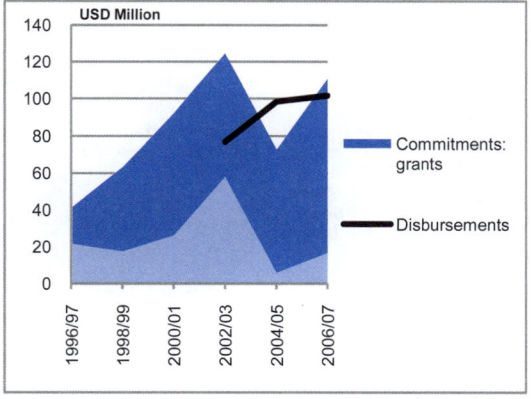

Regional distribution of aid to agriculture

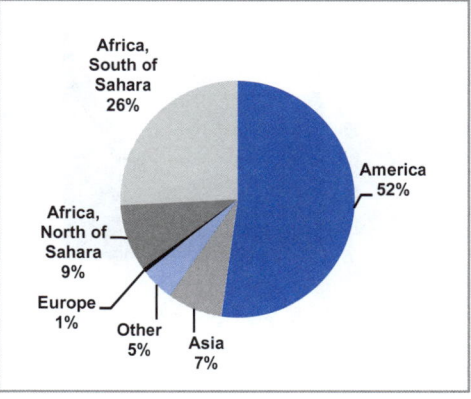

Agriculture aid by subsector

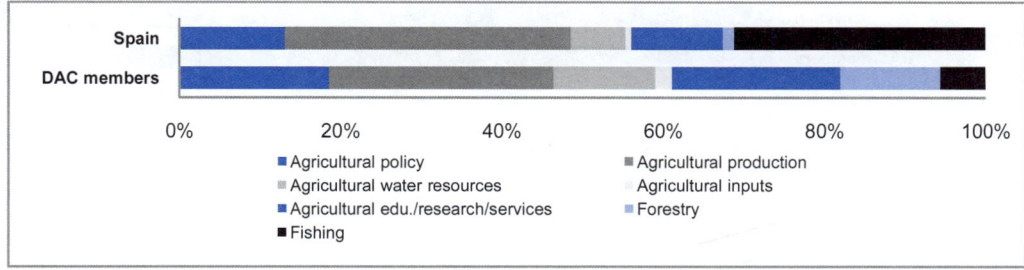

* % of sector allocable aid

SWEDEN
Aid at a glance - Agriculture

Summary statistics
If not otherwise stated, figures refer to 2006-2007 annual average commitments expressed in 2007 constant prices.

	Total aid to agriculture		
	USD million	Aid to agriculture by Sweden as a share of total aid by Sweden*	Aid to agriculture by Sweden as a share of total DAC members' aid to agriculture
Sweden	143.6	8%	3%
For reference, total DAC	4213	6%	100%

Top ten recipients of aid to agriculture			
		Aid to agriculture by Sweden to that recipient as a share of	
	USD million	total aid by Sweden to that recipient*	total DAC members' aid to agriculture to that recipient
Kenya	32.2	49%	45%
Laos	13.9	71%	36%
Mozambique	6.6	7%	9%
Tajikistan	5.4	32%	47%
Bosnia-Herzegovina	4.9	11%	48%
Moldova	4.6	27%	40%
Ethiopia	3.5	21%	8%
Kyrgyz Republic	3.2	24%	34%
Macedonia, FYR	1.9	21%	48%
Croatia	1.3	39%	5%

Aid to all agriculture-related sectors	
	USD million
Agriculture/Forestry/Fishing	143.6
Agro-industries	0.0
Forest-industries	0.0
Rural development	13.2
Food aid	6.9
Emergency food aid	1.1
Total	164.7

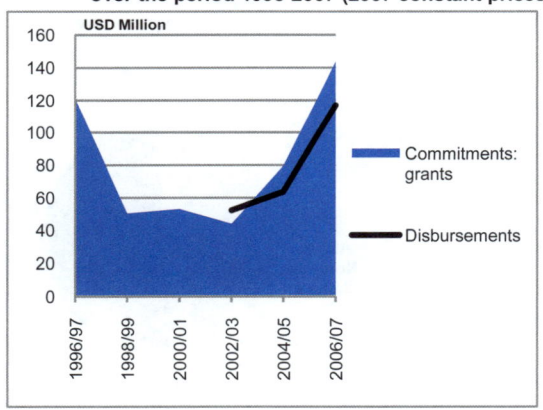

Trend of aid to agriculture over the period 1995-2007 (2007 constant prices)

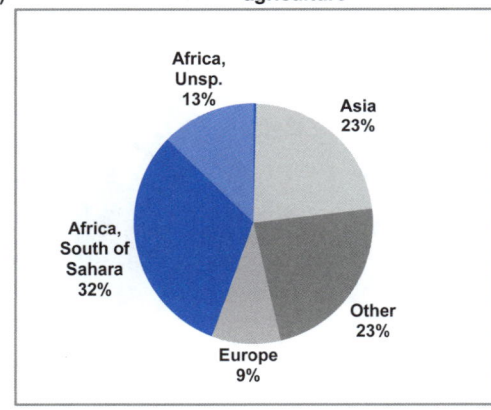

Regional distribution of aid to agriculture

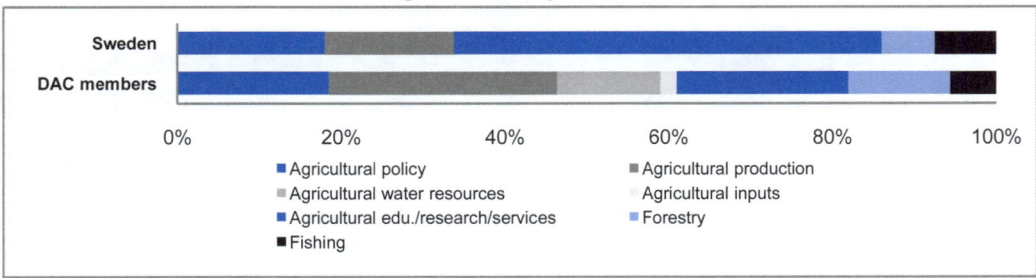

Agriculture aid by subsector

* % of sector allocable aid

SWITZERLAND
Aid at a glance - Agriculture

Summary statistics
If not otherwise stated, figures refer to 2006-2007 annual average commitments expressed in 2007 constant prices.

Total aid to agriculture

	USD million	Aid to agriculture by Switzerland as a share of total aid by Switzerland*	Aid to agriculture by Switzerland as a share of total DAC members' aid to agriculture
Switzerland	68.7	9%	2%
For reference, total DAC	4213	6%	100%

Top ten recipients of aid to agriculture

		Aid to agriculture by Switzerland to that recipient as a share of	
	USD million	total aid by Switzerland to that recipient*	total DAC members' aid to agriculture to that recipient
Madagascar	6.3	79%	13%
Viet Nam	4.5	19%	3%
Nepal	3.3	17%	34%
Laos	3.2	50%	8%
Korea, Dem. Rep.	2.9	65%	88%
India	2.7	23%	1%
Kyrgyz Republic	2.6	22%	28%
Mongolia	1.9	24%	13%
Bhutan	1.6	57%	30%
Serbia	1.4	3%	11%

Aid to all agriculture-related sectors

	USD million
Agriculture/Forestry/Fishing	68.7
Agro-industries	0.0
Forest-industries	0.0
Rural development	28.7
Food aid	0.0
Emergency food aid	58.3
Total	155.7

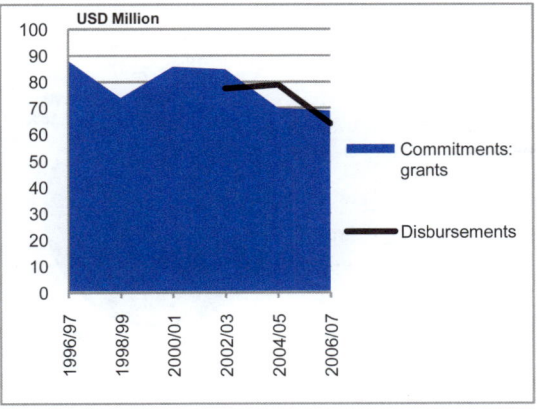
Trend of aid to agriculture over the period 1995-2007 (2007 constant prices)

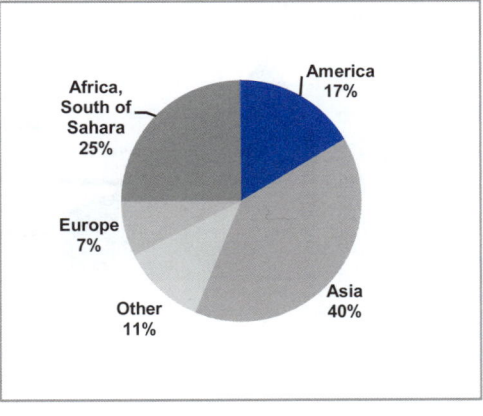
Regional distribution of aid to agriculture

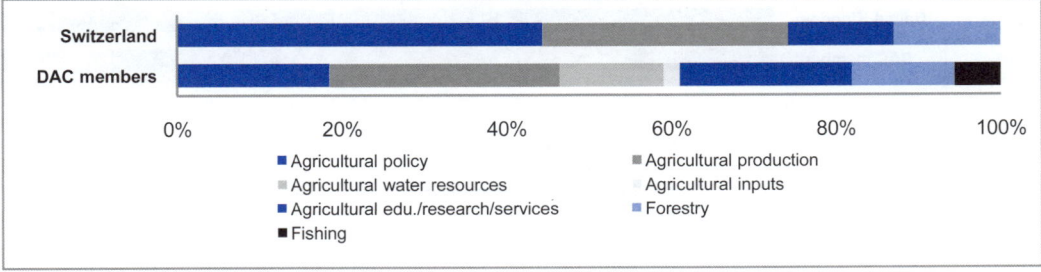
Agriculture aid by subsector

* % of sector allocable aid

UNITED KINGDOM
Aid at a glance - Agriculture

Summary statistics
If not otherwise stated, figures refer to 2006-2007 annual average commitments expressed in 2007 constant prices.

	Total aid to agriculture		
	USD million	Aid to agriculture by United Kingdom as a share of total aid by United Kingdom*	Aid to agriculture by United Kingdom as a share of total DAC members' aid to agriculture
United Kingdom	113.3	2%	3%
For reference, total DAC	*4213*	*6%*	*100%*

Top ten recipients of aid to agriculture			
		Aid to agriculture by United Kingdom to that recipient as a share of	
	USD million	total aid by United Kingdom to that recipient*	total DAC members' aid to agriculture to that recipient
Cambodia	9.3	33%	33%
Ghana	9.0	4%	6%
Afghanistan	7.0	5%	3%
Cameroon	6.7	95%	14%
Mauritius	6.5	99%	35%
Zambia	2.8	26%	7%
Kenya	2.1	1%	3%
Mexico	2.0	66%	22%
Rwanda	2.0	4%	12%
Congo, Dem. Rep.	0.9	1%	5%

Aid to all agriculture-related sectors	
	USD million
Agriculture/Forestry/Fishing	113.3
Agro-industries	0.1
Forest-industries	0.0
Rural development	209.7
Food aid	53.1
Emergency food aid	62.6
Total	*438.9*

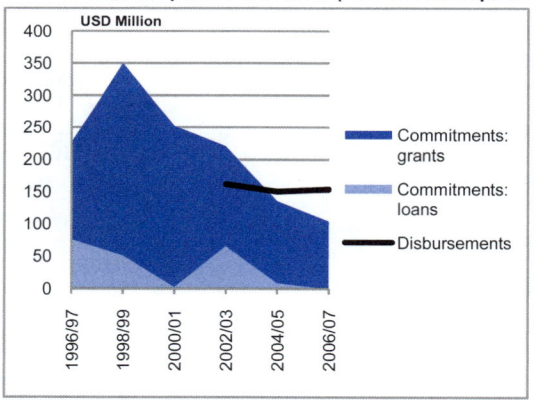

Trend of aid to agriculture over the period 1995-2007 (2007 constant prices)

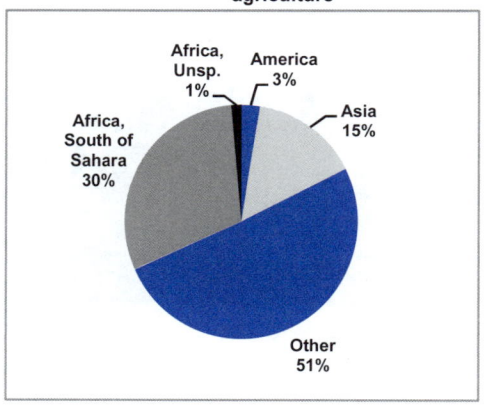

Regional distribution of aid to agriculture

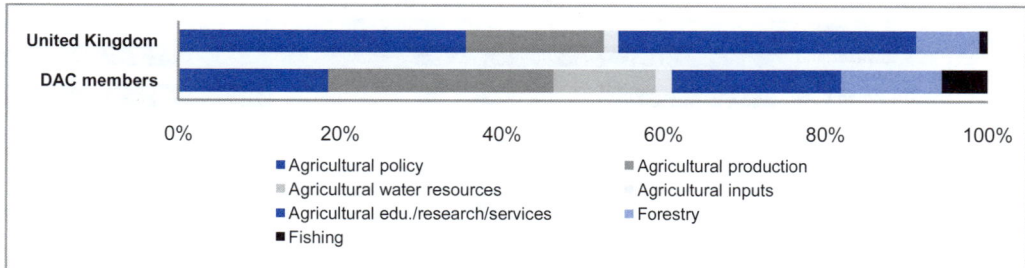

Agriculture aid by subsector

* % of sector allocable aid

UNITED STATES
Aid at a glance - Agriculture

Summary statistics
If not otherwise stated, figures refer to 2006-2007 annual average commitments expressed in 2007 constant prices.

	Total aid to agriculture		
	USD million	Aid to agriculture by United States as a share of total aid by United States*	Aid to agriculture by United States as a share of total DAC members' aid to agriculture
United States	932.0	5%	22%
For reference, total DAC	*4213*	*6%*	*100%*

Top ten recipients of aid to agriculture			
		Aid to agriculture by United States to that recipient as a share of	
	USD million	total aid by United States to that recipient*	total DAC members' aid to agriculture to that recipient
Afghanistan	146.7	9%	70%
Iraq	120.3	3%	99%
Ghana	95.9	32%	59%
Armenia	80.6	42%	95%
Mali	80.3	31%	54%
Colombia	60.3	7%	71%
Peru	45.5	19%	46%
Bolivia	32.5	18%	39%
Mozambique	14.8	9%	21%
Pakistan	12.2	3%	57%

Aid to all agriculture-related sectors	
	USD million
Agriculture/Forestry/Fishing	932.0
Agro-industries	5.9
Forest-industries	0.4
Rural development	55.0
Food aid	647.7
Emergency food aid	1240.7
Total	2881.7

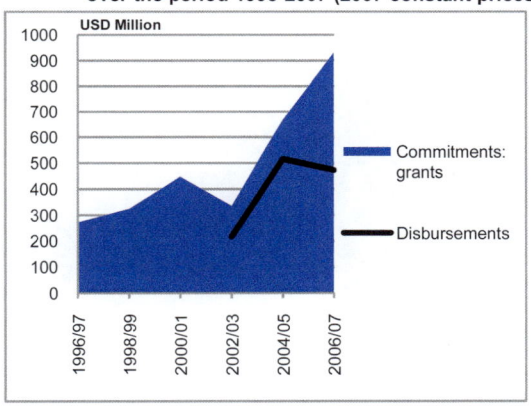

Trend of aid to agriculture over the period 1995-2007 (2007 constant prices)

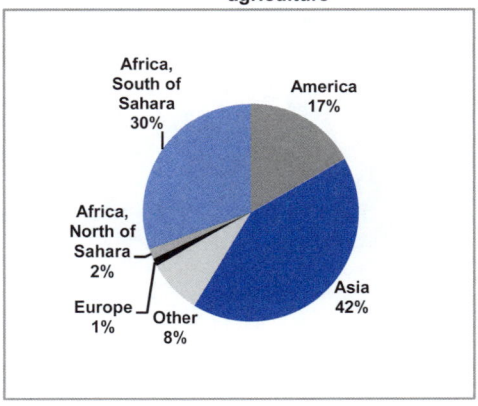

Regional distribution of aid to agriculture

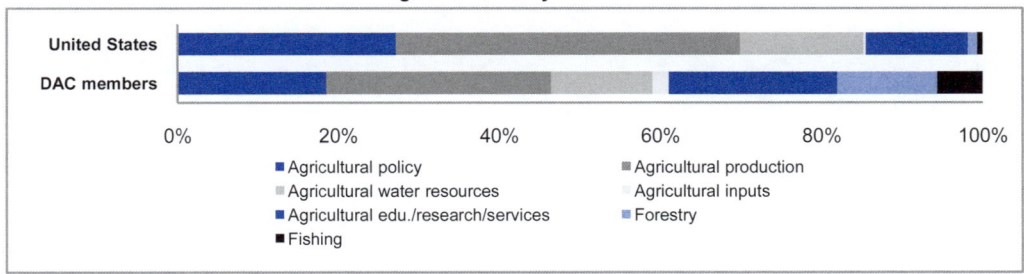

Agriculture aid by subsector

* % of sector allocable aid

PART 3

List of Aid Activities

2007

www.oecd.org/dac/stats/idsonline

(The list of aid activities to the agricultural sector that follows includes transactions over USD 5000)

PARTIE 3

Liste des activités d'aide

2007

www.oecd.org/dac/stats/idsonline

(Cette liste d'activités d'aide dans le secteur de l'agriculture inclue des transactions de montant supérieurs à USD 5000)

AID ACTIVITIES IN SUPPORT OF AGRICULTURE

2007 Commitments, USD thousand

Source: Creditor Reporting system (CRS)

RECIPIENT / Donor	Agency	Sector	Amount USD thousand	Project description	CRS ID Number
Afghanistan					
Australia	AusAID	31182	139	WHEAT AND MAIZE PRODUCTIVITY IMPROVEMENT	2007000840
Canada	CIDA	31161	13963	HORTICULTURE AND LIVESTOCK PROJECT(ARTF)	070059
Canada	CIDA	31161	465	HORTICULTURE AND LIVESTOCK PROJECT(ARTF) - MONITORING EVALUATION	070060
France	AFD	31110	835	MISE A DISPOSITION ASSISTANCE TECHNIQUE	2007152200
Germany	Fed Min	31110	548	MANAGING BIODIVERSITY FOR SUSTAINABLE FOOD SECURITY AND NUTRITION	2007010312
Germany	BMZ	31162	1027	REFORESTATION OF PISTACHIO WOODS IN TAKHAR PROVINCE	2007009533
Germany	Fed Min	31163	486	DEVELOPMENT OF INTEGRATED DAIRY SCHEMES IN AFGHANISTAN	2007010299
Germany	BMZ	31166	70	AGRICULTURAL EXTENSION	2007006713
Germany	Fed Min	31220	876	INITIATING PARTICIPATORY FORESTRY IN SUPPORT TO SUSTAINABLE LIVELIHOOD	2007010314
IDA		31110	1280	AVIAN FLU	2007000736
IDA		31140	25000	IRRIGATION	2007000730
Ireland	DFA	31130	437	SUPPORT TO NGOS - IRISH - MAPS - GENERAL - REFORESTATION & WATERSHED	2007000229
Ireland	DFA	31161	79	FOOD CROP PRODUCTION	2007001291
Japan	MAFF	31110	1505	TC AGGREGATED ACTIVITIES	070009T
Japan	JICA	31110	2253	TC AGGREGATED ACTIVITIES	073062T
Japan	JICA	31130	120	TC AGGREGATED ACTIVITIES	073105T
Japan	JICA	31163	36	TC AGGREGATED ACTIVITIES	073216T
Netherlands	MFA	31110	49	AGRICULTURAL POLICY & ADMIN. MGMT	2007100153
Netherlands	MFA	31161	811	FOOD CROP PRODUCTION	2007000658
Netherlands	MFA	31165	5257	AGRICULTURAL ALTERNATIVE DEVELOPMENT	2007000543
Norway	NORAD	31220	184	FORESTRY DEVELOPMENT	2007002182
UNDP		31130	175	AGRICULTURAL LAND RESOURCES	2007012845
United Kingdom	DFID	31120	14008	HORTICULTURE & LIVESTOCK PROGRAMME	2007000396
United States	AID	31110	263	AGRICULTURAL POLICY & ADMIN. MGMT	2007001664
United States	AID	31110	2519	PROGRAM SUPPORT (AGRICULTURE)	2007013923
United States	AID	31110	8485	AGRICULTURAL ENABLING ENVIRONMENT	2007001658
United States	DOD	31110	2507	COMMANDER'S EMERGENCY RESPONSE PROGRAM (CERP): AGRICULTURE	2007024322
United States	AID	31120	40762	AGRICULTURAL SECTOR PRODUCTIVITY	2007001781
United States	DOD	31140	6681	COMMANDER'S EMERGENCY RESPONSE PROGRAM (CERP): IRRIGATION	2007024330
United States	AID	31165	224380	ALTERNATIVE DEVELOPMENT AND ALTERNATIVE LIVELIHOODS	2007001907

Total aid for water in 2007 for Afghanistan USD thousand   **355201**
And as a share of aid to total recipient countries   **5.29%**

Africa, regional

AfDF		31162	15302	PROJET APPUI FILLIERE COTON - B- FASO	070054
AfDF		31162	7651	PROJET APPUI FILLIERE COTON -TCHAD	070062
AfDF		31162	12242	PROJET APPUI FILLIERE COTON - BENIN	070053
AfDF		31162	3060	PROJET APPUI FILLIERE COTON -UEMOA	070061
AfDF		31162	15302	PROJET APPUI FILLIERE COTON - MALI	070055
Canada	IDRC	31110	47	INNOVATION AS KEY TO THE GREEN REVOLUTION IN AFRICA	070434i

RECIPIENT Donor	Agency	Sector	Amount USD thousand	Project description	CRS ID Number
Canada	IDRC	31110	55	INTERNATIONAL CONFERENCE OF THE AFRICAN ASSOCIATION OF AGR. ECONOMISTS	070690i
Canada	CIDA	31130	71	ARUSHA, TANZANIA / 2007.09.17-21	070300
Canada	IDRC	31130	431	SéCURISATION FONCIÜRE, AMéLIORATION DE LA SéCURITé ALIMENTAIRE	070659i
Canada	IDRC	31130	11	GENDER AND TENURE: PREPARATORY WORK	070561i
Canada	IDRC	31161	28	ATELIER REGIONAL: UNE APPROCHE INTERSECTORIELLE POUR L'ALIMENTATION	070738i
Canada	CIDA	31182	2420	FORUM FOR AGRICULTURAL RESEARCH AFRICA 2	070625
Canada	IDRC	31182	74	REGIONAL IMPACT ASSESSMENT OF THE AFRICAN HIGHLANDS INITIATIVE	070416i
Canada	IDRC	31182	100	IMPACTING AGRICULTURAL RESEARCH IN AFRICA: THE CGIAR REGIONAL PLAN FOR	070692i
Canada	IDRC	31182	8	DEVELOPING A STRATEGIC ENTRY POINT TO AGRA: CONCEPT NOTE DEVELOPMENT	070647i
Canada	IDRC	31210	22	ATELIER RéGIONAL: OPTIONS POLITIQUES ET DE RECHERCHES AGRO-FORESTIÜRES	070677i
France	MISC	31182	137	RECHERCHE AGRICULTURE	2007800361
Germany	BMZ	31110	2738	PAN-AFRICAN REFORM PROCESSES IN AGRIBUSINESS OPERATIONS (NEPAD/CAADP)	2007005918
Germany	DEG	31162	587	EQUITY INVESTMENT	2007001544
Japan	JICA	31110	232	TC AGGREGATED ACTIVITIES	073040T
Japan	MAFF	31181	871	AGRICULTURAL EDUCATION/TRAINING	072161
Japan	MAFF	31194	179	TC AGGREGATED ACTIVITIES	070026T
Norway	NORAD	31110	56	AGRICULTURAL POLICY & ADMIN. MGMT	2007004776
Norway	NORAD	31130	211	CONTRIBUTING TO IMPROVED FOOD SECURITY FOR VULNERABLE HOUSEHOLDS	2007004751
Norway	MFA	31164	512	PHASE 2 REGIONAL COOPERATION LAND REFORM POLICY	2007004746
Norway	MFA	31320	1161	PHASE II OF THE ESTABL. OF AN INFOPÜCHE UNIT IN SOUTHERN A.– INFOSA	2007004855
Portugal	ICP	31182	131	DEVELOPMENT OF INNOVATIVE AND HEALTHFUL MARAMA BEAN (TYLOSEMA ESCULETUM)	070823
Portugal	ICP	31192	112	LISTYPES - DIGITILIZING AFRICAN TYPES AT LISBON HERBARIA (LISC AND LISU)	070591
Sweden	Sida	31110	46	AGRICULTURAL POLICY & ADMIN. MGMT	2007001451
Sweden	Sida	31120	543	AGRICULTURAL DEVELOPMENT	2007002523
Sweden	Sida	31164	3994	273 LAND ADMIN SADC 273 LAND USE SADC	2007005584
Sweden	Sida	31181	29	AGRICULTURAL EDUCATION/TRAINING	2007001420
Sweden	Sida	31182	873	AGRICULTURAL RESEARCH	2007005233
Sweden	Sida	31282	548	FORESTRY RESEARCH	2007005124
Sweden	Sida	31382	6807	FISHERY RESEARCH	2007005264
Switzerland	SDC	31110	83	AGRICULTURAL POLICY & ADMIN. MGMT	2007000170
Switzerland	SDC	31220	161	INTERSESSIONAL ON DECENTRALISATION	2003005532
United Kingdom	DFID	31110	997	FOOD SOVEREIGNTY PROGRAMME CSCF 395	2007000074
United Kingdom	DFID	31110	420	COMPREHENSIVE AFRICA AGRICULTURE DEVELOPMENT PROGRAMME	2007000047

Total aid for water in 2007 for Africa, regional **USD thousand 78251**
And as a share of aid to total recipient countries **1.17%**

Albania

	Agency	Sector	Amount	Project description	CRS ID
Austria	ADA	31120	220	BUSINESS PARTNERSHIP PROGRAM:	2007500110c
Austria	ADA	31181	9	CONTRACTED SERVICES FOR AGRICULTURAL SCHOOL	2007500900
Austria	MISC	31210	12	SMALL-SCALE COMMITMENTS AGGREGATED BY SECTOR AND RECIPIENT COUNTRY	2007980036
Germany	BMZ	31120	246	AGRICULTURAL DEVELOPMENT	2007006540
Germany	BMZ	31166	136	AGRICULTURAL EXTENSION	2007006715
Greece	YPEJ	31140	249	PARTIAL RECONSTRUCTION OF THE IRRIGATION CANAL THOMAS PHILIPEOS	2007000320
Italy	DGCS	31120	373	AGRICULTURAL DEVELOPMENT	070251
Italy	DGCS	31191	7	AGRICULTURAL SERVICES	060871

RECIPIENT Donor	Agency	Sector	Amount USD thousand	Project description	CRS ID Number
Italy	DGCS	31194	21	AGRICULTURAL CO-OPERATIVES	060963
Italy	DGCS	31320	6	FISHERY DEVELOPMENT	060550
Japan	JICA	31110	55	TC AGGREGATED ACTIVITIES	072904T
Spain	MFA	31110	205	DEV. SUPPORT FOR ASSOCIATIONS AND AGRICULTURAL COOPERATIVES IN ALBANIA	075855
Sweden	Sida	31210	562	FORESTRY POLICY & ADMIN. MANAGEMENT	2007002401
Switzerland	SDC	31110	618	PROMOTION LOW-INPUT AGRICULTURE	2001002665
Switzerland	SDC	31110	33	AGRICULTURAL POLICY & ADMIN. MGMT	2001002666
United States	AID	31110	50	AGRICULTURAL POLICY & ADMIN. MGMT	2007013815
United States	TDA	31110	49	AGRICULTURAL POLICY & ADMIN. MGMT	2007028203
United States	AID	31120	3513	AGRICULTURAL SECTOR PRODUCTIVITY	2007001702
United States	AID	31120	50	AGRICULTURAL DEVELOPMENT	2007001705

Total aid for water in 2007 for Albania **USD thousand 6414**
And as a share of aid to total recipient countries **0.1%**

Algeria

Canada	IDRC	31163	362	LA GESTION DURABLE DES PARCOURS STEPPIQUES: APPRENTISSAGE PARTICIPATIF	070558i
France	MISC	31110	13	POLITIQUE AGRICOLE & GESTION ADMINISTRATIVE	2007800766
Italy	DGCS	31120	1588	AGRICULTURAL DEVELOPMENT	070440
Japan	JICA	31110	83	TC AGGREGATED ACTIVITIES	072907T
Japan	JICA	31310	81	TC AGGREGATED ACTIVITIES	073542T
Korea	KOICA	31161	1800	THE PROJECT FOR IMPROVEMENT OF AGRICULTURAL CROP PRODUCTION TECHNIQUES I	2007017561
Spain	MFA	31110	271	VALUE ENHANCEMENT STRATEGY FOR LOCAL KNOWLEDGE OF VITIVINₑCOLA OF WILAYA	076291
Spain	MFA	31120	110	INTRODUCTION OF A RURAL DEVELOPMENT MODEL	074730
Spain	MFA	31120	55	SUPPORT IN THE IDENTIFICATION OF DEVELOPMENT PROJECTS IN THE OLIVE PRODUCTION	073446
Spain	MFA	31120	41	AGRICULTURAL DEVELOPMENT	072967
Spain	MFA	31140	845	TEST LABORATORY FOR TECHNICAL CHARACTERISTICS OF IRRIGATION EQUIPMENT	077687
Spain	AG	31194	82	ORIENTATION AND SUPPORT FOR THE STARTING UP OF A PILOT AGRICULTURAL COOP	074270
Spain	MFA	31195	47	VIGILENCE AND CONTROL SYSTEMS OF THE HIGHLY INFECTIOUS BIRD FLU	073192
Spain	MFA	31210	34	FORESTRY POLICY & ADMIN. MANAGEMENT	072777
Spain	MFA	31320	82	SUPPORT FOR THE MPRH IN THE DEVELOPMENT PROJECTS IN THE FISHING INDUSTRY	074218
Spain	MFA	31381	205	SUPPORT TO THE MPRH IN THE IMPLEMENTATION OF THE TRAINING PROGRAM	075856

Total aid for water in 2007 for Algeria **USD thousand 5699**
And as a share of aid to total recipient countries **0.08%**

America, regional

Canada	IDRC	31181	27	PERMANENT SEMINAR ON AGRICULTURAL RESEARCH (SEPIA XII)	070664i
Canada	CIDA	31281	90	PANAMA, PANAMA / 2008.02.25-29	070747
Japan	JICA	31110	84	TC AGGREGATED ACTIVITIES	072972T
Japan	JICA	31195	588	TC AGGREGATED ACTIVITIES	073267T
Japan	JICA	31210	31	TC AGGREGATED ACTIVITIES	073412T
Japan	JICA	31310	39	TC AGGREGATED ACTIVITIES	073567T
Spain	MFA	31110	1837	TRAINING IN ECONOMICS AND AGRICULTURAL POLICIES FOR RURAL DEVELOPMENT	077933
Spain	AGR	31120	138	COLLABORATION WITH THE MINISTRY OF AGRICULTURE FISHERY AND NUTRITION	075226
Spain	MFA	31120	356	COLLABORATION WITH THE MINISTRY OF AGRICULTURE FISHERY AND NUTRITION	076831
Spain	AG	31161	90	PROMOTION OF FAMILY PRODUCTION SYSTEMS AND TECHNOLOGY TRANSFER	074408

RECIPIENT Donor	Agency	Sector	Amount USD thousand	Project description	CRS ID Number
Spain	AG	31161	322	STRENGTHENING SMALL COFFEE PRODUCERS ORGANISATIONS IN MIDDLE AMERICA	076621
Spain	MFA	31162	963	SUPPORT PLAN TO THE COFFEE PRODUCERS IN CENTRAL AMERICA	077740
Spain	AGR	31181	45	COURSE ABOUT IRRIGATION TECHNIQUES AND THE MANAGEMENT OF IRRIGATED LAND	073147
Spain	MFA	31210	72	COLLABORATION WITH THE INIA OF THE MINISTRY OF EDUCATION AND SCIENCE	074010
Spain	AGR	31320	82	LATIN AMERICA CONGRESS WITH REGARD TO HEALTH AND HYGIENE IN AQUACULTURE	074221
Spain	EMP	31320	178	COMMERCIALIZATION AND INCOME STRATEGIES FOR LATINAMERICAN FISH IN THE EUROPE	075619
Spain	EMP	31320	21	SEMINAR ON MARITIME SUSTAINABILITY	071884
Spain	EMP	31320	21	CAMPAIGN ON THE ANALYSIS OF TRADITIONAL FISHING	071885

Total aid for water in 2007 for America, regional   **USD thousand 4985**
And as a share of aid to total recipient countries   **0.07%**

Angola

Finland	MFA	31120	47	AGRICULTURE DEVELOPMENT	2007000275
Germany	BMZ	31120	516	AGRICULTURAL DEVELOPMENT	2007006322
IFAD		31120	8200	MARKET-ORIENTED SMALLHOLDER AGRICULTURE PROJECT	070009
Ireland	DFA	31120	423	SUPPORT TO NGOS - IRISH - MAPS - GENERAL - REDUCTION OF VULNERABILITY	2007000224
Ireland	DFA	31161	526	SUPPORT TO NGOS - IRISH - MAPS - GENERAL - SUPPORT TO INCREASE AGRIC.	2007000235
Italy	CA	31120	603	AGRICULTURAL DEVELOPMENT	071302
Italy	LA	31166	24	AGRICULTURAL EXTENSION	071041
Japan	JICA	31110	26	TC AGGREGATED ACTIVITIES	072913T
Korea	Kexim	31163	49000	POULTRY PROCESSING PROJECT	2007001414
Norway	NORAD	31310	85	MID-TERM REVIEW OF FISHERY COLLABORATION ANGOLA AGO-2528	2007002219
Norway	NORAD	31320	580	ANALYSIS OF OIL-FISH SAMPLES	2007002213
Norway	MFA	31320	22	FISHERY DEVELOPMENT	2007002570
Portugal	ICP	31110	60	AGRICULTURAL POLICY & ADMIN. MGMT	070657
Portugal	ICP	31110	330	SUPPORT TO NATIONAL PLAN OF RURAL DEVELOPMENT	070658
Portugal	ICP	31130	102	AGRICULTURAL LAND RESOURCES	070594
Portugal	ICP	31182	42	AGRICULTURAL RESEARCH	070818
Portugal	ICP	31195	55	LIVESTOCK/VETERINARY SERVICES	070851
Portugal	ICP	31282	45	FORESTRY RESEARCH	070009
Spain	MFA	31161	411	FOOD INCOME GENERATION GUARANTEES FOR THE RIGHT TO ADEQUATE HOUSING	077026
Spain	MFA	31181	548	TRAINING PROJECT OF COMMUNITIES AND SMALL FISHERIES ABOUT FOOD SECURITY	077404
Spain	MFA	31195	560	RENOVATION OF THE CAMPUS INFRASTRUCTURE AND FORESTRY . STO. ANTONIO	077430
Spain	Misc	31282	11	HIGHER EDUCATION	071235
United States	AID	31110	59	AGRICULTURAL POLICY & ADMIN. MGMT	2007013843

Total aid for water in 2007 for Angola   **USD thousand 62273**
And as a share of aid to total recipient countries   **0.93%**

Antigua and Barbuda

Japan	MAFF	31310	32	TC AGGREGATED ACTIVITIES	070038T
Japan	JICA	31310	123	TC AGGREGATED ACTIVITIES	073562T

Total aid for water in 2007 for Antigua and Barbuda   **USD thousand 155**
And as a share of aid to total recipient countries   **0%**

Donor	Agency	Sector	Amount USD thousand	Project description	CRS ID Number
RECIPIENT					

Argentina

Donor	Agency	Sector	Amount USD thousand	Project description	CRS ID Number
Canada	IDRC	31110	106	CAPACITY FOR INNOVATION, INCREASING PRODUCTIVITY AND ACCESS TO MARKETS	070449i
France	MISC	31110	11	POLITIQUE AGRICOLE & GESTION ADMINISTRATIVE	2007800775
France	MISC	31182	2752	RECHERCHE AGRICULTURE	2007800370
Germany	BMZ	31120	2539	INTEGRATED AGRICULTURAL DEVELOPMENT	2007006251
Germany	BMZ	31120	356	INVESTMENT FUND FOR RURAL AND INDIGENOUS ECONOMIC ACTIVITIES	2007006145
Germany	L G	31182	17	AGRICULTURAL RESEARCH	2007011306
Germany	BMZ	31210	43	FORESTRY POLICY & ADMIN. MANAGEMENT	2007008782
Italy	DGCS	31120	6	AGRICULTURAL DEVELOPMENT	060895
Italy	DGCS	31163	679	LIVESTOCK	070636
Italy	DGCS	31166	785	AGRICULTURAL EXTENSION	070682
Japan	JICA	31110	196	TC AGGREGATED ACTIVITIES	072962T
Japan	PRF	31161	18	TC AGGREGATED ACTIVITIES	075594T
Japan	JICA	31163	16	TC AGGREGATED ACTIVITIES	073208T
Japan	PRF	31181	12	TC AGGREGATED ACTIVITIES	075600T
Japan	JICA	31195	223	TC AGGREGATED ACTIVITIES	073261T
Japan	JICA	31210	801	TC AGGREGATED ACTIVITIES	073398T
Japan	OFCF	31310	8803	FISHING POLICY AND ADMIN. MANAGEMENT	070901
Japan	JICA	31310	94	TC AGGREGATED ACTIVITIES	073564T
Spain	MFA	31110	79	CONSERVATION OF LANDS AND WATERS IN THE BOLIVIAN ALTIPLANO	074146
Spain	MFA	31120	130	CREATION OF A COOPERATIVE OF ARTESIAN OFFICES	075011
Spain	MFA	31120	274	EMPLOYMENT POLICY AND ADMIN. MGMT.	076321
Spain	MFA	31120	137	FIGHT AGAINST POVERTY AND EMPLOYMENT CREATION THROUGH COMPREHENSIVE DEV.	075135
Spain	MFA	31120	41	STRENGTHENING THE SMALL COMMON PRODUCERS	072958
Spain	MFA	31161	137	SOCIAL AND LABOUR INSERTION OF SMALL PRODUCERS OF TYPICAL LOCAL AGRIC.	075124
Spain	AG	31161	104	INTEGRATED STRUCTURE FOR THE CONTROLLED PRODUCTION OF REGIONAL PRODUCTS	074642
Spain	AG	31161	71	STRATEGIES & CAPACITIES IN THE PROD. CONSUMPTION & COMMERCIALIZ. PROCESSES	073995
Spain	AG	31163	41	IMPROVEMENT OF THE QUALITY OF LIFE OF INDIGENOUS FAMILIES (FARMS).	073005
Spain	MIE	31210	136	TOURISM PROGRAM IN S. MIGUEL TUCUMAN	075090
United States	AGR	31191	10	AGRICULTURAL SERVICES	2007000436
United States	IADF	31191	204	AGRICULTURAL SERVICES	2007025060

Total aid for water in 2007 for Argentina   **USD thousand 18822**
And as a share of aid to total recipient countries   **0.28%**

Armenia

Donor	Agency	Sector	Amount USD thousand	Project description	CRS ID Number
Germany	BMZ	31120	821	INTEGRATED COMMUNITY DEVELOPMENT PROGRAM CONTINUATION (2008-2010)	2007006671
Greece	YPESDDA	31120	212	AGRICULTURAL INFRASTRUCTURES: DEV./EXPANSION OF ELECTRICITY SUPPLY LINE	2007000904
Greece	YPEPU	31120	589	AGRICULTURAL DEVELOPMENT PROGRAM IN ARMENIA	2007000900
Greece	YPEJ	31195	2053	PROGRAM FINANCING OF THE CONSTRUCTION OF A SLAUGHTER HOUSE IN ARAGASTOTN	2007000846
IDA		31140	5000	IRRIGATION	2007000699
Japan	JICA	31110	106	TC AGGREGATED ACTIVITIES	072982T
Norway	MFA	31210	341	COMMUNITY FOREST MANAGEMENT	2007001964
Switzerland	SDC	31110	42	AGRICULTURAL POLICY & ADMIN. MGMT	2002002143
Switzerland	SDC	31110	667	COMMUNITY DEVELOPMENT	2002002142
Switzerland	SDC	31120	125	AGRICULTURAL DEVELOPMENT	2006003420

RECIPIENT Donor	Agency	Sector	Amount USD thousand	Project description	CRS ID Number
United States	AGR	31110	4862	CAUCASUS AGRICULTURAL DEVELOPMENT INITIATIVE	2007000444

Total aid for water in 2007 for Armenia — USD thousand 14818
And as a share of aid to total recipient countries — 0.22%

Asia, regional

Recipient	Agency	Sector	Amount	Project description	CRS ID
Canada	IDRC	31310	279	SUSTAINABLE AQUATIC RESOURCE MANAGEMENT INITIATIVE	070473i
Japan	MAFF	31110	190	AGRICULTURAL POLICY & ADMIN. MGMT	072169
Japan	MAFF	31110	202	TC AGGREGATED ACTIVITIES	070012T
Japan	MAFF	31120	1112	AGRICULTURAL DEVELOPMENT	072162
Japan	MAFF	31130	2280	TC AGGREGATED ACTIVITIES	070015T
Japan	MAFF	31140	216	AGRICULTURAL WATER RESOURCES	072163
Japan	MAFF	31140	518	TC AGGREGATED ACTIVITIES	070018T
Japan	JICA	31163	7	TC AGGREGATED ACTIVITIES	073234T
Japan	MOFA	31181	120	TC AGGREGATED ACTIVITIES	070381T
Japan	MAFF	31210	1090	TC AGGREGATED ACTIVITIES	070028T
Japan	MAFF	31210	496	FORESTRY POLICY & ADMIN. MANAGEMENT	072164
Japan	MAFF	31310	36	TC AGGREGATED ACTIVITIES	070044T
Norway	MFA	31192	512	IPM ADDENDUM	2007004797
Norway	MFA	31210	427	TERESTRIAL MONITORING IN CENTRAL ASIA	2007001903
Switzerland	SDC	31182	4167	ICIMOD CORE CONTRIBUTION MOUNTAIN DEVELOPMENT	1983000140
United States	AGR	31181	15	AGRICULTURAL EDUCATION/TRAINING	2007000458
United States	STATE	31310	30	FISHING POLICY AND ADMIN. MANAGEMENT	2007027006

Total aid for water in 2007 for Asia, regional — USD thousand 11697
And as a share of aid to total recipient countries — 0.17%

Azerbaijan

Recipient	Agency	Sector	Amount	Project description	CRS ID
IFAD		31140	17196	RURAL DEVELOPMENT PROJECT FOR THE NORTH-WEST	070017
Japan	JICA	31110	16	TC AGGREGATED ACTIVITIES	072983T
Japan	JICA	31130	56	TC AGGREGATED ACTIVITIES	073101T
Korea	Misc	31181	124	AGRICULTURAL EDUCATION/TRAINING	2007002562
Norway	NORAD	31120	69	AGRICULTURAL DEVELOPMENT	2007003744
Switzerland	SDC	31110	60	AGRICULTURAL POLICY & ADMIN. MGMT	2000002151
Switzerland	SDC	31110	439	AGRICULTURE IDPS	2000002276
United States	AID	31110	160	AGRICULTURAL POLICY & ADMIN. MGMT	2007013943
United States	AID	31110	651	AGRICULTURAL ENABLING ENVIRONMENT	2007001665
United States	AID	31120	14	AGRICULTURAL DEVELOPMENT	2007001784

Total aid for water in 2007 for Azerbaijan — USD thousand 18784
And as a share of aid to total recipient countries — 0.28%

Bangladesh

Recipient	Agency	Sector	Amount	Project description	CRS ID
Australia	AusAID	31150	6	DEVELOPMENT OF CONSERVATION FARMING IMPLEMENTS FOR TWO-WHEELER TRACTORS	2007002130
Australia	AusAID	31161	9	SCOPING STUDY TO ASSESSE THE TECHNICAL & ECONOMIC FEASABILITY OF WHEAT PROD	2007002029
Australia	AusAID	31161	58	EXPANDING THE AREA FOR RABI-SEASON CROPPING IN SOUTHERN BANGLADESH	2007001766

RECIPIENT Donor	Agency	Sector	Amount USD thousand	Project description	CRS ID Number
Australia	AusAID	31161	210	FOOD CROP PRODUCTION	2007000978
Australia	AusAID	31182	210	AGRICULTURAL RESEARCH	2007001229
Australia	AusAID	31182	36	PLANT HEALTH MANAGEMENT FOR FABA BEAN, CHICKPEA AND LENTILS	2007001543
Canada	IDRC	31310	344	IMPROVING FLOOD PLAIN MANAGEMENT THROUGH ADAPTIVE LEARNING NETWORKS	070277i
Denmark	MFA	31120	401	AGRICULTURAL DEVELOPMENT	071496
Germany	BMZ	31120	313	AGRICULTURAL DEVELOPMENT	2007006349
Germany	BMZ	31120	1975	DEVELOPMENT EXTENSION EDUCATION SERVICES, PEOPLE'S ORGANISATIONS' INITIATIVE	2007006222
Germany	BMZ	31181	359	INTEGRATED CONVEYANCE OF FAMILIES OF SMALL FARMERS, BANGLADESH	2007005396
Germany	L G	31191	14	AGRICULTURAL SERVICES	2007011564
IDA		31110	2720	AVIAN FLU	2007000862
IDA		31140	27610	WATER MANAGEMENT IMPROVEMENT PROJECT	2007000881
IFAD		31182	19450	NATIONAL AGRICULTURAL TECHNOLOGY PROJECT	070020
Ireland	DFA	31130	714	SUPPORT TO NGOS - IRISH - MAPS - GENERAL - SUPPORT SMALL ENTERPRISES	2007000226
Japan	JICA	31110	1531	TC AGGREGATED ACTIVITIES	072997T
Japan	JICA	31130	323	TC AGGREGATED ACTIVITIES	073111T
Japan	JICA	31163	258	TC AGGREGATED ACTIVITIES	073223T
Japan	JICA	31195	48	TC AGGREGATED ACTIVITIES	073273T
Korea	KOICA	31120	77	AGRICULTURAL DEVELOPMENT	2007017887
Korea	KOICA	31130	15	AGRICULTURAL LAND RESOURCES	2007016127
Korea	KOICA	31140	30	AGRICULTURAL WATER RESOURCES	2007016066
Korea	KOICA	31163	14	LIVESTOCK	2007016960
Korea	KOICA	31181	27	AGRICULTURAL EDUCATION/TRAINING	2007017137
Norway	NORAD	31120	51	INTEGRATED COMMUNITY DEVELOPMENT PROGRAM (ICDP), AGRICULTURE PROGRAM	2007003650
Switzerland	SDC	31182	521	AFIP AGROFORESTRY IMPROVEM PARTNERSHIP	2004004632
UNDP		31161	11	FOOD CROP PRODUCTION	2007013391
UNDP		31163	9	LIVESTOCK	2007012418
UNDP		31210	38	FORESTRY POLICY & ADMIN. MANAGEMENT	2007012414
United States	AID	31110	525	AGRICULTURAL POLICY & ADMIN. MGMT	2007013898
United States	AID	31120	3209	GROWTH OF AGRIBUSINESS AND SMALL BUSINESS	2007008446

Total aid for water in 2007 for Bangladesh   **USD thousand 61117**
And as a share of aid to total recipient countries   **0.91%**

Barbados

EU institutions	EDF	31162	15240	ACCOMPANYING MEASURES 2007 FOR SUGAR PROTOCOL COUNTRIES	2007201267

Total aid for water in 2007 for Barbados   **USD thousand 15240**
And as a share of aid to total recipient countries   **0.23%**

Belarus

Germany	BMZ	31166	64	AGRICULTURAL EXTENSION	2007006801
Germany	Fed Min	31181	456	TRAINING OF CONSULTANTS AND MANAGERS	2007010323

Total aid for water in 2007 for Belarus   **USD thousand 520**
And as a share of aid to total recipient countries   **0.01%**

Belize

EU institutions	CEC	31120	272	AGRICULTURAL DEVELOPMENT	2005100108

RECIPIENT / Donor	Agency	Sector	Amount USD thousand	Project description	CRS ID Number
EU institutions	EDF	31181	2464	AGRICULTURAL EDUCATION/TRAINING	2007201280
Japan	JICA	31110	162	TC AGGREGATED ACTIVITIES	072955T

Total aid for water in 2007 for Belize — USD thousand 2898
And as a share of aid to total recipient countries — 0.04%

Benin

Donor	Agency	Sector	Amount	Project description	CRS ID
Belgium	DGCD	31120	1863	APPUI MONDE RURAL DEPARTEMENTS ATACORA & DONGA (PAMRAD)	2004001133
Belgium	DGCD	31120	455	RENFORCEMENT DES CAPACITéS DES COMMUNAUTéS RURALES	2006003973
Belgium	DGCD	31140	477	PROGRAMME D'AMéNAGEMENTS HYDROAGRICOLES	2007002923
Belgium	DGCD	31161	232	DEVELOPPEMENT DE LA FILIERE MANIOC DANS LE DEPARTEMENT DU ZOU	2004003885
Belgium	MPRF	31161	58	FOOD CROP PRODUCTION	2005002058
Belgium	DGCD	31163	62	LIVESTOCK	2004005341
Belgium	DGCD	31320	3422	APPUI AU DéVELOPPEMENT DES FILIèRES HALIEUTIQUES DU BéNIN (ADEFIH)	2007903204
EU institutions	CEC	31164	13689	APPUI A LA FILIERE COTON BENINOISE	2007100005
France	AFD	31161	13689	PROJET D'APPUI AUX DYNAMIQ.PROD DU BéNIN	2007151200
France	MISC	31182	6201	RECHERCHE AGRICULTURE	2007800343
Germany	BMZ	31110	74	AGRICULTURAL POLICY & ADMIN. MGMT	2007008835
Germany	BMZ	31120	50	AGRICULTURAL DEVELOPMENT	2007008836
Germany	BMZ	31140	545	IMPROVEMENT OF HYDRAULIC STRUCTURES	2007006459
Germany	BMZ	31166	166	AGRICULTURAL EXTENSION	2007008970
Germany	BMZ	31181	67	AGRICULTURAL EDUCATION/TRAINING	2007008837
Ireland	DFA	31120	20	AGRICULTURAL DEVELOPMENT	2007002149
Italy	DGCS	31120	136	AGRICULTURAL DEVELOPMENT	060815
Italy	LA	31162	134	INDUSTRIAL CROPS/EXPORT CROPS	071498
Italy	LA	31181	144	AGRICULTURAL EDUCATION/TRAINING	070930
Japan	MAFF	31110	69	TC AGGREGATED ACTIVITIES	070001T
Japan	JICA	31110	405	TC AGGREGATED ACTIVITIES	072918T
Japan	JICA	31310	962	TC AGGREGATED ACTIVITIES	073547T
Netherlands	MFA	31110	525	AGRIC.: SUBS. TO PSOM 2003-2009	2007100129
Netherlands	MFA	31120	66	AGRICULTURAL DEVELOPMENT	2007000829
Netherlands	MFA	31194	6790	COT PROCOTON	2007000838
Spain	MFA	31310	46	CONTINENTAL AQUACULTURE DEVELOPMENT PROGRAM.	073162

Total aid for water in 2007 for Benin — USD thousand 50348
And as a share of aid to total recipient countries — 0.75%

Bhutan

Donor	Agency	Sector	Amount	Project description	CRS ID
Australia	AusAID	31110	20	OPPORTUNITIES TO IMPROVE LAND AND WATER MANAGEMENT IN BHUTAN	2007001779
Australia	AusAID	31161	85	FOOD CROP PRODUCTION	2007001070
Japan	JICA	31110	1237	TC AGGREGATED ACTIVITIES	072990T
Japan	JICA	31130	122	TC AGGREGATED ACTIVITIES	073106T
Japan	MOFA	31150	2037	THE ASSISTANCE FOR UNDERPRIVILEGED FARMERS	070015
Japan	MOFA	31150	1783	GRANT ASSISTANCE FOR UNDERPRIVILEGED FARMERS	070295
Japan	JICA	31150	253	TC AGGREGATED ACTIVITIES	073163T
Japan	JICA	31163	5	TC AGGREGATED ACTIVITIES	073217T
Japan	JICA	31195	181	TC AGGREGATED ACTIVITIES	073269T

Donor	Agency	Sector	Amount USD thousand	Project description	CRS ID Number
RECIPIENT					
Japan	JICA	31210	34	TC AGGREGATED ACTIVITIES	073414T
Japan	JICA	31310	12	TC AGGREGATED ACTIVITIES	073517T
Korea	Misc	31181	5	AGRICULTURAL EDUCATION/TRAINING	2007002565
Switzerland	SDC	31220	3042	PARTICIPATORY FORESTRY MANAGEMENT PROGRAMME	2002002257
Switzerland	SDC	31220	58	FORESTRY DEVELOPMENT	2002002258

Total aid for water in 2007 for Bhutan — **USD thousand 8874**
And as a share of aid to total recipient countries — **0.13%**

Bilateral, unspecified

Donor	Agency	Sector	Amount	Project description	CRS ID
Australia	AusAID	31110	1321	INTERNATIONAL AGRICULTURAL ASSISTANCE PROGRAM	2007000410
Australia	AusAID	31110	833	INTERNATIONAL AGRICULTURAL ASSISTANCE- LIVE ANIMAL	2007000626
Australia	AusAID	31181	2905	FT- ADMINISTRATION	2007000224
Australia	AusAID	31182	2410	NON PROJECT SPECIFIC ACTIVITIES (R & D RESEARCH OTHER ACTIVITIES)	2007000142
Australia	AusAID	31182	55	AGRICULTURAL RESEARCH	2007001325
Australia	AusAID	31195	34	SHARING EXPERIENCES WITH THE MANAGEMENT & CONTROL OF AVIAN INFLUENZA	2007002107
Australia	AusAID	31220	1673	FOREST MONITORING AND ASSESSMENT SYSTEM	2007000196
Australia	AusAID	31220	49	GIFC PROGRAM DEVELOPMENT AND LIAISON	2007001366
Australia	AusAID	31220	20221	KALIMANTAN FORESTS AND CLIMATE PARTNERSHIP	2007000019
Australia	AusAID	31220	2510	PEATLAND FIRE MANAGEMENT	2007000140
Australia	AusAID	31220	3347	REDD WORKING GROUP	2007000119
Australia	AusAID	31282	4880	KALIMANTAN FORESTS AND CLIMATE PARTNERSHIP	2007000088
Australia	AusAID	31282	837	REDUCING DEFORESTATION; IMPROVE LOCAL FORESTRY GOV.	2007000278
Austria	BMLFUW	31210	21	CONTRIBUTION TO FACILITATE PARTICIPATION OF DEV. COUNTRIES IN FORUM ON FOREST	2007818448
Austria	BMLFUW	31210	34	CONTRIBUTION TO NATIONAL FOREST PROGRAMM FACILITY	2007818447
Austria	BMLFUW	31210	30	ANALYSIS OF INTERVENTION FIELDS FOR AUSTRIAN FORESTRY DEVELOPMENT POLICY	2007818457
Austria	MISC	31210	6	SMALL-SCALE COMMITMENTS AGGREGATED BY SECTOR AND RECIPIENT COUNTRY	2007980249
Austria	BMLFUW	31282	103	TRAINING FOR FOREST RESEARCHERS AND PRACTITIONERS IN THE USE OF GFIS	2007818446
Belgium	DGCD	31110	280	AGRICULTURAL POLICY & ADMIN. MGMT	2005001713
Belgium	MPRF	31110	143	AGRICULTURAL POLICY & ADMIN. MGMT	2007004795
Belgium	DGCD	31181	455	IMT - BOURSES MSTAH	2005001154
Belgium	DGCD	31181	438	IMT - CURSUS MASTER OF SCIENCE IN TROPICAL ANIMAL HEALTH	2005001143
Belgium	DGCD	31182	159	ILRI PROGRAMME PERSONNEL COOPERATION MULTILATERALE	2004003732
Belgium	DGCD	31191	841	FAO- HORTIVAR - PROGRAMME 2001-2007	2006002434
Belgium	DGCD	31195	86	LIVESTOCK/VETERINARY SERVICES	2006003553
Belgium	MPRF	31310	97	FISHING POLICY AND ADMIN. MANAGEMENT	2007004796
Canada	IDRC	31110	42	SUPPORT FOR AGROPOLIS II PUBLICATION	070443i
Canada	IDRC	31110	192	RESOURCES ON GENDER AND URBAN AGRICULTURE	070189i
Canada	IDRC	31161	8	SUPPORT FOR CLOSING THE LOOP AND DISSEMINATION ACTIVITIES	070020i
Canada	CIDA	31181	92	DCFRN/DEV.COUN.FARM RADIO NET. 2005-2008	050828
Canada	CIDA	31181	92	CANADIAN LUTHERAN WORLD RELIEF	050830
Canada	CIDA	31182	419	CGIAR LINKAGE FUND CCLF - CGIAR LINKAGE FUND CCLF 2006/07	070181
Canada	CIDA	31182	419	CGIAR LINKAGE FUND - FUNDING FOR 2007/08	070933
Canada	CIDA	31210	22	INTERNATIONAL COMMUNITY FORESTRY WORKSHO - TORONTO, CANADA / 2007.09.30	070756
Canada	CIDA	31310	61	VICTORIA, CANADA / 2007.05.15-18	070314
Canada	IDRC	31310	7	GLOBAL AQUACULTURE & AQUATIC GENETIC RESOURCES: SOUTH-SOUTH PARTNERSHIP	070117i
Denmark	MFA	31282	915	TROPICAL FORESTS FOR POVERTY ALLEVIATION	071363

RECIPIENT Donor	Agency	Sector	Amount USD thousand	Project description	CRS ID Number
EU institutions	EDF	31110	342	AGRICULTURAL POLICY & ADMIN. MGMT	2007201473
EU institutions	CEC	31120	42437	AGRICULTURAL DEVELOPMENT	2007100366
EU institutions	EDF	31162	48	INDUSTRIAL CROPS/EXPORT CROPS	2007201278
EU institutions	CEC	31182	17796	ACP SUGAR RESEARCH PROGRAMME	2007100386
EU institutions	CEC	31210	13689	FOREST LAW ENFORCEMENT, GOVERNANCE AND TRADE SUPPORT PROJECT	2007100330
EU institutions	CEC	31310	41068	STRENGTHENING FISHERIES MANAGEMENT IN ACP COUNTRIES	2007100329
Finland	MFA	31120	1164	INT. FEDERATION OF AGRICULTURAL PRODUCERS (IFAP) / AGRICORD	2002003058
Finland	MFA	31181	1916	SUPPORT TO CONSULTATIVE GROUP ON INT. AGRICULTURAL RESEARCH (CGIAR)	1999003132
Finland	MFA	31210	411	FORESTRY POLICY AND ADMINISTRATIVE MANAGEMENT - SUPPORT TO UNFF/CPF/NFP	2001003041
Finland	MFA	31210	411	PROFOR PROJECT (WORLD BANK)	2007000037
Finland	MFA	31220	4791	FOREST CARBON PARTNERSHIP FACILITY - FCPF	2007000114
Finland	MFA	31282	137	SUPPORT TO INTERNATIONAL UNION OF FOREST RESEARCH ORGANIZATIONS (IUFRO)	2003003065
France	MAE	31110	2738	RENF. DE L'APPROCHE REGIONALE DANS LE PARTENARIAT EUROPE - AFR SUR LE COTON	2007901188
France	MISC	31110	192	POLITIQUE AGRICOLE & GESTION ADMINISTRATIVE	2007800787
France	MISC	31181	397	ÉDUCATION & FORMATION DANS LE DOMAINE AGRICOLE	2007800795
Germany	BMZ	31110	1643	SUSTAINABLE USE OF BIOMASS FOCUSSING ON BIOENERGY	2007005883
Germany	BMZ	31110	23	AGRICULTURAL POLICY & ADMIN. MGMT	2007008452
Germany	Fed Min	31110	41	AGRICULTURAL POLICY & ADMIN. MGMT	2007010321
Germany	BMZ	31130	2053	SUSTAINABLE AGRICULTURE INFORMATION NETWORK - SUSTAINET	2007005906
Germany	KFW	31140	1095	EDUCATION AND TRAINING	2007000010
Germany	KFW	31181	1335	EDUCATION AND TRAINING	2007000011
Germany	BMZ	31182	17645	INTERNATIONAL AGRICULTURAL RESEARCH	2007009359
Germany	Fed Min	31195	1232	AVIAN INFLUENZA HPAI STRATEGIES FOR SMALLHOLDER LIVELIHOODS AND BIODIVERSITY	2007010309
Germany	BMZ	31210	4107	SUPPORT TO INTERNATIONAL FOREST-RELATED PROGRAMS (IWRP)	2007002131
Ireland	DFA	31120	97	AGRICULTURAL DEVELOPMENT	2007001201
Ireland	DFA	31150	185	AGRICULTURAL INPUTS	2007000833
Italy	DGCS	31120	1368	AGRICULTURAL DEVELOPMENT	070408
Japan	MAFF	31110	1230	TC AGGREGATED ACTIVITIES	070008T
Japan	MOFA	31110	1952	ASSISTANCE FOR UNDERPRIVILEGED FARMERS	070196
Japan	JICA	31110	973	TC AGGREGATED ACTIVITIES	073074T
Japan	MAFF	31140	185	TC AGGREGATED ACTIVITIES	070017T
Japan	JICA	31150	207	TC AGGREGATED ACTIVITIES	073174T
Japan	MAFF	31181	68	TC AGGREGATED ACTIVITIES	070023T
Japan	MAFF	31191	149	TC AGGREGATED ACTIVITIES	070025T
Japan	MAFF	31210	1052	TC AGGREGATED ACTIVITIES	070030T
Japan	JICA	31210	38	TC AGGREGATED ACTIVITIES	073461T
Japan	MAFF	31281	170	TC AGGREGATED ACTIVITIES	070032T
Japan	MAFF	31291	689	TC AGGREGATED ACTIVITIES	070033T
Japan	MAFF	31310	418	TC AGGREGATED ACTIVITIES	070047T
Japan	MAFF	31381	1099	TC AGGREGATED ACTIVITIES	070078T
Netherlands	MFA	31110	96	AGRICULTURAL POLICY & ADMIN. MGMT	2007000402
Netherlands	MFA	31110	68	AGRIC.: DSI FAO	2003012969
Netherlands	MFA	31120	205	AGRICULTURAL DEVELOPMENT	2007000400
Netherlands	MFA	31130	241	AGRIC. LAND RES.: DDE IFDC	2003012602
Netherlands	MFA	31130	1441	AGRIC. LAND RES.: INTERNAT. CONTRIBUTIONS/ISRIC	2007100159
Netherlands	MFA	31162	69	INDUSTRIAL CROPS/EXPORT CROPS	2007000050

RECIPIENT Donor	Agency	Sector	Amount USD thousand	Project description	CRS ID Number
Netherlands	MFA	31165	6571	DMW TECHN. AND MARKET DEVELOP.	2007000735
Netherlands	MFA	31182	616	AGRIC. RESEARCH: DSI IITA	2003012976
Netherlands	MFA	31182	3787	AGRIC. RESEARCH: ILAC	2007000692
Netherlands	MFA	31182	369	AGRIC. RESEARCH: DSI ICRISAT	2003013022
Netherlands	MFA	31182	3789	AGRIC. RESEARCH: CAS-IP INNOVATION SYSTEMS	2007000689
Netherlands	MFA	31210	164	DMW IUCN LIVELIHOOD	2006000947
Netherlands	MFA	31281	1717	KYOTO TGAL	2007000728
Netherlands	MFA	31282	15195	FORESTRY: DMW TROPENBOS INT. 2006	2006000877
Norway	NORAD	31110	43	TC TO PREPARE AND ARRANGE A CONFERENCE IN OSLO ON AGR. DVPT/ FOOD SECURITY	2007003754
Norway	MFA	31110	85	STANDING COMMITTEE ON NUTRITION (SCN) 6TH REPORT	2007001229
Norway	MFA	31110	373	AGRICULTURAL POLICY & ADMIN. MGMT	2007002080
Norway	MFA	31110	5974	FAO NORWEGIAN EXTRA BUDGETARY CONTRIBUTION 2005	2007000100
Norway	NORAD	31150	85	CONSULTING SERVICES IN RELATION TO PPP WORK AGRI. DEV, FERTILIZER STRATEGY	2007003875
Norway	MFA	31161	853	LESS WATER MORE RICE	2007001336
Norway	MFA	31182	51	AGRICULTURAL RESEARCH	2007001245
Norway	MFA	31182	5015	THE GLOBAL SEED VAULT, SPITSBERGEN	2007001356
Norway	MFA	31182	885	BIOS (BIOLOGICAL INNOVATION FOR OPEN SOCIETY)	2007000099
Norway	MFA	31182	4182	GCDT MATCHING GATES FOUNDATION	2007001357
Norway	MFA	31210	45	PROJECT SUPPORT FOR TROPICAL TIMBER ORG.	2007001329
Norway	NORAD	31220	341	RRI - ACCELERATING REFORMS IN FOREST TENURE AND GOVERNANCE	2007003915
Norway	NORAD	31310	290	ICSF PROGRAMME FUNDING 2007	2007003753
Norway	NORAD	31310	307	FRAME AGREEMENT NORWEGIAN COLLEGE OF FISHERY SCIENCE 2007	2007003763
Norway	NORAD	31310	19	ASSISTANCE WITH FISHERY-PRIVATE SECTOR DOCUMENT	2007003906
Norway	MFA	31310	82	FISHING POLICY AND ADMIN. MANAGEMENT	2007000293
Norway	MFA	31310	55	FAO. DATABASE ON PORT STATE MEASURES	2007000296
Norway	MFA	31310	171	FAO TRUST FUND FOR PORT STATE MEASURES	2007000297
Norway	MFA	31310	95	FAO DEEP SEAS FISHERIES PROGRAMME	2007001348
Norway	MFA	31310	40	AGREEM. ON PORT STATE CONTRL. EXPERT CONSULTATIONS	2007001333
Norway	NORAD	31320	5121	EAF MARINE, NEW NANSEN PROGRAMME-ADDITIONAL GRANT	2007003831
Norway	NORAD	31320	546	FRAME AGREEMENT DIRECTORATE OF FISHERIES & INST. OF MARINE RESEARCH 2007	2007003764
Norway	NORAD	31391	5	FISHERY SERVICES	2007003994
Portugal	ICP	31182	537	RESEA.&PRODUCTION OF STRAIN COFFE PLANTS RESISTANTS TO FRUITS ANTHRACNOSES	070008
Spain	AGR	31110	108	AGRICULTURAL POLICY & ADMIN. MGMT	074694
Spain	MFA	31110	6845	SPANISH FUND	078156
Spain	AG	31120	146	START OF A CO-DEVELOPMENT PLAN BETWEEN THE VALENCIA COMMUNITY AND COLOMB.	075296
Spain	AGR	31150	15	AGRARIAN CODES AND SYSTEM /TRACTORS	071549
Spain	AGR	31150	8	AGRARIAN CODES AND SYSTEM /FOREST	070874
Spain	AGR	31150	15	AGRARIAN CODES AND SYSTEM /SEEDS	071520
Spain	AGR	31150	11	FORUM ON PESTICIDES	071216
Spain	AG	31181	25	TRAINING AND TECHNICAL ASSISTANCE TO SOCIAL FIRMS OF THE RURAL WORKERS	072271
Spain	Misc	31181	22	EDUC./TRNG:WATER SUPPLY & SANITATION	072093
Spain	AGR	31182	14	EUROPEAN COORDINATOR ANIMAL GENETIC RESOURCES	071406
Spain	AGR	31182	20	FIGHT AGAINST THE FOOT AND MOUTH DISEASE	071854
Spain	AGR	31182	41	AGREMEENT ABOUT PLANT GENETIC RESOURCES (TIRFAA)	072945
Spain	AGR	31182	101	AGRICULTURAL RESEARCH	074582

RECIPIENT Donor	Agency	Sector	Amount USD thousand	Project description	CRS ID Number
Spain	AG	31194	21	EXCHANGE RESEARCH AND TECHNICAL ASSISTANCE TO COOPERATIVES	071980
Spain	MFA	31210	80	INTERNATIONAL MEETING: RESPONSIBLE PRODUCTION AND USE OF WOOD	074173
Spain	AGR	31310	85	FISHING POLICY AND ADMIN. MANAGEMENT	073101
Sweden	Sida	31110	4500	AGRICULTURAL POLICY & ADMIN. MGMT	2007002491
Sweden	Sida	31120	307	AGRICULTURAL DEVELOPMENT	2007002492
Sweden	Sida	31181	2175	PLANT BREEDING	2006008462
Sweden	Sida	31182	20363	AGRICULTURAL RESEARCH	2007005253
Sweden	Sida	31194	4484	AGRICULTURAL CO-OPERATIVES	2007002522
Sweden	Sida	31210	5179	FORESTRY POLICY & ADMIN. MANAGEMENT	2007002472
Sweden	Sida	31220	30	FORESTRY DEVELOPMENT	2007002420
Sweden	Sida	31310	966	FISHING POLICY AND ADMIN. MANAGEMENT	2007002452
Switzerland	FA	31110	146	AGRICULTURE DURABLE ET DéVELOPPEMENT RURAL DANS DES RéGIONS DE MONTAGNE	2007001403
Switzerland	SDC	31110	129	GLOBAL DONOR PLATFORM FOR RURAL DEVELOPMENT	1999001600
Switzerland	SDC	31110	192	INTERNATIONAL YEAR OF POTATO (IYP2008)	2007000438
Switzerland	SDC	31110	350	GREENFACTS PROGRAMMBEITRAG	2007000075
Switzerland	SDC	31110	2215	CDE DEVELOPMENT AND ENVIRONMENT BACKSTOPPING	2002002118
Switzerland	SDC	31110	759	AGRICULTURAL POLICY & ADMIN. MGMT	2007000440
Switzerland	SDC	31110	1440	WOCAT - SOIL AND WATER CONSERVATION	2007000418
Switzerland	seco	31110	27	AGRICULTURAL POLICY & ADMIN. MGMT	2007001883
Switzerland	seco	31110	1167	PLATFORM RENEWABLE ENERGIES REPIC II	2007001943
Switzerland	seco	31110	667	BETTER COTTON INITITAVE (BCI)	2007001920
Switzerland	FA	31120	104	AGRICULTURAL DEVELOPMENT	2007001401
Switzerland	SDC	31120	292	THE GOOD SEED INITIATIVE	2004001778
Switzerland	SDC	31120	74	AGRICULTURAL DEVELOPMENT	2004004580
Switzerland	SDC	31120	1917	GLOBAL CROP DIVERSITY TRUST	2004004537
Switzerland	SDC	31120	266	IUED DROIT AUX L'ALIMENTATION	2001002636
Switzerland	SDC	31182	833	AGRICULTURAL PRODUCTION SYSTEMS	2007000042
Switzerland	SDC	31182	175	YPARD YOUNG PROFESSIONAL PLATFORM	2006000333
Switzerland	SDC	31220	375	FAO NATIONAL FOREST PROGRAMME	2007000273
Switzerland	SDC	31220	128	NATURAL RESOURCE GOUVERNANCE EVENTS	2007000383
Switzerland	SDC	31220	21	LANDSCAPE MOSAICS: INTEGRATING LIVELIHOOD	2006000640
Switzerland	SDC	31220	46	FORESTRY DEVELOPMENT	1992000427
United Kingdom	DFID	31110	5	TRADITIONAL KNOWLEDGE PAPER	2007000792
United Kingdom	DFID	31110	532	SADC MAPP DEVELOPMENT	2007000797
United Kingdom	DFID	31110	17503	SCARDA PROGRAMME	2007000798
United Kingdom	DFID	31110	1501	STANDARD AND TRADE DEVELOPMENT	2007000662
United Kingdom	DFID	31110	160	EUREPGAP AFRICA OBSERVER INITIATIVE	2007000664
United Kingdom	DFID	31110	200	2ND EUROPEAN FORUM FOR SUSTAINABLE RURAL DEVELOPMENT	2007000659
United Kingdom	DFID	31110	26	AGRICULTURAL POLICY & ADMIN. MGMT	2007000852
United Kingdom	DFID	31130	30	EVALUATION OF LAND POTENTIAL 1956 - 2001	2007000795
United Kingdom	DFID	31163	7805	AVIAN FLU CONTROL POLICIES	2007000796
United Kingdom	DFID	31210	1001	PROGRAMME FOR FORESTS (PROFOR) TRUST FUND	2007000927
United Kingdom	DFID	31210	166	MEGAFLORESTAIS 0708	2007000860
United Kingdom	DFID	31210	86	INTERNATIONAL TROPICAL TIMBER ORGANISATION	2007000693
United Kingdom	DFID	31310	1601	PROFISH	2007000663

RECIPIENT Donor	Agency	Sector	Amount USD thousand	Project description	CRS ID Number
United States	AID	31110	500	AGRICULTURAL ENABLING ENVIRONMENT	2007001654
United States	AID	31110	55386	AGRICULTURE	2007001817
United States	AID	31110	2743	AGRICULTURAL POLICY & ADMIN. MGMT	2007013895
United States	AGR	31110	6550	COCHRAN FELLOWSHIP PROGRAM	2007000447
United States	AGR	31110	334	AGRICULTURAL POLICY & ADMIN. MGMT	2007000470
United States	AID	31120	2470	AGRICULTURAL SECTOR PRODUCTIVITY	2007001758
United States	AID	31120	371	AGRICULTURAL DEVELOPMENT	2007001757
United States	STATE	31162	96	INDUSTRIAL CROPS/EXPORT CROPS	2007026987
United States	AGR	31163	95	LIVESTOCK	2007000452
United States	AGR	31166	987	INTERNATIONAL SCIENCE AND EDUCATION COMPETITIVE GRANTS PROGRAM	2007000460
United States	AGR	31182	276	AGRICULTURAL RESEARCH	2007000440
United States	Misc	31192	80	PLANT/POST-HARVEST PROT. & PEST CTRL	2007024541
United States	AGR	31210	20	FORESTRY POLICY & ADMIN. MANAGEMENT	2007000450
United States	STATE	31210	100	FORESTRY POLICY & ADMIN. MANAGEMENT	2007027029
United States	STATE	31320	70	FISHERY DEVELOPMENT	2007027007

Total aid for water in 2007 for Bilateral, unspecified   **USD thousand 429736**
And as a share of aid to total recipient countries   **6.4%**

Bolivia

	Agency	Sector	Amount	Project description	CRS ID
Austria	MISC	31110	7	SMALL-SCALE COMMITMENTS AGGREGATED BY SECTOR AND RECIPIENT COUNTRY	2007980162
Belgium	DGCD	31120	35	AGRICULTURAL DEVELOPMENT	2005000616
Belgium	DGCD	31161	225	FOOD CROP PRODUCTION	2007004220
Belgium	DGCD	31165	57	AGRICULTURAL ALTERNATIVE DEVELOPMENT	2005001239
Belgium	DGCD	31166	51	AGRICULTURAL EXTENSION	2005001233
Belgium	DGCD	31181	167	AGRICULTURAL EDUCATION/TRAINING	2007005087
Belgium	DGCD	31194	61	AGRICULTURAL CO-OPERATIVES	2004003571
Belgium	DGCD	31194	153	RENFORCEMENT DE LA POSITION DES PETITS AGRICULTEURS ET LES INDIGENES EN AGR.	2004003231
Belgium	DGCD	31210	1538	UTILISATION INTEGRALE ET DURABLE DES RESSOURCES FORESTIERES	2005000593
Belgium	MPRF	31210	134	FORESTRY POLICY & ADMIN. MANAGEMENT	2007004669
Belgium	MPRF	31220	8	FORESTRY DEVELOPMENT	2006003620
Belgium	DGCD	31382	30	FISHERY RESEARCH	2007900050
Canada	IDRC	31130	242	INF. & COM. TECHNOLOGIES FOR BUILDING DEMOCRATIC DIALOGUE: THE AGRARIAN REV.	070095i
Canada	IDRC	31130	19	AGRARIAN REVOLUTION OBSERVATORY	070300i
Canada	CIDA	31181	287	AGRICULTURE ÉCOLOGIQUE ET RENTABLE	070743
EU institutions	EDF	31165	15058	AGRICULTURAL ALTERNATIVE DEVELOPMENT	2007200577
Finland	MFA	31161	67	CONSTRUCTING GREENHOUSES AND WATER PUMPS IN ANDEAN HIGHLAND	2007000276
Germany	BMZ	31120	816	AGRICULTURAL DEVELOPMENT	2007008749
Germany	BMZ	31120	684	ENHANCEMENT OF AGRIC. SELF HELP COOPERATIVES	2007005597
Germany	BMZ	31130	495	AGRICULTURAL LAND RESOURCES	2007006764
Germany	BMZ	31140	11	AGRICULTURAL WATER RESOURCES	2007008751
Germany	KFW	31140	10951	NATIONAL IRRIGATION PROGRAMME	2007000344
Germany	BMZ	31182	309	AGRICULTURAL RESEARCH	2007007644
Germany	BMZ	31210	95	FORESTRY POLICY & ADMIN. MANAGEMENT	2007008780
IDA		31110	1000	PARTICIPATORY RURAL INVESTMENT	2007000603
IDA		31110	12000	LAND FOR AGRICULTURAL DEVELOPMENT	2007000600

RECIPIENT Donor	Agency	Sector	Amount USD thousand	Project description	CRS ID Number
IDA		31130	3000	LAND FOR AGRICULTURAL DEVELOPMENT	2007000601
IDA		31140	6000	PARTICIPATORY RURAL INVESTMENT	2007000602
Italy	DGCS	31120	844	AGRICULTURAL DEVELOPMENT	070281
Italy	CA	31181	214	AGRICULTURAL EDUCATION/TRAINING	071316
Japan	JICA	31110	2115	TC AGGREGATED ACTIVITIES	072963T
Japan	JICA	31130	456	TC AGGREGATED ACTIVITIES	073093T
Japan	MOFA	31140	3175	THE REHABILITATION OF THE IRRIGATION SYSTEM	070152
Japan	JICA	31163	628	TC AGGREGATED ACTIVITIES	073209T
Japan	JICA	31195	87	TC AGGREGATED ACTIVITIES	073262T
Japan	JICA	31310	8	TC AGGREGATED ACTIVITIES	073507T
Luxembourg	MFA	31181	71	AGRICULTURAL EDUCATION/TRAINING	070383
Luxembourg	MFA	31194	35	AGRICULTURAL CO-OPERATIVES	070092
Netherlands	MFA	31164	597	LAP NATIONAL LAND REFORM PLAN	2003014269
Norway	NORAD	31120	128	INTEGRATED DEV., INTERANDEAN VALLEYS PRODUCTION/AGRICULTURE	2007003552
Norway	NORAD	31120	232	REGIONAL DEVELOPMENT ALCOCHE PRODUCTION/AGRICULTURE	2007003534
Norway	NORAD	31195	14	REGIONAL DEVELOPMENT, ALCOCHE, LIVESTOCK	2007003535
Spain	MFA	31110	639	IMPROVEMENT OF THE QUALITY OF INDIGENOUS AND RURAL PEOPLE	077500
Spain	AG	31120	397	IMPROVEMENT OF SUSTAINABLE FAMILY-BASED AGRI. PRODUCTION IN THE VALLEYS	076951
Spain	AG	31120	158	SUSTAINABLE MANAGEMENT OF PRODUCTION RESOURCES BY FARMERS	075401
Spain	AG	31120	304	AGRICULTURAL DEVELOPMENT	076493
Spain	AG	31120	68	FOOD SECURITY IN THE MUNICIPALITY OF CARIPUYO	073945
Spain	AG	31140	289	RETAINING WALL AND RECOVERY OF ARABLE LANDS PHASE 4	076454
Spain	AG	31161	383	FOOD CROP PRODUCTION	076909
Spain	AG	31161	41	IMPROVING THE CONDITIONS OF FOOD SECURITY OF FIVE COMMUNITIES	073047
Spain	AG	31161	423	SUSTAINABLE RECOVERY OF THE PRODUCTIVE & DIVERSIFICATED PRODUCTION OF FARMS	077142
Spain	AG	31163	55	CONSOLIDATION OF THE LIVESTOCK FARMERS' PRODUCTION	073502
Spain	AG	31163	27	GENETIC AND HEALTH IMPROVEMENT OF THE INDIGENOUS POPULATION OF COCHABAMBA	072487
Spain	AG	31164	130	IMPROVEMENT OF THE ALIMENTATION IN THE TAYA CAJA PROVINCE	075010
Spain	MFA	31181	82	EMPOWERMENT AND PROFESSIONAL TRAINING ACTIVITIES FOR YOUNG RURAL WOMEN	074198
Spain	MFA	31181	377	DISASTER PREVENTION PLAN AND AGRICULTURAL RENOVATION IN THE BOLIVIAN AMAZON	076888
Spain	AG	31181	113	RURAL DEVELOPMENT	074797
Spain	AG	31181	15	CONSTRUCTION OF SOLAR PANEL IN QUINTAPAMPA	071510
Spain	Misc	31182	7	FOOD CULTURE IN THE ANDEAN ECO REGION OF BOLIVIA	070823
Spain	MFA	31194	274	RURAL DEVELOPMENT	076315
Spain	AG	31195	38	SERVICES OF VETERINARIAN MEDICINE FROM FARMER TO FARMER (BOLIVIA).	072892
Spain	MFA	31210	1027	FORESTRY DEVELOPMENT PROJECT IN LA CHIQUITANIA	077756
Spain	MFA	31220	93	PROJECT FOR THE PRESERVATION OF THE DRY FOREST CHIQUITANO	074440
Spain	EMP	31220	25	FORESTRY DEVELOPMENT	072283
Spain	AG	31320	205	CONSOLIDATION OF THE SMALL SCALE RURAL FISH FARMING	075906
Sweden	Sida	31110	148	AGRICULTURAL POLICY & ADMIN. MGMT	2007001854
Switzerland	SDC	31110	500	FONDO ESPECIES ANDINAS	2007000291
Switzerland	SDC	31110	42	AGRICULTURAL POLICY & ADMIN. MGMT	2007000302
Switzerland	seco	31110	42	AGRICULTURAL POLICY & ADMIN. MGMT	2007001875
United States	AID	31110	90	AGRICULTURAL POLICY & ADMIN. MGMT	2007013885
United States	AID	31120	154	AGRICULTURAL DEVELOPMENT	2007001743

RECIPIENT Donor	Agency	Sector	Amount USD thousand	Project description	CRS ID Number
United States	AID	31120	2361	AGRICULTURAL SECTOR PRODUCTIVITY	2007001746
United States	AID	31165	2859	LICIT ECONOMIC GROWTH IN COCA-GROWING AND ASSOCIATED AREAS	2007011025
United States	AID	31165	25864	ALTERNATIVE DEVELOPMENT AND ALTERNATIVE LIVELIHOODS	2007001897

Total aid for water in 2007 for Bolivia   **USD thousand 99332**
And as a share of aid to total recipient countries   **1.48%**

Bosnia-Herzegovina

Donor	Agency	Sector	Amount	Project description	CRS ID
EU institutions	EDF	31110	2053	AGRICULTURAL POLICY & ADMIN. MGMT	2007202334
Germany	BMZ	31120	548	AFSC COMMUNITY GARDENS ASSOCIATION OF BOSNIA ANDHERZEGOVINA PROGRAMME	2007006575
Germany	BMZ	31194	123	AGRICULTURAL CO-OPERATIVES	2007009470
IDA		31110	18900	AGRICULTURE/RURAL DEVELOPMENT	2007000058
IDA		31163	100	AVIAN FLU	2007000060
IDA		31166	1050	AVIAN FLU	2007000059
IDA		31166	2100	AGRICULTURE/RURAL DEVELOPMENT	2007000055
IDA		31210	3351	FOREST	2007000071
Italy	DGCS	31120	27	AGRICULTURAL DEVELOPMENT	060924
Italy	DGCS	31161	759	FOOD CROP PRODUCTION	070653
Italy	LA	31161	11	FOOD CROP PRODUCTION	070971
Italy	DGCS	31194	17	AGRICULTURAL CO-OPERATIVES	060823
Japan	JICA	31110	248	TC AGGREGATED ACTIVITIES	072902T
Japan	JICA	31163	14	TC AGGREGATED ACTIVITIES	073186T
Netherlands	MFA	31191	633	AGRICULTURAL SERVICES	2007000513
Norway	MFA	31120	896	AGRICULTURAL DEVELOPMENT	2007001541
Spain	MFA	31110	248	CREATION AND CONSOLIDATION OF THE AGRICULTURAL AN RURAL DEVELOPMENT CENTER	076166
Spain	EMP	31120	25	AGRICULTURAL DEVELOPMENT	072285
Sweden	Sida	31120	6539	AGRICULTURAL DEVELOPMENT	2007002486
Switzerland	SDC	31110	618	SME PROMOTION FRUITS AND VEGETABLES	2000002135

Total aid for water in 2007 for Bosnia-Herzegovina   **USD thousand 38261**
And as a share of aid to total recipient countries   **0.57%**

Botswana

Donor	Agency	Sector	Amount	Project description	CRS ID
France	MISC	31182	7009	RECHERCHE AGRICULTURE	2007800338
Germany	BMZ	31182	60	AGRICULTURAL RESEARCH	2007008985
Germany	BMZ	31210	77	FORESTRY POLICY & ADMIN. MANAGEMENT	2007008981
Japan	JICA	31110	225	TC AGGREGATED ACTIVITIES	072914T
Japan	JICA	31130	59	TC AGGREGATED ACTIVITIES	073078T
Japan	JICA	31150	17	TC AGGREGATED ACTIVITIES	073139T
Japan	JICA	31163	11	TC AGGREGATED ACTIVITIES	073189T
Japan	JICA	31210	73	TC AGGREGATED ACTIVITIES	073363T
United Kingdom	DFID	31110	79	AGRICULTURAL POLICY & ADMIN. MGMT	2007000791
United States	ADF	31163	76	LIVESTOCK	2007000114

Total aid for water in 2007 for Botswana   **USD thousand 7687**
And as a share of aid to total recipient countries   **0.11%**

Brazil

RECIPIENT Donor	Agency	Sector	Amount USD thousand	Project description	CRS ID Number
Australia	AusAID	31191	12	INTERCONNECTING AGRI-FOOD TRADE INTELLIGENCE	2007001958
Austria	ADA	31120	58	NGO-VOLUNTEERS: ADVISORY SERVICES FOR SMALL SCALE FARMERS	2007500160d
Austria	ADA	31120	37	NGO-VOLUNTEERS: AGRICULTURAL PROD. & MARKETING IN THREE MUNICIPALITIES	2007500160a
Austria	Reg	31161	14	IMPROVING OFFER IN FOOD PRODUCTS BY AGRICULTURAL SUPPORT	2007828562
Austria	Reg	31162	17	SUSTAINABLE INTRODUCTION OF BIO-FUEL PRODUCTION	2007828606
Austria	ADA	31164	49	NGO-VOLUNTEERS: SUPPORT FOR AGRARIAN REFORM IN BARREIRAS	2007500160a
Austria	Reg	31181	16	TRAINING FOR SMALL-SCALE AND LANDLESS FARMERS	2007828603
Austria	Reg	31181	24	TRAINING TO IMPROVE FOOD SECURITY FOR 180 FAMILIES	2007828601
Austria	ADA	31191	49	NGO-VOLUNTEERS: MARKETING PROMOTION - ORGANIC PRODUCTS OF IRPAA PARTNERS	2007500160bj
Austria	ADA	31220	49	NGO-VOLUNTEERS: AFFORESTATION - REGION OF THE INDIGEN. MAXACALI PEOPLE	2007500160as
Belgium	DGCD	31164	52	AGRARIAN REFORM	2004003572
Belgium	DGCD	31194	23	AGRICULTURAL CO-OPERATIVES	2004000139
Belgium	MPRF	31210	18	FORESTRY POLICY & ADMIN. MANAGEMENT	2006003623
Canada	CIDA	31210	60	VANCOUVER, CANADA / 2007.05.07-09	070359
France	MISC	31110	17	POLITIQUE AGRICOLE & GESTION ADMINISTRATIVE	2007800776
France	MISC	31182	71034	RECHERCHE AGRICULTURE	2007800371
Germany	BMZ	31120	3054	AGRICULTURAL DEVELOPMENT	2007006791
Germany	BMZ	31120	378	INTEGRATED RURAL DEVELOPMENT	2007006132
Germany	BMZ	31130	342	AGRICULTURAL LAND RESOURCES	2007006303
Germany	BMZ	31140	854	ENHANCING CIVIC PART. & ADVOCACY FOR CONSERVING LANDSCAPE & WATERSHED	2007006258
Germany	BMZ	31140	44	AGRICULTURAL WATER RESOURCES	2007005670
Germany	BMZ	31164	329	AGRARIAN REFORM	2007006363
Germany	BMZ	31165	48	AGRICULTURAL ALTERNATIVE DEVELOPMENT	2007005698
Germany	BMZ	31166	1231	AGRICULTURAL EXTENSION	2007006286
Germany	BMZ	31166	465	AGRO-ECOLOGICAL TRAINING PROGRAMME FOR SMALL FARMERS	2007006414
Germany	BMZ	31182	58	AGRICULTURAL RESEARCH	2007007642
Germany	BMZ	31210	67	FORESTRY POLICY & ADMIN. MANAGEMENT	2007008778
Germany	BMZ	31220	434	FORESTRY DEVELOPMENT	2007008987
Germany	BMZ	31291	7	FORESTRY SERVICES	2007006789
Italy	DGCS	31110	1518	AGRICULTURAL POLICY & ADMIN. MGMT	060385
Italy	LA	31110	11	AGRICULTURAL POLICY & ADMIN. MGMT	071331
Italy	DGCS	31120	20	AGRICULTURAL DEVELOPMENT	060776
Italy	CA	31120	358	AGRICULTURAL DEVELOPMENT	071303
Italy	LA	31163	31	LIVESTOCK	071328
Italy	LA	31165	53	AGRICULTURAL ALTERNATIVE DEVELOPMENT	071075
Italy	CA	31181	31	AGRICULTURAL EDUCATION/TRAINING	071310
Italy	LA	31191	68	AGRICULTURAL SERVICES	071557
Italy	DGCS	31281	1314	FORESTRY EDUCATION/TRAINING	070132
Japan	JICA	31110	398	TC AGGREGATED ACTIVITIES	073048T
Japan	JICA	31130	33	TC AGGREGATED ACTIVITIES	073094T
Japan	JICA	31163	11	TC AGGREGATED ACTIVITIES	073210T
Japan	JICA	31195	167	TC AGGREGATED ACTIVITIES	073263T
Japan	JICA	31210	822	TC AGGREGATED ACTIVITIES	073404T
Japan	JICA	31310	76	TC AGGREGATED ACTIVITIES	073565T
Luxembourg	MFA	31194	27	AGRICULTURAL CO-OPERATIVES	070096

RECIPIENT Donor	Agency	Sector	Amount USD thousand	Project description	CRS ID Number
Norway	NORAD	31140	66	BASIC DRINKING WATER SUPPLY AND BASIC SANITATION	2007003014
Spain	AG	31120	27	ALTERNATIVE HEALTH ECOLOGICAL GARDEN AND MEDICINAL HERBS	072448
Spain	AG	31120	118	DEVELOPMENT OF ECONOMIC SOCIAL CULTURAL AND HUMAN RIGHTS IN 3 AREAS	074865
Spain	AG	31164	237	EXPANDING THE RURAL TEACHING MODEL OF THE ENFF	076093
Spain	AG	31164	29	SOLIDARITY AND FORMATION IN AREAS OF AGRARIAN REFORM BRASIL	072559
Spain	AG	31164	49	IMPLEMENTATION OF AN ENERGY SOVEREIGNTY MODEL FOR SMALL COMMUNITIES	073284
Spain	AG	31164	172	ECONOMIC AND DEVELOPMENT POLICY/PLANNING	075581
Spain	AG	31165	71	FIGHT AGAINST POVERTY IN AGRARIAN REFORM SETTLEMENTS	074005
Spain	AG	31181	127	TRAINING OF NATIONAL TRAINERS FOR THE STRENGTHENING OF AGRARIAN REFORM	074988
Spain	MFA	31310	27	SYSTEM OF MANAGEMENT FOR THE COSTAL FISHERY IN THE NORTHEAST OF BRASIL	072440
Spain	MFA	31320	315	INSHORE FISHERY IN NORTH-EAST BRAZIL	076574
Spain	MFA	31382	31	STUDY ON THE FATTENING OF OCTOPUS USING DIFFERENT TYPES OF ARTIFICIAL DIETS	072635
Sweden	Sida	31194	888	UBV/BRAZIL 2007-2008	2007001747
United Kingdom	MISC	31210	67	CAPACITY BUILDING OF SUSTAINABLE DEVT AND EXTRACTIVE RESERVES IN THE AMAZON	2007800607
United Kingdom	MISC	31210	88	AMAZONIAN DEFORESTATION	2007800603
United Kingdom	MISC	31210	247	ESTABLISH RESPONSIBLE BUSINESS PRACTICE IN ATLANTIC RAINFOREST CORRIDOR	2007800599
United Kingdom	MISC	31210	16	BRAZILIAN EXPERIENCES IN SUSTAINABLE RESERVES	2007800606
United States	IADF	31181	92	AGRICULTURAL EDUCATION/TRAINING	2007024976
United States	AGR	31210	158	FORESTRY POLICY & ADMIN. MANAGEMENT	2007000451

Total aid for water in 2007 for Brazil   **USD thousand 86605**
And as a share of aid to total recipient countries   **1.29%**

Burkina Faso

Austria	Reg	31140	62	CONSTRUCTION OF AN AGRICULTURAL WATER RESERVOIR	2007829027
Austria	ADA	31191	138	IMPROVEMENT OF NUTRITION FOR 24 VILLAGES IN BURKINA FASO	2007500145
Belgium	DGCD	31110	217	COOPERANTS	2001008681
Belgium	DGCD	31110	23	AGRICULTURAL POLICY & ADMIN. MGMT	2004003265
Belgium	DGCD	31120	138	AGRICULTURAL DEVELOPMENT	2005000791
Belgium	DGCD	31120	383	DéVELOPPEMENT DU TERROIR DE TENSOBENTENGA	2007005611
Belgium	DGCD	31140	310	AGRICULTURAL WATER RESOURCES	2005001074
Belgium	DGCD	31163	186	AMELIORATION DE L'ELEVAGE TRADITIONNEL	2004003368
Belgium	DGCD	31163	363	DéVELOPPEMENT DU ZéBU PEULH AU SAHEL	2004003879
Belgium	DGCD	31163	10	LIVESTOCK	2001001070
Belgium	DGCD	31163	155	SéLECTION ET MULTIPLICATION DU ZEBU AZAWAK	2004001117
Belgium	MPRW	31166	594	FORMATION PROFESSIONNELLE, AGRICULTURE	2001008261
Belgium	DGCD	31194	61	AGRICULTURAL CO-OPERATIVES	2004003367
Belgium	DGCD	31194	307	RENFORCEMENT GRENIERS ALIMENTAIRES GROUPEMENTS NAAM - SOS FAIM	2002000574
EU institutions	CEC	31162	13689	INDUSTRIAL CROPS/EXPORT CROPS	2007100012
France	MISC	31110	31	POLITIQUE AGRICOLE & GESTION ADMINISTRATIVE	2007800770
France	MISC	31182	18001	RECHERCHE AGRICULTURE	2007800360
Germany	BMZ	31110	46	AGRICULTURAL POLICY & ADMIN. MGMT	2007008808
Germany	BMZ	31120	1315	AGRICULTURAL DEVELOPMENT	2007009006
Germany	BMZ	31140	693	INTEGRATED SUPPORT FOR SUSTAINABLE CARE OF HYDRAULIC STRUCTURES	2007006141
Germany	BMZ	31181	40	AGRICULTURAL EDUCATION/TRAINING	2007008810
Germany	BMZ	31192	108	PLANT/POST-HARVEST PROT. & PEST CTRL	2007006167

RECIPIENT	Agency	Sector	Amount USD thousand	Project description	CRS ID Number
Donor					
IDA		31110	9620	RURAL DEVELOPMENT	2007000450
IFAD		31140	400	AGRICULTURAL WATER RESOURCES	070010a
IFAD		31140	11038	PROJET D'IRRIGATION ET DE GESTION DE L'EAU ◦ PETITE ÉCHELLE (PIGEPE)	070010
Italy	LA	31110	57	AGRICULTURAL POLICY & ADMIN. MGMT	071069
Italy	DGCS	31120	636	AGRICULTURAL DEVELOPMENT	070348
Italy	LA	31120	110	AGRICULTURAL DEVELOPMENT	070918
Italy	LA	31150	274	AGRICULTURAL INPUTS	071545
Italy	DGCS	31194	494	AGRICULTURAL CO-OPERATIVES	070514
Japan	JICA	31110	1197	TC AGGREGATED ACTIVITIES	073037T
Japan	MOFA	31150	2801	THE ASSISTANCE FOR UNDERPRIVILEGED FARMERS	070023
Japan	JICA	31150	21	TC AGGREGATED ACTIVITIES	073152T
Japan	JICA	31163	60	TC AGGREGATED ACTIVITIES	073200T
Japan	JICA	31210	640	TC AGGREGATED ACTIVITIES	073381T
Japan	JICA	31310	112	TC AGGREGATED ACTIVITIES	073557T
Korea	KOICA	31320	156	TRAINING PROGRAM	2007017472
Luxembourg	MFA	31110	28	AGRICULTURAL POLICY & ADMIN. MGMT	071701
Luxembourg	MFA	31120	444	AGRICULTURAL DEVELOPMENT	071655
Luxembourg	MFA	31130	14	AGRICULTURAL LAND RESOURCES	070158
Luxembourg	MFA	31150	937	RECONSTITUTION DES STOCKS DE SEMENCES-BURKINA FASO	071767
Netherlands	MFA	31110	605	AGRIC.: SUBS. TO PSOM 2003-2009	2007100156
Spain	MFA	31140	1154	HELP TO IMPROVE THE FOOD & NUTRITION	077804
Spain	Misc	31161	8	CONSTRUCTION OF A SMALL CHICKEN FARM FOR THE WOMEN	070993
Spain	AG	31181	62	SUPPORT PROGRAM FOR THE IMPROVEMENT OF LIVING CONDITIONS OF WOMEN	073651
Spain	Misc	31194	11	AGRICULTURAL DEVELOPMENT AT THE WOUOL COOPERATIVE	071237
Spain	MFA	31310	46	CONTINENTAL AQUACULTURE DEVELOPMENT PROGRAM.	073163
Sweden	Sida	31210	2427	FORESTRY POLICY & ADMIN. MANAGEMENT	2007002421
Switzerland	SDC	31120	2750	APPUI AUX LA PRODUCTION RURALE	2007000360
United States	TDA	31192	320	PLANT/POST-HARVEST PROT. & PEST CTRL	2007028846

Total aid for water in 2007 for Burkina Faso   **USD thousand 73294**
And as a share of aid to total recipient countries   **1.09%**

Burundi

RECIPIENT	Agency	Sector	Amount USD thousand	Project description	CRS ID Number
Belgium	DGCD	31110	1369	APPUI INSTITUTIONNEL AU MINISTÈRE DE L'AGRICULTURE ET DE L'ÉLEVAGE	2007004237
Belgium	DGCD	31120	116	AGRICULTURAL DEVELOPMENT	2007004694
Belgium	DGCD	31165	171	ONG LOCALE ASSOCIATION ABAKENYEZI DUHAGURUKIRE ITERAMBERE	2007003143
Belgium	DGCD	31166	86	AGRICULTURAL EXTENSION	2007004428
Belgium	DGCD	31194	793	LUTTE CONTRE LA PAUVRETÉ DANS LA PROVINCE DE NGOZI - LPPN	2005007004
IFAD		31163	13978	LIVESTOCK SECTOR REHABILITATION SUPPORT PROJECT (PARSE)	070004
Ireland	DFA	31130	225	AGRICULTURAL LAND RESOURCES	2007000726
Ireland	DFA	31381	120	FISHERY EDUCATION/TRAINING	2007001084
Italy	LA	31120	274	AGRICULTURAL DEVELOPMENT	071524
Italy	DGCS	31166	5	AGRICULTURAL EXTENSION	060208
Japan	JICA	31210	8	TC AGGREGATED ACTIVITIES	073297T
Japan	JICA	31310	22	TC AGGREGATED ACTIVITIES	073468T
Spain	AG	31161	221	REINSERTION AND SOCIAL & ECO. REINTEGRATION OF THE VULNERABLE POPULATION	076000

RECIPIENT / Donor	Agency	Sector	Amount USD thousand	Project description	CRS ID Number
Spain	AG	31163	315	SOCIAL AND ECONOMIC REINTEGRATION OF RETURNING REFUGEES	076591
Spain	AG	31181	52	ALL TOGETHER FOR AUTONOMY AND GENDER EQUALITY IN BURUNDI	073363
Spain	AG	31181	104	A NEW STEP: FROM PRODUCTION TO COMMERCIALIZATION	074644
United States	AID	31120	623	AGRICULTURAL DEVELOPMENT	2007001763

Total aid for water in 2007 for Burundi USD thousand **18482**
And as a share of aid to total recipient countries **0.28%**

Cambodia

Recipient	Agency	Sector	Amount	Project description	CRS ID
Australia	AusAID	31120	209	COMMUNITY DEVELOPMENT FUND (CDF)	2007000656
Australia	AusAID	31130	46	AGRICULTURAL LAND RESOURCES	2007001410
Australia	AusAID	31150	8	DEVELOPMENT OF CONSERVATION FARMING IMPLEMENTS FOR TWO-WHEELER TRACTORS	2007002041
Australia	AusAID	31161	224	FOOD CROP PRODUCTION	2007000943
Australia	AusAID	31161	222	DIVERSIFICATION AND INTENSIFICATION OF RAINFED LOWLAND CROPPING SYSTEMS	2007000976
Australia	AusAID	31163	74	DEVELOPING BEST PRACTICE CATTLE AND BUFFALO HEALTH AND HUSBANDRY SYSTEMS	2007001161
Australia	AusAID	31182	37	IMPROVED RISK ASSESSMENT FOR TRANSBOUNDARY DISEASE IN LAOS AND CAMBODIA	2007001528
Australia	AusAID	31182	60	AGRICULTURAL RESEARCH	2007001273
Australia	AusAID	31191	48	AGRICULTURAL SERVICES	2007001373
Australia	AusAID	31192	329	PLANT/POST-HARVEST PROT. & PEST CTRL	2007001349
Australia	AusAID	31195	16	LIVESTOCK/VETERINARY SERVICES	2007001852
Australia	AusAID	31195	25	IDENTIFYING RESEARCH PRIORITIES FOR THE DEVELOPMENT OF THE BEEF INDUSTRY	2007001679
Australia	AusAID	31195	13	LIVESTOCK HEALTH AND VACCINES: SCOPING STUDY AND ECONOMIC ASSESSMENT	2007001924
Australia	AusAID	31320	33	FISHERY DEVELOPMENT	2007001583
Australia	AusAID	31382	10	FISHERY RESEARCH	2007002006
Belgium	DGCD	31120	6	AGRICULTURAL DEVELOPMENT	2004008734
Belgium	DGCD	31162	31	INDUSTRIAL CROPS/EXPORT CROPS	2004008727
Belgium	DGCD	31195	215	LE DIAGNOSE, L'EPIDEM. ET LE CONTROLE DES INFECTIONS PARASITIQUES DU BETAIL	2007004752
Canada	CIDA	31110	465	KNOWLEDGE BUILDING PROJECT -II	070651
France	AFD	31161	1150	DEVELOPPEMENT HEVEACULTURE	2007160500
France	MISC	31182	1848	RECHERCHE AGRICULTURE	2007800377
Germany	BMZ	31110	169	AGRICULTURAL POLICY & ADMIN. MGMT	2007007129
Germany	BMZ	31120	78	AGRICULTURAL DEVELOPMENT	2007005703
Germany	BMZ	31166	702	AGRICULTURAL EXTENSION	2007009295
Germany	BMZ	31181	64	AGRICULTURAL EDUCATION/TRAINING	2007008864
Germany	Fed Min	31195	1232	AVIAN INFLUENZA CONTROL (CAMBODIA)	2007010305
Germany	BMZ	31281	572	FORESTRY EDUCATION/TRAINING	2007009046
IDA		31140	7250	RURAL INVESTMENT AND LOCAL GOVERNMENT	2007000886
IFAD		31120	9514	RURAL LIVELIHOODS IMPROVEMENT PROJECT	070001
Ireland	DFA	31130	1456	SUPPORT TO NGOS - IRISH - MAPS - GENERAL - MULTI LEVEL SUPPORT	2007000225
Ireland	DFA	31310	97	FISHING POLICY AND ADMIN. MANAGEMENT	2007001199
Japan	MAFF	31110	89	TC AGGREGATED ACTIVITIES	070004T
Japan	JICA	31110	2419	TC AGGREGATED ACTIVITIES	073066T
Japan	JICA	31130	2116	TC AGGREGATED ACTIVITIES	073131T
Japan	JICA	31163	93	TC AGGREGATED ACTIVITIES	073224T
Japan	JICA	31195	54	TC AGGREGATED ACTIVITIES	073274T
Japan	JICA	31210	1037	TC AGGREGATED ACTIVITIES	073422T

RECIPIENT Donor	Agency	Sector	Amount USD thousand	Project description	CRS ID Number
Japan	MAFF	31310	78	TC AGGREGATED ACTIVITIES	070040T
Japan	JICA	31310	1412	TC AGGREGATED ACTIVITIES	073572T
Korea	KOICA	31110	1800	DEVELOPMENT STUDY FOR FORMULATION OF RURAL DEVELOPMENT POLICY AND STRATE	2007017592
Korea	Misc	31110	9	AGRICULTURAL POLICY & ADMIN. MGMT	2007002555
Korea	KOICA	31120	99	TRAINING PROGRAM	2007017399
Korea	KOICA	31120	36	AGRICULTURAL DEVELOPMENT	2007016243
Korea	Misc	31120	43	AGRICULTURAL DEVELOPMENT	2007002563
Korea	KOICA	31130	14	AGRICULTURAL LAND RESOURCES	2007016623
Korea	KOICA	31130	106	TRAINING PROGRAM	2007017401
Korea	KOICA	31140	17	AGRICULTURAL WATER RESOURCES	2007016222
Korea	KOICA	31162	150	TRAINING PROGRAM	2007017361
Korea	KOICA	31163	92	LIVESTOCK	2007017907
Korea	KOICA	31166	22	AGRICULTURAL EXTENSION	2007016895
Korea	Misc	31181	64	AGRICULTURAL EDUCATION/TRAINING	2007002566
Norway	MFA	31195	36	LIVESTOCK/VETERINARY SERVICES	2007002652
Spain	MFA	31320	353	SUSTAINABLE POVERTY REDUCTION AQUACULTURE PROMOTION AND DEVELOPMENT	076820

Total aid for water in 2007 for Cambodia   **USD thousand 36341**
And as a share of aid to total recipient countries   **0.54%**

Cameroon

Belgium	DGCD	31150	204	DEV. ECONOMIQUE ET SOCIAL DE LA COMMUNAUTE DES PYGMEES BAKA (PADES-BAKA)	2005007222
Belgium	DGCD	31194	228	RENFORCEMENT DES CAPACITéS D'ACTIONS D'ORGANISATIONS PAYSANNES	2004003369
EU institutions	CEC	31110	2987	AGRICULTURAL POLICY & ADMIN. MGMT	2007100034
EU institutions	EDF	31162	5832	INDUSTRIAL CROPS/EXPORT CROPS	2007200091
France	MISC	31182	35866	RECHERCHE AGRICULTURE	2007800339
Germany	BMZ	31166	75	AGRICULTURAL EXTENSION	2007006168
Germany	L G	31191	12	AGRICULTURAL SERVICES	2007011685
Italy	LA	31110	24	AGRICULTURAL POLICY & ADMIN. MGMT	071059
Italy	LA	31120	5	AGRICULTURAL DEVELOPMENT	070919
Japan	JICA	31110	200	TC AGGREGATED ACTIVITIES	072915T
Japan	JICA	31310	183	TC AGGREGATED ACTIVITIES	073546T
Spain	AG	31181	69	CONSTRUCTION OF A FARMING SCHOOL	073968
Spain	ICO	31310	8716	FISHING AND NAUTICAL INSTITUTE	070015
Switzerland	SDC	31110	750	COOPERATION JURA-CAMEROUN: RURAL AUX LèKIè	2002002324
United Kingdom	MISC	31210	219	ENABLING INDEPENDENT MONITORING OF FOREST RR BY LOCAL AND INDIGENOUS	2007800677

Total aid for water in 2007 for Cameroon   **USD thousand 55370**
And as a share of aid to total recipient countries   **0.83%**

Cape Verde

EU institutions	EDF	31140	684	AGRICULTURAL WATER RESOURCES	2007201281
Italy	DGCS	31195	411	LIVESTOCK/VETERINARY SERVICES	070439
Japan	JICA	31110	132	TC AGGREGATED ACTIVITIES	072916T
Japan	JICA	31195	6	TC AGGREGATED ACTIVITIES	073244T
Japan	JICA	31310	180	TC AGGREGATED ACTIVITIES	073470T

RECIPIENT Donor	Agency	Sector	Amount USD thousand	Project description	CRS ID Number
Japan	MOFA	31391	2598	PROJET D EXTENSION DES INSTALLATIONS DU PORT	070230
Portugal	ICP	31120	61	AGRICULTURAL DEVELOPMENT	070865
Portugal	ICP	31192	51	PLANT/POST-HARVEST PROT. & PEST CTRL	070593
Portugal	ICP	31195	11	LIVESTOCK/VETERINARY SERVICES	070750
Spain	MFA	31161	541	STRENGTHENING AND DIVERSIFICATION OF FOOD PRODUCTION IN CAPE VERDE	077374
Spain	MFA	31310	68	DIAGNOSIS FOR THE STRENGTHENING OF SHIPS INSPECTION CAPACITIES & SEA RESCUE	073863
Spain	MFA	31320	252	SUSTAINABLE DEVELOPMENT OF TRADITIONAL FISHERIES	076184
Spain	MFA	31381	123	FORMATION OF FISHERY AS TOOL FOR SUSTAINABLE DEVELOPMENT	074919
Spain	MFA	31391	34	DELIVERY OF EQUIPMENT FOR SURVEILLANCE AND FISHERIES INSPECTION	072742

Total aid for water in 2007 for Cape Verde   **USD thousand 5152**
And as a share of aid to total recipient countries   **0.08%**

Central African Rep.

France	MISC	31182	18344	RECHERCHE AGRICULTURE	2007800340

Total aid for water in 2007 for Central African Rep.   **USD thousand 18344**
And as a share of aid to total recipient countries   **0.27%**

Central Asia, regional

Germany	BMZ	31192	411	SUSTAINABLE LOCUST CONTROL IN CENTRALASIA	2007002643
Switzerland	SDC	31110	532	MOUNTAINS	2001002642
Switzerland	SDC	31110	158	AGRICULTURAL POLICY & ADMIN. MGMT	2007000459
Switzerland	SDC	31120	233	ORGANISATION AGRICOLE	2002002300
Switzerland	SDC	31120	50	AGRICULTURAL DEVELOPMENT	2002002301
United States	AGR	31181	250	AGRICULTURAL EDUCATION/TRAINING	2007000454
United States	AGR	31182	2959	FORMER SOVIET UNION COOPERATIVE RESEARCH PROGRAM	2007000459

Total aid for water in 2007 for Central Asia, regional   **USD thousand 4592**
And as a share of aid to total recipient countries   **0.07%**

Chad

France	MISC	31182	151	RECHERCHE AGRICULTURE	2007800341
Germany	BMZ	31120	240	AGRICULTURAL DEVELOPMENT	2007006150
Germany	BMZ	31120	471	INTEGRATED SUPPORT FOR SELF RELIANCE, SUSTAINABLE AGRIC.	2007006139
Germany	BMZ	31130	89	AGRICULTURAL LAND RESOURCES	2007006314
Italy	DGCS	31181	10	AGRICULTURAL EDUCATION/TRAINING	061097
Japan	JICA	31110	40	TC AGGREGATED ACTIVITIES	072917T
Switzerland	SDC	31110	133	AGRICULTURAL POLICY & ADMIN. MGMT	1992000440
Switzerland	SDC	31120	1363	APESS GAROUA	1999001709
Switzerland	SDC	31120	42	AGRICULTURAL DEVELOPMENT	1999001710

Total aid for water in 2007 for Chad   **USD thousand 2537**
And as a share of aid to total recipient countries   **0.04%**

Chile

Belgium	MPRF	31162	103	INDUSTRIAL CROPS/EXPORT CROPS	2007004725

RECIPIENT Donor	Agency	Sector	Amount USD thousand	Project description	CRS ID Number
Belgium	MPRF	31210	111	FORESTRY POLICY & ADMIN. MANAGEMENT	2007004729
France	MISC	31110	43	POLITIQUE AGRICOLE & GESTION ADMINISTRATIVE	2007800777
Germany	BMZ	31210	15	FORESTRY POLICY & ADMIN. MANAGEMENT	2007008779
Germany	BMZ	31220	159	FORESTRY DEVELOPMENT	2007006816
Germany	BMZ	31261	64	FUELWOOD/CHARCOAL	2007006820
Italy	LA	31140	5	AGRICULTURAL WATER RESOURCES	070041
Japan	JICA	31110	515	TC AGGREGATED ACTIVITIES	073050T
Japan	JICA	31150	31	TC AGGREGATED ACTIVITIES	073179T
Japan	JICA	31163	18	TC AGGREGATED ACTIVITIES	073211T
Japan	JICA	31210	604	TC AGGREGATED ACTIVITIES	073407T
Japan	JICA	31310	44	TC AGGREGATED ACTIVITIES	073509T
Switzerland	FA	31182	67	AGRICULTURAL RESEARCH	2007001402

Total aid for water in 2007 for Chile   **USD thousand 1779**
And as a share of aid to total recipient countries   **0.03%**

China

Donor	Agency	Sector	Amount	Project description	CRS ID
Australia	AusAID	31110	512	AUSTRALIA CHINA AGRICULTURAL TECHNICAL COOPERATION	2007001337
Australia	AusAID	31110	9	AUSTRALIA-CHINA LINKAGE FOR IMPROVED RICE COLD TOLERANCE	2007002031
Australia	AusAID	31120	49	ECONOMIC ANALYSIS OF TECHNICAL BARRIERS LIMITING AGRICULTURAL TRADE OF CHINA	2007001368
Australia	AusAID	31120	175	AGRICULTURAL DEVELOPMENT	2007001390
Australia	AusAID	31130	83	SUSTAINABLE LAND USE CHANGE IN NORTH WEST PROVINCE	2007001102
Australia	AusAID	31130	13	MORE EFFECTIVE WATER USE BY RAINFED WHEAT IN CHINA AND AUSTRALIA	2007001922
Australia	AusAID	31130	19	AGRICULTURAL LAND RESOURCES	2007002042
Australia	AusAID	31130	53	IMPACT OF RE-VEGETATION ON WATER RESOURCES	2007001335
Australia	AusAID	31140	21	EVALUATION OF CATCHMENT FILTER PILOT STUDY IN SHANXI, CHINA	2007001743
Australia	AusAID	31140	113	AGRICULTURAL WATER RESOURCES	2007001324
Australia	AusAID	31163	33	ESTABLISHMENT OF BEEF INDUSTRIES IN AN ADDITIONAL 10 RED SOIL PROVINCES	2007001569
Australia	AusAID	31181	21	TRAVEL TO CHINA FOR LUCERNE TRAINING WORKSHOPS	2007001756
Australia	AusAID	31182	178	OILSEED BRASSICA IMPROVEMENT	2007001670
Australia	AusAID	31182	63	SUSTAINABLE CONTROL OF YELLOW RUST OF WHEAT IN ASIA	2007001752
Australia	AusAID	31182	59	DIISR MOLECULAR MECHANISMS OF DISEASE BIO-CONTROL. (LY88)	2007001285
Australia	AusAID	31182	412	AGRICULTURAL RESEARCH	2007001350
Australia	AusAID	31192	157	PLANT/POST-HARVEST PROT. & PEST CTRL	2007001532
Australia	AusAID	31192	7	POSTHARVEST HANDLING AND DISEASE CONTROL IN MELONS	2007002093
Australia	AusAID	31192	82	TO REDUCE SPOILAGE AND CONTAMINATION RISKS OF FRESH VEGETABLES	2007001113
Australia	AusAID	31192	20	ACIAR - MANAGEMENT OF SMALL MAMMALS IN TIBET	2007001774
Australia	AusAID	31192	13	FINALISE UPTAKE OF RESULTS FOR IMPROVED MANAGEMENT OF RODENTS	2007001932
Australia	AusAID	31195	64	LIVESTOCK/VETERINARY SERVICES	2007001680
Australia	AusAID	31282	246	FORESTRY RESEARCH	2007001652
Austria	BMF	31140	1730	INTEREST SUBSIDY COMMITMENT: WATER SAVING IRRIGATION	2007604023
Belgium	DGCD	31192	172	NEMATODES ENTOMOPATHOGENIQUES POUR UN CONTROLE DURABLE DES CHIVE MIDGE	2007004746
Canada	IDRC	31110	31	BIOSAFETY MANAGEMENT OF GENETICALLY MODIFIED CROPS (CHINA)	070185i
Canada	CIDA	31161	99	VS/DEHYDRATED ALFALFA PRODUCTION - VS - DEHYDRATED ALFALFA PRODUCTION	070164
Canada	CIDA	31163	197	PS/PEAK SWINE/HSAHTRAINING PROGRAM	070867
Finland	MFA	31150	6079	HEILONGJIANG TRACTOR PROJECT	2007002002

RECIPIENT Donor	Agency	Sector	Amount USD thousand	Project description	CRS ID Number
Finland	MFA	31161	2005	SHIHEZI COLD STORAGE PROJECT	2007002003
Finland	MFA	31166	259	TRAIN TIBETAN FARMERS IN GONGHE	2007000256
Finland	MFA	31210	192	MASTER PLAN FOR FOREST PLANTATIONS IN CHINA	2007006016
France	MISC	31110	74	POLITIQUE AGRICOLE & GESTION ADMINISTRATIVE	2007800783
France	MISC	31182	411	RECHERCHE AGRICULTURE	2007800378
Germany	KFW	31120	6160	SUST. DEVEL. OF DISADVANT. AREAS, YUNNAN	2007000493
Germany	L G	31181	21	AGRICULTURAL EDUCATION/TRAINING	2007011011
Germany	BMZ	31182	549	MANAGING THE GLOBAL GENETIC RESOURCES	2007007651
Germany	BMZ	31182	102	AGRICULTURAL RESEARCH	2007007650
Germany	KFW	31210	6160	PROG. SUSTAIN. FORESTRY MANAG. (GUIZHOU)	2007000492
Germany	BMZ	31220	55	FORESTRY DEVELOPMENT	2007002369
Germany	Fed Min	31282	1859	RESEARCH FOR A FOREWARD-LOOKING FORESTRY INDUSTRY	2007010538
IFAD		31110	30001	INNER MONGOLIA AUTONOMOUS REGION RURAL ADVANCEMENT PROGRAMME	070011
Italy	Art.	31130	15058	POVERTY REDUCTION AND RURAL DEVELOPMENT IN HETIAN PREFECTURE XINJIANG	070005
Japan	JICA	31110	898	TC AGGREGATED ACTIVITIES	073069T
Japan	Oth. MIN	31110	5	TC AGGREGATED ACTIVITIES	075148T
Japan	PRF	31110	19	TC AGGREGATED ACTIVITIES	075590T
Japan	JICA	31130	967	TC AGGREGATED ACTIVITIES	073133T
Japan	PRF	31130	70	TC AGGREGATED ACTIVITIES	075591T
Japan	JICA	31150	119	TC AGGREGATED ACTIVITIES	073183T
Japan	PRF	31161	45	TC AGGREGATED ACTIVITIES	075596T
Japan	JICA	31163	576	TC AGGREGATED ACTIVITIES	073242T
Japan	PRF	31163	12	TC AGGREGATED ACTIVITIES	075597T
Japan	PRF	31166	21	TC AGGREGATED ACTIVITIES	075599T
Japan	PRF	31181	28	TC AGGREGATED ACTIVITIES	075601T
Japan	PRF	31182	17	TC AGGREGATED ACTIVITIES	075607T
Japan	PRF	31192	14	TC AGGREGATED ACTIVITIES	075608T
Japan	JICA	31195	27	TC AGGREGATED ACTIVITIES	073275T
Japan	JICA	31210	4826	TC AGGREGATED ACTIVITIES	073430T
Japan	JBIC	31220	80645	JILIN AFFORESTATION PROJECT	073034
Japan	JBIC	31220	53480	QINGHAI ECOLOGICAL ENVIRONMENTAL IMPROVEMENT PROJECT	073088
Japan	PRF	31281	24	TC AGGREGATED ACTIVITIES	075612T
Japan	JICA	31310	50	TC AGGREGATED ACTIVITIES	073523T
Japan	MAFF	31320	48	TC AGGREGATED ACTIVITIES	070057T
Japan	PRF	31320	20	TC AGGREGATED ACTIVITIES	075613T
Korea	Misc	31110	108	CONSULTING FOR RURAL DEVELOPMENT MODEL IN CHINA	2007002550
Korea	KOICA	31130	110	SUPPORT TO NGO	2007017905
Korea	KOICA	31162	22	INDUSTRIAL CROPS/EXPORT CROPS	2007017903
Korea	KOICA	31166	21	AGRICULTURAL EXTENSION	2007016334
Korea	Misc	31181	5	AGRICULTURAL EDUCATION/TRAINING	2007002564
Korea	Misc	31210	323	THE PROJECT OF REHABILITATION OF ECOLOGY AND ENVIRONMENT IN CHINA	2007010937
Korea	KOICA	31281	23	FORESTRY EDUCATION/TRAINING	2007015733
Netherlands	MFA	31110	2798	AGRIC.: SUBS. TC PSOM 2003-2009	2007100148
New Zealand	NZAid	31120	11	YINCHUAN FARMING	070366
New Zealand	NZAid	31161	57	SAFE VEGETABLES	070527
New Zealand	NZAid	31161	46	TIBET SHANNAN ZHANANG COUNTY GREENHOUSE BUILDING	070532
New Zealand	NZAid	31162	45	PEPPERCORN PRODUCTION TRG	070531

RECIPIENT Donor	Agency	Sector	Amount USD thousand	Project description	CRS ID Number
New Zealand	NZAid	31163	406	GUIZHOU: KARST MOUNTAIN PROJECT	070422
Norway	NORAD	31140	10	DEVELOPMENT PROJECT IN TIBET AGRICULTURAL AND WATER SOURCES	2007003554
Norway	NORAD	31140	43	AGRICULTURAL WATER RESOURCES	2007003618
Norway	NORAD	31220	9	DEVELOPMENT PROJECT IN TIBET- AGRICULTURAL AND FORESTRY	2007003667
Norway	NORAD	31320	147	PILOT PROJECT FOR CRAYFISH CULTURE	2007002328
Norway	MFA	31320	14	Training PARTNERS ON TILAPIA HATCHERY/GROW-OUT MANAGEMENT AND OPERATIONS	2007002681
Norway	MFA	31320	11	DEVELOPING A HIGH STD TILAPIA HATCHERY AND GROW-OUT MANAGEMENT PROTOCOL	2007002571
Spain	MIE	31110	267	MUSHROOM FARMING DEVELOPMENT IN CHINA	076268
UNDP		31163	22	LIVESTOCK	2007014852
United Kingdom	MISC	31210	96	FACILITATING LEGALITY IN THE GLOBAL COMMUNITY CHAIN OF FOREST PRODUCTS	2007800718
United States	AGR	31210	173	FORESTRY POLICY & ADMIN. MANAGEMENT	2007000457

Total aid for water in 2007 for China   **USD thousand 220295**
And as a share of aid to total recipient countries   **3.28%**

Colombia

Belgium	MPRF	31162	7	INDUSTRIAL CROPS/EXPORT CROPS	2007004733
Belgium	DGCD	31181	175	DéVELOPPEMENT HORTICULTURE BOGOTA SAVANNE	2007008600
France	MISC	31182	7543	RECHERCHE AGRICULTURE	2007800372
Germany	BMZ	31110	222	AGRICULTURAL POLICY & ADMIN. MGMT	2007006293
Germany	BMZ	31120	657	AGRICULTURAL DEVELOPMENT	2007008685
Germany	BMZ	31130	25	AGRICULTURAL LAND RESOURCES	2007008691
Germany	BMZ	31165	48	AGRICULTURAL ALTERNATIVE DEVELOPMENT	2007006882
Germany	BMZ	31182	75	AGRICULTURAL RESEARCH	2007007636
Germany	BMZ	31195	64	LIVESTOCK/VETERINARY SERVICES	2007006879
Germany	BMZ	31210	63	FORESTRY POLICY & ADMIN. MANAGEMENT	2007008775
Germany	KFW	31220	19304	SILVICULT SUSTAINABLE IN CAFETERA ZONE	2007000817
Italy	DGCS	31120	1316	CONTRIBUTION TO FAO	070626
Japan	JICA	31110	504	TC AGGREGATED ACTIVITIES	073051T
Japan	JICA	31210	371	TC AGGREGATED ACTIVITIES	073409T
Japan	JICA	31310	26	TC AGGREGATED ACTIVITIES	073510T
Luxembourg	MFA	31120	118	AGRICULTURAL DEVELOPMENT	070388
Netherlands	MFA	31110	723	AGRIC.: SUBS. TO PSOM 2003-2009	2007100157
Spain	AG	31120	242	PROGRAM FOR SOCIAL MANAGEMENT AND ENVIRONMENTAL PARTICIPATION	076123
Spain	AG	31120	38	SOCIAL TECHNICAL AND AGRO ENTERPRISE DEVELOPMENT OF SMALL FARMERS	072860
Spain	AG	31120	435	PRODUCTION OF ORGANIC COFFEE FOR SUSTAINABILITY IN THE PAEZ INDIGENOUS COM.	077173
Spain	AG	31120	41	ASOCAL: A COMMUNITY INITIATIVE FOR LOCAL DEVELOPMENT	072947
Spain	AG	31120	51	COMPREHENSIVE PROJECT ON FOOD SOVEREIGNTY	073351
Spain	AG	31120	67	COMPREHENSIVE PROJECT ON PRODUCTION DEV. AGRICULTURE AND CATTLE FARMING	073818
Spain	AG	31161	425	PROJECT ON FOOD SOVEREIGNTY & HUMAN RIGHTS WITH FARMING COMMUNITIES	077146
Spain	AG	31161	96	FOOD CROP PRODUCTION: STRENGTHENING OF INDIGENOUS IDENTITY	074510
Spain	AG	31161	193	SUPPORT FOR THE SOCIOECONOMIC REESTABLISHMENT AND FOOD SECURITY	075718
Spain	AG	31163	110	PROJECT ON FOOD SOVEREIGNTY PROMOTION OF HUMAN RIGHTS	074743
Spain	AG	31165	38	EDUCATIONAL PROPOSALS ON BIO CONSTRUCTION AND ECOLOGICAL TOURISM	072864
Spain	AG	31165	52	IMPLEMENTATION OF A TECHNICAL SCHOOL ON AGRO ECOLOGY	073374
Spain	AG	31320	583	INT. COOP. IN THE FIELD OF TECHNICAL EXCHANGE REGARDING THE FARMING OF MERO	077455
United States	AID	31165	46000	ALTERNATIVE DEVELOPMENT AND ALTERNATIVE LIVELIHOODS	2007001898

RECIPIENT / Donor	Agency	Sector	Amount USD thousand	Project description	CRS ID Number
Total aid for water in 2007 for Colombia And as a share of aid to total recipient countries					**USD thousand 79609** 1.19%
Comoros					
Belgium	DGCD	31163	119	LIVESTOCK	2004008413
Belgium	DGCD	31166	58	AGRICULTURAL EXTENSION	2004008421
Belgium	DGCD	31382	26	FISHERY RESEARCH	2005008508
EU institutions	CEC	31164	1287	AGRARIAN REFORM	2007100444
IFAD		31110	4654	NATIONAL PROGRAMME FOR SUSTAINABLE HUMAN DEVELOPMENT	070002
UNDP		31310	85	FISHING POLICY AND ADMIN. MANAGEMENT	2007014408
Total aid for water in 2007 for Comoros And as a share of aid to total recipient countries					**USD thousand 6228** 0.09%
Congo, Dem. Rep.					
Belgium	DGCD	31110	3422	RESTRUCTURATION DES SERVICES DU MINISTÈRE DE L'AGRICULTURE	2007004383
Belgium	DGCD	31110	38	AGRICULTURAL POLICY & ADMIN. MGMT	2004003659
Belgium	DGCD	31110	375	APPUI A LA DEFINITION DES POLITIQUES DE DEVELOPPEMENT AGRICOLE RDC 2004-2007	2005008406
Belgium	DGCD	31120	36	AGRICULTURAL DEVELOPMENT	2005008424
Belgium	DGCD	31120	159	DÉVELOPPEMENT AGRICOLE DANS LA REGION YAKA	2004008813
Belgium	DGCD	31120	1605	PROGRAMME DE RELANCE AGRICOLE DANS LA PROVINCE DE L'EQUATEUR	2004003904
Belgium	DGCD	31120	189	SÉCURITÉ ALIMENTAIRE BASANKUSU	2006001421
Belgium	DGCD	31120	510	SÉCURITÉ ALIMENTAIRE KANANGA	2004003878
Belgium	DGCD	31120	185	SÉCURITÉ ALIMENTAIRE PLATEAU DES BATEKES	2005000416
Belgium	DGCD	31161	247	FOOD CROP PRODUCTION	2007004684
Belgium	DGCD	31161	1361	INERA RECHERCHE PRODUCTION VÉGÉTALE	2005001584
Belgium	MPRF	31162	22	INDUSTRIAL CROPS/EXPORT CROPS	2007004738
Belgium	DGCD	31165	392	FAO - HUP RDC - PROGRAMME 2001-2007	2005002432
Belgium	MPRF	31181	55	AGRICULTURAL EDUCATION/TRAINING	2007004737
Belgium	DGCD	31193	66	AGRICULTURAL FINANCIAL SERVICES	2004005353
Belgium	DGCD	31194	231	AMELIORER LA SITUATION SANITAIRE ET NUTRITIONNELLE DE LA POPULATION	2004005317
Belgium	DGCD	31194	188	RENFORCEMENT DES ORGANISATIONS DES ONG'S	2006003975
Belgium	MPRF	31194	30	AGRICULTURAL CO-OPERATIVES	2005002010
Belgium	DGCD	31210	526	AIDE DIVERS	2005008498
Belgium	DGCD	31210	956	FAO- FORESTERIE COMMUNAUTAIRE RDC - 2004-2007	2006002209
Belgium	DGCD	31210	1369	FONDS FIDUCIAIRE SECTEUR FORESTIER	2007908524a
Germany	BMZ	31120	411	SUSTAINABLE AGRICULTURE WITH LAND RESOURCE MANAGEMENT FOR SMALL GROUPS	2007006193
Germany	BMZ	31120	890	SUPPORT OF REGIONAL DEVELOPMENT WITH EMPHASIS ON AGRICULTURE AND TRAINING	2007006642
Germany	BMZ	31210	4107	PROGRAMM CONSERVATION OF BIODIVERSITY AND FOREST MANAGEMENT	2007003907
Greece	YPEJ	31163	66	CREATION OF A CATTLE BREEDING UNIT	2007000218
Ireland	DFA	31120	29	AGRICULTURAL DEVELOPMENT	2007001770
Ireland	DFA	31150	251	AGRICULTURAL INPUTS	2007000683
Ireland	DFA	31161	542	SUPPORT TO NGOS - IRISH - MAPS - GENERAL - AGRICULTURE & LIVESTOCK SUPPORT	2007000234
Italy	CA	31181	694	AGRICULTURAL EDUCATION/TRAINING	071319
Italy	LA	31210	67	FORESTRY POLICY & ADMIN. MANAGEMENT	071064
Italy	LA	31220	67	FORESTRY DEVELOPMENT	071049

RECIPIENT Donor	Agency	Sector	Amount USD thousand	Project description	CRS ID Number
Luxembourg	MFA	31161	55	FOOD CROP PRODUCTION	070416
Luxembourg	MFA	31210	274	FORESTRY POLICY & ADMIN. MANAGEMENT	071651
Norway	NORAD	31120	42	LIKATI AGRICULTURAL PROGRAM	2007003557
Norway	NORAD	31120	145	BANDUNDU FOREST & BEE KEEPING PROGRAMME	2007003704
Spain	AG	31120	328	FOOD SOVEREIGNTY OF FARMING FAMILIES IN THE REGION OF CIYAMBA	076648
Spain	AG	31120	193	AGRICULTURAL PILOT OPERATION IN KINSHASA	075720
Spain	AG	31161	181	IMPROVEMENT OF THE SOCIOECONOMIC DEVELOPMENT OF COMMUNITIES	075639
Spain	AG	31161	27	NUTRITIONAL IMPROVEMENT OF ONE HUNDRED RURAL FAMILIES OF KINDELE	072411
Spain	AG	31163	150	STRENGTHENING OF THE ALIMANTARY SOVEREIGNTY IN THE RURAL AREAS	075322
Switzerland	SDC	31110	12	AGRICULTURAL POLICY & ADMIN. MGMT	1980001039
United States	AID	31110	289	AGRICULTURAL POLICY & ADMIN. MGMT	2007013850
United States	AID	31120	882	AGRICULTURAL SECTOR PRODUCTIVITY	2007001712
United States	AID	31120	150	AGRICULTURAL DEVELOPMENT	2007001721
United States	STATE	31210	25	FORESTRY POLICY & ADMIN. MANAGEMENT	2007025833

Total aid for water in 2007 for Congo, Dem. Rep. **USD thousand 21838**
And as a share of aid to total recipient countries **0.33%**

Congo, Rep.

Donor	Agency	Sector	Amount	Project description	CRS ID
EU institutions	EDF	31162	958	INDUSTRIAL CROPS/EXPORT CROPS	2007201272
France	MISC	31182	7050	RECHERCHE AGRICULTURE	2007800342
IDA		31110	5000	AGRICULTURAL REHABILITATION	2007000179
IDA		31161	5400	AGRICULTURAL REHABILITATION	2007000175
IDA		31163	1600	AGRICULTURAL REHABILITATION	2007000176
Spain	MFA	31120	660	RURAL DEVELOPMENT	077519

Total aid for water in 2007 for Congo, Rep. **USD thousand 20669**
And as a share of aid to total recipient countries **0.31%**

Costa Rica

Donor	Agency	Sector	Amount	Project description	CRS ID
France	MISC	31182	18891	RECHERCHE AGRICULTURE	2007800362
Germany	BMZ	31120	890	PEASANT SELF HELP GROUPS FOR SUSTAINABILITY OF AGRICULTURAL PLANTS	2007005584
Germany	BMZ	31191	51	AGRICULTURAL SERVICES	2007006883
Germany	BMZ	31210	5	FORESTRY POLICY & ADMIN. MANAGEMENT	2007008772
Japan	JICA	31110	542	TC AGGREGATED ACTIVITIES	072948T
Japan	JICA	31210	64	TC AGGREGATED ACTIVITIES	073318T
Japan	JICA	31310	567	TC AGGREGATED ACTIVITIES	073558T
Spain	AG	31120	21	AGRICULTURAL DEVELOPMENT	071997

Total aid for water in 2007 for Costa Rica **USD thousand 21032**
And as a share of aid to total recipient countries **0.31%**

Cote d'Ivoire

Donor	Agency	Sector	Amount	Project description	CRS ID
Belgium	DGCD	31163	519	DéVELOPPEMENT LAITIER DANS LE SUD	2002000175
Belgium	DGCD	31320	231	INITIATIVE PROPRE KUL	2007002401
EU institutions	CEC	31120	29062	AGRICULTURAL DEVELOPMENT	2007100445
EU institutions	EDF	31162	5900	INDUSTRIAL CROPS/EXPORT CROPS	2007201279
France	MISC	31182	329	RECHERCHE AGRICULTURE	2007800348

RECIPIENT Donor	Agency	Sector	Amount USD thousand	Project description	CRS ID Number
Greece	YPEJ	31181	424	AGRICULTURAL EDUCATION/TRAINING	2007000387
Italy	LA	31120	5	AGRICULTURAL DEVELOPMENT	071695
Italy	LA	31191	14	AGRICULTURAL SERVICES	070039
Italy	LA	31194	7	AGRICULTURAL CO-OPERATIVES	071730
Japan	JICA	31110	100	TC AGGREGATED ACTIVITIES	072924T
Japan	JICA	31150	180	TC AGGREGATED ACTIVITIES	073142T
Japan	JICA	31310	62	TC AGGREGATED ACTIVITIES	073478T
Spain	MFA	31120	110	SUPPORT PROJECT FOR THE DEVELOPMENT OF AGRO PASTORALIST ACTIVITIES	074728
Spain	MFA	31310	46	CONTINENTAL AQUACULTURE DEVELOPMENT PROGRAM.	073164

Total aid for water in 2007 for Cote d'Ivoire USD thousand   **36988**
And as a share of aid to total recipient countries   **0.55%**

Croatia

RECIPIENT Donor	Agency	Sector	Amount USD thousand	Project description	CRS ID Number
EU institutions	EDF	31110	8385	NATIONAL PROGRAMME FOR CROATIA UNDER THE IPA-TRANSITION ASSISTANCE AND INSTITUTION BUILDING COMPONENT FOR 2007	2007202288
Norway	MFA	31194	28	AGRICULTURAL CO-OPERATIVES	2007001655
United States	AID	31110	133	AGRICULTURAL POLICY & ADMIN. MGMT	2007013820
United States	AID	31120	750	AGRICULTURAL SECTOR PRODUCTIVITY	2007001703
United States	TDA	31140	496	METEOROLOGICAL AND HYDROLOGICAL SERVICE MODERNIZATION - FEASIBILITY STUDY	2007028830

Total aid for water in 2007 for Croatia   USD thousand   **9791**
And as a share of aid to total recipient countries   **0.15%**

Cuba

RECIPIENT Donor	Agency	Sector	Amount USD thousand	Project description	CRS ID Number
Belgium	DGCD	31110	96	AGRICULTURAL POLICY & ADMIN. MGMT	2005000976
Japan	JICA	31110	69	TC AGGREGATED ACTIVITIES	073041T
Japan	JICA	31150	43	TC AGGREGATED ACTIVITIES	073178T
Japan	JICA	31310	44	TC AGGREGATED ACTIVITIES	073559T
Netherlands	MFA	31110	84	AGRICULTURAL POLICY & ADMIN. MGMT	2007100127
Spain	AG	31110	339	LOCAL CAPACITIES TO IMPLEMENT THE PROGR. FOR THE INNOVATION OF LOCAL AGR.	076695
Spain	AG	31120	56	IMPROVEMENT OF THE ELECTRICAL CAPACITY AND PRODUCTIVE INFRASTRUCTURE	073528
Spain	AG	31120	72	ORGANISATIONAL CAPACITY BUILDING FOR SELF MANAGEMENT	074016
Spain	AG	31120	152	SOCIAL AND PRODUCTIVE DEVELOPMENT OF THE COOPERATIVE AND FARMING SECTOR	075359
Spain	AG	31120	157	SUPPORT FOR THE NATIONAL MEDICINAL PLANTS PROG. IN THE EASTERN REGION	075386
Spain	AG	31120	63	TRANSFORMATION OF A SUGARCANE UBPC	073711
Spain	AG	31120	93	SOCIAL AND PRODUCTIVE DEVELOPMENT OF THE CREDIT AND SERVICES COOPERATIVE	074446
Spain	AG	31140	66	MITIGATING THE EFFECTS OF DROUGHT	073788
Spain	AG	31161	1187	STRENGTHENING URBAN AGRICULTURE IN THE MUNICIPALITIES OF JAG-EY GRANDE PEDRO	077813
Spain	EDUC	31163	14	OBTAINING AND EVALUATION OF PROBIOTICS FOR RUMINANTS' FEEDING	071405
Spain	AG	31163	341	SUPPORT FOR THE RECOVERY OF FARMING (LIVESTOCK) PRODUCTION	076697
Spain	AG	31164	44	AGRARIAN REFORM	073134
Spain	Misc	31182	14	OBTAINING AND EVALUATION OF PROBIOTICS FOR RUMINANTS' FEEDING	071467
Spain	AG	31194	163	ENHANCE IN A SUSTAINABLE WAY THE DIVERSIFICATION OF PRODUCTION	075435
Spain	AG	31194	42	AGRICULTURAL CO-OPERATIVES	073075
Spain	AG	31194	35	EMPOWERMENT OF THE COOPERATION'S OF SMALL OWNERS	072796
Switzerland	SDC	31220	108	FORESTRY DEVELOPMENT	2005000443

RECIPIENT Donor	Agency	Sector	Amount USD thousand	Project description	CRS ID Number
		Total aid for water in 2007 for Cuba			**USD thousand 3283**
		And as a share of aid to total recipient countries			0.05%

Djibouti

Japan	JICA	31110	250	TC AGGREGATED ACTIVITIES	072939T
Japan	JICA	31210	16	TC AGGREGATED ACTIVITIES	073310T
		Total aid for water in 2007 for Djibouti			**USD thousand 265**
		And as a share of aid to total recipient countries			0%

Dominica

Japan	JICA	31110	34	TC AGGREGATED ACTIVITIES	072960T
Japan	JICA	31310	155	TC AGGREGATED ACTIVITIES	073504T
Japan	MAFF	31381	34	TC AGGREGATED ACTIVITIES	070071T
		Total aid for water in 2007 for Dominica			**USD thousand 223**
		And as a share of aid to total recipient countries			0%

Dominican Republic

Belgium	DGCD	31161	43	FOOD CROP PRODUCTION	2005000338
France	MISC	31182	55	RECHERCHE AGRICULTURE	2007800363
Germany	BMZ	31130	8	AGRICULTURAL LAND RESOURCES	2007009309
Germany	BMZ	31162	26	INDUSTRIAL CROPS/EXPORT CROPS	2007006897
Germany	BMZ	31182	65	AGRICULTURAL RESEARCH	2007006896
Japan	JICA	31110	941	TC AGGREGATED ACTIVITIES	073042T
Japan	JICA	31130	635	TC AGGREGATED ACTIVITIES	073091T
Japan	JICA	31163	43	TC AGGREGATED ACTIVITIES	073202T
Japan	JICA	31210	793	TC AGGREGATED ACTIVITIES	073385T
Korea	KOICA	31120	72	AGRICULTURAL DEVELOPMENT	2007016802
Spain	MFA	31110	342	FOOD SOVEREIGNTY THROUGH SUPPORT TO SMALL PRODUCERS	076740
Spain	MFA	31120	583	IRRIGATION AGRICULTURAL PRODUCTION MARKETING AND SOCIAL DEVELOPMENT	077454
Spain	MFA	31120	14	PRODUCTION OF BIO BEAN FERTILIZERS	071403
Spain	AG	31120	44	IMPROVEMENT OF THE COFFEE PRODUCTION INFRASTRUCTURE. DOMINICAN REPUBLIC	073140
Spain	AG	31120	205	A DECENT LIFE FOR THE FAMILIES OF SMALL PRODUCERS OF COFFEE	075821
Spain	AG	31120	44	ENCOURAGEMENT OF THE PROCESSING OF THE AGRICULTURAL PRODUCTION IN POLO	073124
Spain	AG	31120	21	AGRICULTURAL DEVELOPMENT	072015
Spain	Misc	31120	14	PRODUCTION OF BIO BEAN FERTILIZERS	071466
Spain	AG	31140	102	TRANSFERENCE OF INTERNATIONAL INNOVATION FOR THE USE & MGMT OF WATER	074592
Spain	AG	31161	376	CAPACITY BUILDING FOR FOOD SOVEREIGNTY DEFENCE IN THE COFFEE GROWING SECT.	076875
Spain	AG	31161	224	ENCOURAGEMENT OF GENERATION RELIEF AND GENDER EQUALITY IN THE FAMILY COFFE	076006
Spain	AG	31161	55	FOOD CROP PRODUCTION	073466
Spain	AG	31161	269	PRODUCTION OF COFFEE FRUIT TREES AND TIMBER WITH AN AGROFORESTRY APPROACH	076284
Spain	AG	31161	153	PROMOTING GENDER EQUALITY AND FOOD SOVEREIGNTY IN THE EASTERN REGION	075367
Spain	AG	31163	235	THE CACHIMBO BREEDING OF COWS FOR THE PRODUCTION OF CHEESE PASTEURIZED MILK	076090
Spain	MFA	31182	753	PROGRAM ON AGRICULTURAL TECHNOLOGY OF THE SOUTH (PROTESUR)	077633
Spain	MFA	31194	274	RURAL DEVELOPMENT	076317
United States	AID	31110	92	AGRICULTURAL POLICY & ADMIN. MGMT	2007013886

RECIPIENT Donor	Agency	Sector	Amount USD thousand	Project description	CRS ID Number
United States	AID	31120	4815	AGRICULTURAL SECTOR PRODUCTIVITY	2007001747

Total aid for water in 2007 for Dominican Republic   **USD thousand 11296**
And as a share of aid to total recipient countries   **0.17%**

Ecuador

Donor	Agency	Sector	Amount	Project description	CRS ID
Austria	ADA	31140	70	BUSINESS PARTNERSHIP PROGRAM: IRRIGATION FOR HIGH QUALITY COFFEE	2007500108a
Belgium	DGCD	31110	75	AGRICULTURAL POLICY & ADMIN. MGMT	2007003896
Belgium	DGCD	31120	675	RENFORCEMENT DE L'AGRICULTURE DURABLE	2006003979
Belgium	DGCD	31120	439	RENFORCEMENT DU DEV. ECO. ET SOCIAL DES MICRO ET PETIT PRODUCTEURS	2004003461
Belgium	DGCD	31120	136	AGRICULTURAL DEVELOPMENT	2007004651
Belgium	DGCD	31120	19165	DÉVELOPPEMENT RURAL ORIENTÉ VERS LA GÉNÉRATION DE REVENUS	2007004412
Belgium	DGCD	31163	977	PRODUCTION DE PRODUITS CARNÉS SAINES DANS LE NORD DE L'EQUATEUR PROCANOR	2004001147
Belgium	DGCD	31182	248	IMT- CENTRE INTERNATIONAL DE ZOONOSE	2005000047
Belgium	DGCD	31194	136	AGRICULTURAL CO-OPERATIVES	2005001105
Belgium	DGCD	31194	1354	APPUI AUX FAMILLES DE PETITS CAFICULTEURS DE MANABÍ	2004001152
Belgium	MPRF	31210	189	FORESTRY POLICY & ADMIN. MANAGEMENT	2007004739
Canada	CIDA	31120	172	DÉVELOPPEMENT DURABLE DES AGRICULTEURS	070742
Canada	IDRC	31162	163	ENVIRONMENTAL AND HEALTH IMPACTS OF FLORICULTURE IN ECUADOR - PHASE II	070160i
France	MISC	31182	4162	RECHERCHE AGRICULTURE	2007800373
Germany	BMZ	31120	1754	AGRICULTURAL DEVELOPMENT	2007008705
Germany	BMZ	31130	29	AGRICULTURAL LAND RESOURCES	2007008711
Germany	BMZ	31182	167	AGRICULTURAL RESEARCH	2007007639
Germany	BMZ	31191	85	AGRICULTURAL SERVICES	2007006920
Germany	BMZ	31210	179	FORESTRY POLICY & ADMIN. MANAGEMENT	2007008776
Germany	BMZ	31220	65	FORESTRY DEVELOPMENT	2007006909
Germany	BMZ	31291	82	FORESTRY SERVICES	2007006905
Italy	DGCS	31161	354	FOOD CROP PRODUCTION	070317
Italy	DGCS	31191	19	AGRICULTURAL SERVICES	060536
Italy	LA	31210	27	FORESTRY POLICY & ADMIN. MANAGEMENT	071566
Japan	JICA	31110	283	TC AGGREGATED ACTIVITIES	073052T
Japan	MOFA	31150	3820	THE ASSISTANCE FOR UNDERPRIVILEGED FARMERS	070096
Japan	JICA	31150	37	TC AGGREGATED ACTIVITIES	073159T
Japan	JICA	31163	64	TC AGGREGATED ACTIVITIES	073212T
Japan	JICA	31210	74	TC AGGREGATED ACTIVITIES	073410T
Japan	JICA	31310	7	TC AGGREGATED ACTIVITIES	073511T
Japan	MAFF	31320	300	TC AGGREGATED ACTIVITIES	070056T
Korea	KOICA	31120	23	AGRICULTURAL DEVELOPMENT	2007016538
Korea	KOICA	31181	23	AGRICULTURAL EDUCATION/TRAINING	2007016736
Korea	KOICA	31381	21	FISHERY EDUCATION/TRAINING	2007016157
Luxembourg	MFA	31181	99	AGRICULTURAL EDUCATION/TRAINING	070172
Spain	MFA	31110	753	PROMOTE FOOD SOVEREIGNTY: GUARANTEE ACCESS TO ENOUGH FOOD NUTRITIVE	077636
Spain	AG	31110	821	BOLIVAR PARTNERSHIPS FOR DEVELOPMENT	077679
Spain	MFA	31120	252	IMPLEMENTATION OF AGRO ECOLOGICAL PRODUCTION SYSTEMS IN SEVILLA DE ORO	076178
Spain	MFA	31120	319	AGRICULTURAL DEVELOPMENT	076602
Spain	AG	31120	165	IMPLEMENTATION OF A PROD. SS CENTRE IN THE NORTHERN REGION OF MANABI	075507

RECIPIENT Donor	Agency	Sector	Amount USD thousand	Project description	CRS ID Number
Spain	AG	31120	171	IMPROVEMENT OF THE COMMUNITY RISK SYSTEM OF MALAL GAZA AND ZHUYA	075575
Spain	AG	31120	42	PROMOTING A SUSTAINABLE HUMAN DEV. IN CHIMBORAZO AND COTOPAXI ECUADOR	073079
Spain	AG	31120	82	SYSTEM OF AGRICULTURAL PRODUCTION; PROCESSING AND COMERCIALIZATION	074258
Spain	AG	31120	156	SUPPORT FOR INDIGENOUS ORG. PEASANTS & ECUADORIAN AFRO-AMERICANS	075385
Spain	AG	31120	188	PROMOTING THE DEVELOPMENT OF THE SMALL WHEAT PRODUCER RURAL COMMUNITIES	075678
Spain	AG	31130	172	PROJECT ON SUSTAINABLE MANAGEMENT OF NATURAL RESOURCES	075583
Spain	AG	31140	190	REHABILITATION OF THE RISKSYSTEM OF SUNFO LAIGUA	075687
Spain	AG	31161	127	IMPROVEMENT OF COMMERCIAL AND PRODUCTIVE DEVELOPMENT OF CACAO PRODUCERS	074985
Spain	AG	31161	411	ENCOURAGEMENT OF FOOD SOVEREIGNTY	077097
Spain	AG	31161	298	FOOD SOVEREIGNTY AND IMPROVEMENT OF THE INCOME	076502
Spain	AG	31163	46	IMPLEMENTATION AND FUNCTIONING OF A POULTRY FARM	073178
Spain	AG	31165	49	PRODUCTION AND MARKETING OF PRODUCTS OF MEDICINAL PLANTS	073285
Spain	AG	31166	88	AGRO ECOLOGICAL DEVELOPMENT OF THE INDIGENOUS COMMUNITIES	074369
Spain	AG	31182	136	AGRICULTURAL RESEARCH	075082
Spain	AG	31182	316	SUSTAINABLE HUMAN DEVELOPMENT OF SMALL CACAO PRODUCERS	076594
Spain	AG	31191	314	STRENGTHENING THE PRODUCTIVE AND COMMERCIAL ASSOCIATIVE SYSTEM	076567
Spain	MFA	31194	177	PROMOTING SUSTAINABLE HUMAN DEVELOPMENT OF 6 RURAL ORGANISATIONS	075613
Spain	AG	31194	162	ORGANISATIONS OF CACAO PRODUCERS OF THE ECUADORIAN COAST	075421
Spain	AG	31194	98	STRENGTHENING CIVIL SOCIETY	074537
Spain	Misc	31220	6	FORESTRY DEVELOPMENT	070715
Spain	AG	31282	6	BIOLOGY OF MAMMALS' REPRODUCTION	070693
Switzerland	SDC	31110	583	AGRICULTURAL POLICY & ADMIN. MGMT	1998000975
Switzerland	SDC	31110	1917	CENACAM II - CAMAREN RECURSOS NATURALES RENOVABLES	1996001426
Switzerland	SDC	31110	167	AGRIC: CENACAM II	1996001511
Switzerland	SDC	31220	93	FORESTRY DEVELOPMENT	1991000295
United States	IADF	31110	348	AGRICULTURAL POLICY & ADMIN. MGMT	2007025065
United States	AID	31165	8192	ALTERNATIVE DEVELOPMENT AND ALTERNATIVE LIVELIHOODS	2007001904

Total aid for water in 2007 for Ecuador USD thousand 52317
And as a share of aid to total recipient countries 0.78%

Egypt

Austria	ADA	31110	137	NGO/EU COFINANCING - SUPPORT TO AGRICULTURE NETWORKS (SAN)	2007500101
Japan	JICA	31110	415	TC AGGREGATED ACTIVITIES	073020T
Japan	JICA	31130	455	TC AGGREGATED ACTIVITIES	073077T
Japan	MOFA	31140	645	AGRICULTURAL WATER RESOURCES	070278
Japan	MOFA	31140	2615	THE PROJECT FOR REHABILITATION OF FLOATING PUMP STATIONS	070181
Japan	MOFA	31150	2462	THE ASSISTANCE FOR UNDERPRIVILEGED FARMERS	070029
Japan	MOFA	31150	6723	MODERNIZATION OF AGRICULTURAL MECHANIZATION CENTER	070140
Japan	MOFA	31150	2462	GRANT ASSISTANCE FOR UNDERPRIVILEGED FARMERS	070279
Japan	MOFA	31150	390	AGRICULTURAL INPUTS	070028
Japan	JICA	31150	34	TC AGGREGATED ACTIVITIES	073138T
Japan	JICA	31163	37	TC AGGREGATED ACTIVITIES	073188T
Japan	JICA	31210	40	TC AGGREGATED ACTIVITIES	073294T
Korea	Kexim	31150	30000	MODERNIZATION AND MECHANIZATION PROCESS OF RICE CROP	2007001415
Netherlands	MFA	31110	339	AGRICULTURAL POLICY & ADMIN. MGMT	2007100138

RECIPIENT	Agency	Sector	Amount USD thousand	Project description	CRS ID Number
Donor					
Norway	MFA	31110	100	AGRICULTURAL POLICY & ADMIN. MGMT	2007000108
Spain	MFA	31182	254	CONCLUSION PROJECT OF BOVINE LIVESTOCK'S ARTIFICIAL INSEMINATION	076193
Spain 073193	MFA	31195	47	REINFORCEMENT OF THE VIGILENCE AND CONTROL SYSTEMS OF THE HIGHLY INFECTIOUS BIRD FLU	
United States	AID	31110	500	PROGRAM SUPPORT (AGRICULTURE)	2007013921
United States	AID	31110	100	AGRICULTURAL POLICY & ADMIN. MGMT	2007013914
United States	AID	31120	10900	AGRICULTURAL SECTOR PRODUCTIVITY	2007001777
United States	AID	31310	250	FISHING POLICY AND ADMIN. MANAGEMENT	2007011369

Total aid for water in 2007 for Egypt   **USD thousand 58905**
And as a share of aid to total recipient countries   **0.88%**

El Salvador

Austria	ADA	31120	100	NGO COFINANCED PROJECT: FOOD SECURITY PROGR. WITH GENDER VIEW	2006000022b
Belgium	DGCD	31194	51	AGRICULTURAL CO-OPERATIVES	2004003219
France	MISC	31182	1903	RECHERCHE AGRICULTURE	2007800364
Germany	BMZ	31120	337	AGRICULTURAL DEVELOPMENT	2007006240
Germany	BMZ	31120	454	INTEGRATED RURAL DEVELOPMENT PROGRAMME IN THE DIOCESE OF CHALATENANGO	2007006423
Germany	BMZ	31166	392	AGRICULTURAL EXTENSION	2007006338
Ireland	DFA	31181	83	AGRICULTURAL EDUCATION/TRAINING	2007001268
Japan	JICA	31110	668	TC AGGREGATED ACTIVITIES	073043T
Japan	JICA	31210	7	TC AGGREGATED ACTIVITIES	073321T
Japan	JICA	31310	872	TC AGGREGATED ACTIVITIES	073560T
Spain	MIE	31110	27	AGROPLASTICULTURAL EXTENSION IN EL SALVADOR	072499
Spain	AG	31120	116	STRENGTHENING OF THE PRODICTIVE ORGANIZATIONAL AND ECONOMIC CAPACITIES	074842
Spain	AG	31120	69	IMPROVEMENT OF THE FAMILY ECONOMY IN 8 RURAL COMMUNITIES IN EL SALVADOR	073965
Spain	AG	31120	34	DELIVERY OF EQUIPMENT FOR SURVEILLANCE AND FISHERIES INSPECTION	072770
Spain	AG	31140	57	IMPLEMENTATION OF A COMPREHENSIVE SYSTEM FOR THE TREATMENT OF WASTE WATER	073546
Spain	AG	31161	245	BOOSTING FOOD SOVEREIGNTY IN THE COOPERATIVES OF CONFRAS	076137
Spain	AG	31161	83	FOOD CROP PRODUCTION: STRENGTHENING TOURISM COMPETITIVENESS IN HONDURAS	074302
Spain	AG	31161	84	FOOD CROP PRODUCTION: STRENGTHENING OF THE ORG. AND THE LOCAL ECONOMY	074306
Spain	AG	31161	120	STRENGTHENING OF DIVERSIFIED AGRO-ECOLOGIC PRODUCTION TO IMPROVE NUTRITION	074876
Spain	AG	31161	21	FOOD CROP PRODUCTION	072014
Spain	AG	31162	242	BOOSTING THE INCREASE IN PRODUCTION CAPACITIES OF THE PEOPLE OF TECOLUCA	076125
Spain	AG	31163	123	IMPROVING THE PRODUCTION OF DAIRY COWS THROUGH MODERNISATION	074934
Spain	AG	31163	100	ESTABLISHMENT OF THREE LIVESTOCK MICRO-ENTERPRISES	074573
Spain	AG	31181	66	IMPROVEMENT OF THE SOCIO-ECONOMIC CONDITIONS OF THE POPULATION	073787
Spain	AG	31181	146	PROCESS OF SUSTAINABLE ORGANIC AGRICULTURE IN SAN JUAN TEPEZONTES	075293
Spain	AG	31310	164	PROMOTION PLAN FOR 4 ECONOMIC SOLIDARITY INITIATIVES IN THE FISHERY SECTOR	075438
Spain	MFA	31320	72	EMPOWERING THE LOCAL ARTESANIAL FISHERMEN OF THE CONCHAGUITA	074014
Spain	AG	31320	205	CONSERVATION OF MARINE FISHING RR & DEV. OF THE INSHORE FISHERY SECTOR	075816
Spain	AG	31381	114	STRENGTHENING OF THE SUSTAINABILITY PROCESS	074805
Spain	AG	31391	432	SEMI-ARTISANAL PRODUCTION OF SHRIMPS AND ENVIRONMENTAL CONSERVATION	077160
United States	AID	31110	642	AGRICULTURAL POLICY & ADMIN. MGMT	2007013905
United States	AID	31120	2677	AGRICULTURAL SECTOR PRODUCTIVITY	2007001764

RECIPIENT / Donor	Agency	Sector	Amount USD thousand	Project description	CRS ID Number
		Total aid for water in 2007 for El Salvador			**USD thousand 10708**
		And as a share of aid to total recipient countries			**0.16%**

Equatorial Guinea

Spain	MFA	31120	71	AGRICULTURAL FARM IN EOKO	074009
Spain	MFA	31120	49	AGRICULTURAL FARM MOKA	073307
Spain	MFA	31120	33	AGRICULTURAL DEVELOPMENT	072675
Spain	MFA	31120	82	AGRICULTURAL FARM IN ONDENG	074199
Spain	MFA	31120	77	AGRICULTURAL FARM IN SIPOLO	074103
Spain	AG	31120	23	AGRICULTURAL DEVELOPMENT	072106
Spain	AG	31120	74	IMPLEMENTATION OF AGROCULTURAL ACTIVITIES OF THE DEVELOPMENT AGENCY C.I.P.A	074058
Spain	AG	31161	180	INTRODUCTION TO AGRICULTURAL DEVELOPMENT BASED ON PRESERVING AGRICULTURE	075634

Total aid for water in 2007 for Equatorial Guinea   **USD thousand 589**
And as a share of aid to total recipient countries   **0.01%**

Eritrea

Belgium	DGCD	31120	1139	DéVELOPPEMENT DES CULTURES VIVRIèRES IRRIGUéES » GASH BARKA - PHASE II	2007004259
Ireland	DFA	31120	418	AGRICULTURAL DEVELOPMENT	2007000986
Ireland	DFA	31181	270	AGRICULTURAL EDUCATION/TRAINING	2007000642
Italy	DGCS	31181	120	AGRICULTURAL EDUCATION/TRAINING	061084
Italy	LA	31195	49	LIVESTOCK/VETERINARY SERVICES	071073
Japan	MOFA	31150	2632	THE ASSISTANCE FOR UNDERPRIVILEGED FARMERS	070095
Japan	JICA	31150	25	TC AGGREGATED ACTIVITIES	073150T
Norway	NORAD	31120	39	Mid term Review of Program Funded Through Contract Agreement	2007002359
Norway	NORAD	31130	122	AGRICULTURAL LAND RESOURCES	2007004747

Total aid for water in 2007 for Eritrea   **USD thousand 4814**
And as a share of aid to total recipient countries   **0.07%**

Ethiopia

Austria	ADA	31120	342	FOOD SECURITY PROJECT IN GAMO GOFA	2007500147
Austria	ADA	31120	134	IMPROVEMENT OF THE FOOD SECURITY STATUS OF VULNERABLE HOUSEHOLDS	2007500146
Belgium	DGCD	31181	33	AGRICULTURAL EDUCATION/TRAINING	2005001833
Belgium	DGCD	31195	69	LIVESTOCK/VETERINARY SERVICES	2006003914
Canada	CIDA	31120	2793	POST-HARVEST MANAGEMENT TO IMPROVE LIVEL	070971
Canada	IDRC	31120	98	MANAGING RISK, REDUCING VULNERABILITY AND ENHANCING PRODUCTIVITY	070311i
Canada	CIDA	31166	18151	RURAL CAPACITY BUILDING PROJECT	070027
Canada	CIDA	31166	279	TECH ASST/MON/EVAL - RURAL CAP BLDG PROJ	070028
Finland	MFA	31181	30	FRAME AGREEMENT WITH NGO (SUOMEN LäHETYSSEURA RY)	2007000294
France	AFD	31162	2327	APPUI EXPORTATIONS PRODUITS HORTICOLES	2007153800
France	MISC	31182	5476	RECHERCHE AGRICULTURE	2007800344
Germany	BMZ	31120	595	INTEGRATED RURAL FOOD SECURITY	2007006552
Germany	BMZ	31120	238	AGRICULTURAL DEVELOPMENT	2007006441
Germany	BMZ	31120	705	INTEGRATED FOOD SECURITY FOR PASTORALISTS AND AGROPASTORALISTS	2007006579
Germany	BMZ	31130	690	AGRICULTURAL LAND RESOURCES	2007008959
Germany	BMZ	31140	548	FOOD SECURITY PROJECT IN IROB DISTRICT 2007-2010	2007006259

RECIPIENT Donor	Agency	Sector	Amount USD thousand	Project description	CRS ID Number
Germany	BMZ	31181	9	AGRICULTURAL EDUCATION/TRAINING	2007008358
IDA		31140	85000	IRRIGATION AND DRAINAGE	2007000206
IDA		31166	21500	PROTECTION BASIC SERVICES	2007000214
IDA		31166	5000	IRRIGATION AND DRAINAGE	2007000203
IDA		31191	10000	IRRIGATION AND DRAINAGE	2007000207
IFAD		31140	19999	AGRICULTURAL WATER RESOURCES	070028a
IFAD		31140	19998	PARTICIPATORY SMALL-SCALE IRRIGATION DEVELOPMENT PROGRAMME	070028
Ireland	DFA	31110	41	AGRICULTURAL POLICY & ADMIN. MGMT	2007001598
Ireland	DFA	31120	821	SUPPORT TO NGOS - IRISH - MAPS - GENERAL - SUPPORT FOR INNOVATION	2007000220
Ireland	DFA	31120	110	AGRICULTURAL DEVELOPMENT	2007001848
Ireland	DFA	31130	439	SUPPORT TO NGOS - IRISH - MAPS - GENERAL - SUPPORT IRRIGATION & WATER	2007000228
Ireland	DFA	31150	715	SUPPORT TO NGOS - IRISH - MAPS - GENERAL - RURAL DEV. PROG.	2007000230
Ireland	DFA	31161	368	FOOD CROP PRODUCTION	2007000456
Ireland	DFA	31182	412	AGRICULTURAL RESEARCH	2007001883
Ireland	DFA	31194	750	AGRICULTURAL CO-OPERATIVES	2007000451
Italy	DGCS	31120	102	AGRICULTURAL DEVELOPMENT	060239
Japan	JICA	31110	2254	TC AGGREGATED ACTIVITIES	072919T
Japan	JICA	31130	181	TC AGGREGATED ACTIVITIES	073121T
Japan	MOFA	31150	3820	GRANT ASSISTANCE FOR UNDERPRIVILEGED FARMERS	070270
Japan	JICA	31163	190	TC AGGREGATED ACTIVITIES	073190T
Japan	JICA	31210	661	TC AGGREGATED ACTIVITIES	073365T
Japan	JICA	31310	74	TC AGGREGATED ACTIVITIES	073548T
Korea	KOICA	31110	87	DISPATCH OF EXPERT	2007017736
Korea	KOICA	31162	126	TRAINING PROGRAM	2007017364
Korea	KOICA	31163	23	LIVESTOCK	2007017178
Luxembourg	MFA	31130	110	AGRICULTURAL LAND RESOURCES	070311
Luxembourg	MFA	31140	74	AGRICULTURAL WATER RESOURCES	070418
Netherlands	MFA	31110	732	AGRIC.: SUBS. TO PSOM 2003-2009	2007100154
Netherlands	MFA	31191	519	AGRICULTURAL SERVICES	2007000419
Norway	MFA	31120	107	AGRICULTURAL DEVELOPMENT	2007002616
Norway	NORAD	31130	45	BIKILAL INTEGRATED RURAL DEV. (BIRDP). -NATURAL RESOURCE CONSERVATION	2007003611
Norway	NORAD	31130	19	INTEGRATED RURAL DEVELOPMENT, NONNO. NATURAL RESOURCES CONSERVATION	2007003706
Norway	NORAD	31130	301	DCG ETHIOPIA	2007004748
Norway	MFA	31150	4097	FAO SEED SECURITY PHASE II (2007-2011) OROMIA	2007002373
Norway	NORAD	31161	9	INTEGRATED DEVELOPMENT, BEGI-GIDAMI- FOOD CROP PRODUCTION	2007003710
Norway	NORAD	31161	21	BIKILAL INTEGRATED RURAL DEV. (BIRDP). - FOOD CROP PRODUCTION	2007003612
Norway	NORAD	31161	20	INTEGRATED RURAL DEVELOPMENT, NONNO. FOOD PRODUCTION	2007003705
Norway	NORAD	31163	21	BIKILAL INTEGRATED RURAL DEV. (BIRDP). - LIFESTOCK	2007003614
Norway	NORAD	31195	36	RAYTU COMM. DEV. PROJECT (RCDP), PASTORAL DEVELOPMENT	2007003693
Norway	NORAD	31195	6	INTEGRATED DEVELOPMENT, BEGI-GIDAMI. - VETERINARY POST	2007003713
Norway	MFA	31220	122	FORESTRY DEVELOPMENT	2007002382
Norway	NORAD	31282	522	TISSUE CULTURE AND MODERN NURSERY TECHNOLOGY	2007002388
Spain	MFA	31120	1127	FOOD SOVEREIGNTY DIVERSIFICATION OF SOURCES OF INCOME	077794
Spain	AG	31120	329	AGRICULTURAL PRODUCTION AND EDUCATION CENTRE OF MEKI	076649
Spain	MFA	31191	238	MARKET ACCESS FOR FAMILY BASED AGRICULTURE	076098

RECIPIENT Donor	Agency	Sector	Amount USD thousand	Project description	CRS ID Number
Sweden	Sida	31164	5475	AGRARIAN REFORM	2007001364
UNDP		31110	6	AGRICULTURAL POLICY & ADMIN. MGMT	2007011649
United Kingdom	DFID	31110	1670	ETHIOPIA STRATEGIC SUPPORT PROGRAMME	2007000097
United States	AID	31110	321	AGRICULTURAL POLICY & ADMIN. MGMT	2007013863
United States	AID	31110	1500	AGRICULTURAL ENABLING ENVIRONMENT	2007001642
United States	AID	31120	3380	AGRICULTURAL SECTOR PRODUCTIVITY	2007001728
United States	AID	31181	24	AGRICULTURAL EDUCATION/TRAINING	2007001014
United States	AID	31191	325	AGRICULTURAL SERVICES	2007001699
United States	AID	31191	1695	AGRICULTURAL SAFETY NETS AND LIVELIHOOD SERVICES	2007001698

Total aid for water in 2007 for Ethiopia **USD thousand 228038**
And as a share of aid to total recipient countries **3.4%**

Europe, regional

EU institutions	EDF	31210	8214	IMPROVING FOREST LAW ENFORCEMENT AND GOVERNANCE (FLEG)	2007200545
France	MISC	31181	48	ƒDUCATION & FORMATION DANS LE DOMAINE AGRICOLE	2007800789
Norway	MFA	31110	683	INSTIT COOPERATION BETWEEN ACADEMIC	2007001532
Sweden	Sida	31110	44	AGRICULTURAL POLICY & ADMIN. MGMT	2007002455
Switzerland	seco	31210	1100	BOIS TROPICAL 2005-2007 - CONSULTANT	2007001933

Total aid for water in 2007 for Europe, regional **USD thousand 10088**
And as a share of aid to total recipient countries **0.15%**

Far East Asia, regional

France	MISC	31181	336	ƒDUCATION & FORMATION DANS LE DOMAINE AGRICOLE	2007800794
Japan	MAFF	31110	324	TC AGGREGATED ACTIVITIES	070011T
Japan	MAFF	31140	278	TC AGGREGATED ACTIVITIES	070016T
Japan	MAFF	31220	139	TC AGGREGATED ACTIVITIES	070031T
Netherlands	MFA	31220	494	DMW ASIA FOREST PARTNERSHIP	2003015579
Spain	MFA	31320	192	SEMINAR FISHERY-AGRICULTURE	075700

Total aid for water in 2007 for Far East Asia, regional **USD thousand 1762**
And as a share of aid to total recipient countries **0.03%**

Fiji

Australia	AusAID	31110	10	A REVIEW OF THE POLICY & ECO. ENVIRONMENT IN THE SOUTH PACIFIC	2007001990
Australia	AusAID	31110	189	AGRICULTURAL POLICY & ADMIN. MGMT	2007001747
Australia	AusAID	31182	63	AGRICULTURAL RESEARCH	2007001237
Australia	AusAID	31182	38	DEVELOPING THE ORNAMENTALS INDUSTRY IN THE PACIFIC	2007001519
Australia	AusAID	31192	21	TARO BEETLE MANAGEMENT	2007001745
Australia	AusAID	31192	8	DEVELOP SENSITIVE TESTS TO DETECT TARO VIRUSES	2007002091
Australia	AusAID	31192	298	PLANT/POST-HARVEST PROT. & PEST CTRL	2007001992
Australia	AusAID	31195	31	RISKS ASSOCIATED WITH ZOONOTIC INFECTION	2007001606
Australia	AusAID	31220	142	FORESTRY DEVELOPMENT	2007001422
Australia	AusAID	31320	6	AQUACULTURE RESEARCH PRIORITIES	2007002140
Germany	BMZ	31181	694	CAPACITY BUILDING PROGRAM FOR THE TUTU RURAL TRAINING CENTRE IN VAUNA LEVU	2007006328
Japan	JICA	31110	218	TC AGGREGATED ACTIVITIES	073009T
Japan	JICA	31150	37	TC AGGREGATED ACTIVITIES	073172T

RECIPIENT Donor	Agency	Sector	Amount USD thousand	Project description	CRS ID Number
Japan	JICA	31163	35	TC AGGREGATED ACTIVITIES	073235T
Japan	JICA	31195	25	TC AGGREGATED ACTIVITIES	073292T
Japan	JICA	31210	35	TC AGGREGATED ACTIVITIES	073458T
Japan	JICA	31310	249	TC AGGREGATED ACTIVITIES	073583T
Japan	MAFF	31320	333	TC AGGREGATED ACTIVITIES	070058T
Korea	KOICA	31161	77	FOOD CROP PRODUCTION	2007017514
Korea	KOICA	31181	25	AGRICULTURAL EDUCATION/TRAINING	2007016411
New Zealand	NZAid	31130	191	SUSTAINABLE LAND USE IN SUGAR CANE BELT	070437

Total aid for water in 2007 for Fiji **USD thousand 2723**
And as a share of aid to total recipient countries **0.04%**

Gabon

France	MISC	31182	13730	RECHERCHE AGRICULTURE	2007800345
Japan	JICA	31110	55	TC AGGREGATED ACTIVITIES	072920T
Japan	JICA	31310	1043	TC AGGREGATED ACTIVITIES	073549T
Japan	MAFF	31320	247	TC AGGREGATED ACTIVITIES	070049T
Japan	MAFF	31381	23	TC AGGREGATED ACTIVITIES	070067T

Total aid for water in 2007 for Gabon **USD thousand 15098**
And as a share of aid to total recipient countries **0.22%**

Gambia

Belgium	DGCD	31161	125	FOOD CROP PRODUCTION	2005000881
Belgium	DGCD	31195	97	LIVESTOCK/VETERINARY SERVICES	2006003922
Germany	BMZ	31220	10	FORESTRY DEVELOPMENT	2007001837
Germany	BMZ	31282	98	FORESTRY RESEARCH	2007007023
Japan	MOFA	31150	1188	AGRICULTURAL INPUTS	070085
Japan	JICA	31150	40	TC AGGREGATED ACTIVITIES	073140T
Japan	JICA	31310	321	TC AGGREGATED ACTIVITIES	073474T
Netherlands	MFA	31110	11	AGRICULTURAL POLICY & ADMIN. MGMT	2007100149
Norway	NORAD	31194	66	AGRICULTURAL CO-OPERATIVES	2007003786

Total aid for water in 2007 for Gambia **USD thousand 1957**
And as a share of aid to total recipient countries **0.03%**

Georgia

Germany	BMZ	31210	104	FORESTRY POLICY & ADMIN. MANAGEMENT	2007006988
Greece	YPEPU	31161	205	COMPLETION OF A PROGRAM FOR A POTATO SEED PRODUCTION CENTER IN GEORGIA	2007000899
Japan	JICA	31110	50	TC AGGREGATED ACTIVITIES	072984T
Netherlands	MFA	31110	305	AGRICULTURAL POLICY & ADMIN. MGMT	2007100158
Sweden	Sida	31120	1361	AGRICULTURAL DEVELOPMENT	2007005426
Switzerland	SDC	31110	46	AGRICULTURAL POLICY & ADMIN. MGMT	2006003364
Switzerland	SDC	31110	129	TOURISM & RURAL DEVELOPMENT	2006003362
United States	AID	31110	48	AGRICULTURAL POLICY & ADMIN. MGMT	2007013934
United States	AID	31110	890	AGRICULTURAL ENABLING ENVIRONMENT	2007001668
United States	AGR	31110	1413	CAUCASUS AGRICULTURAL DEVELOPMENT INITIATIVE	2007000445
United States	AID	31120	2411	AGRICULTURAL SECTOR PRODUCTIVITY	2007001785

RECIPIENT / Donor	Agency	Sector	Amount USD thousand	Project description	CRS ID Number
		Total aid for water in 2007 for Georgia			**USD thousand 6964**
		And as a share of aid to total recipient countries			**0.1%**

Ghana

Donor	Agency	Sector	Amount USD thousand	Project description	CRS ID Number
Belgium	DGCD	31120	199	CONDITIONS DE BASE DE MOYENS D'EXISTENCE DANS LA RéGION DU SUPéRIEUR-EST	2004003466
Denmark	MFA	31220	220	FORESTRY DEVELOPMENT	071147
EU institutions	CEC	31162	6845	COCOA SECTOR SUPPORT PROGRAMME - PHASE 2	2007100084
France	AFD	31120	1780	APPUI A LA FILIERE DE RIZ DS 4 REGIONS	2007164800
France	AFD	31120	17112	APPUI A LA FILIERE RIZ DANS 4 REGIONS	2007110200
France	MISC	31182	8419	RECHERCHE AGRICULTURE	2007800346
Germany	BMZ	31110	10951	PROMOTION OF MARKET ORIENTED AGRICULTURE	2007005985
Germany	BMZ	31120	516	AGRICULTURAL DEVELOPMENT	2007009020
Germany	BMZ	31166	108	AGRICULTURAL EXTENSION	2007007002
Germany	BMZ	31191	31	AGRICULTURAL SERVICES	2007007022
Germany	BMZ	31210	104	FORESTRY POLICY & ADMIN. MANAGEMENT	2007007005
Germany	BMZ	31220	718	FORESTRY DEVELOPMENT	2007009022
Japan	MAFF	31110	69	TC AGGREGATED ACTIVITIES	070002T
Japan	JICA	31110	1627	TC AGGREGATED ACTIVITIES	073021T
Japan	JICA	31130	113	TC AGGREGATED ACTIVITIES	073080T
Japan	MOFA	31150	3141	GRANT ASSISTANCE FOR UNDERPRIVILEGED FARMERS	070281
Japan	JICA	31163	33	TC AGGREGATED ACTIVITIES	073191T
Japan	JICA	31195	47	TC AGGREGATED ACTIVITIES	073245T
Japan	JICA	31210	689	TC AGGREGATED ACTIVITIES	073366T
Netherlands	MFA	31110	2092	AGRIC.: SUBS. TO PSOM 2003-2009	2007100145
Netherlands	MFA	31181	1369	AGRIC. TRAINING: ACCCACAO	2007000093
Netherlands	MFA	31320	128	FISHERY DEVELOPMENT	2006000052
Norway	MFA	31110	32	AGRICULTURAL POLICY & ADMIN. MGMT	2007000107
Spain	MFA	31310	46	CONTINENTAL AQUACULTURE DEVELOPMENT PROGRAM.	073165
Spain	ICO	31391	9582	ARTISANAL FISHING SECTOR REFRIGERATION AND FREEZING.	070024
United Kingdom	DFID	31110	7230	SUPPORT TO FOOD & AGRICULTURE SECTOR HARMONISATION	2007000122
United Kingdom	DFID	31210	30	Ghana: Lesson Learning Review	2007000658
United States	AID	31110	925	AGRICULTURAL POLICY & ADMIN. MGMT	2007013864
United States	AID	31120	4426	AGRICULTURAL SECTOR PRODUCTIVITY	2007001729
United States	ADF	31120	75	AGRICULTURAL DEVELOPMENT	2007000203
United States	MCC	31120	10688	AGRICULTURAL DEVELOPMENT COMPACT ACTIVITY	2007025217
United States	MCC	31140	27608	AGRICULTURAL WATER RESOURCES COMPACT ACTIVITY	2007025221
United States	MCC	31181	67384	AGRICULTURAL EDUCATION/TRAINING COMPACT ACTIVITY	2007025224
United States	MCC	31192	20376	PLANT AND POST-HARVEST PROTECTION AND PEST CONTROL COMPACT ACTIVITY	2007025229
United States	MCC	31193	58400	AGRICULTURAL FINANCIAL SERVICES COMPACT ACTIVITY	2007025230
		Total aid for water in 2007 for Ghana			**USD thousand 263113**
		And as a share of aid to total recipient countries			**3.92%**

Grenada

Donor	Agency	Sector	Amount USD thousand	Project description	CRS ID Number
Japan	JICA	31310	54	TC AGGREGATED ACTIVITIES	073563T

RECIPIENT / Donor	Agency	Sector	Amount USD thousand	Project description	CRS ID Number

Total aid for water in 2007 for Grenada — **USD thousand 54**
And as a share of aid to total recipient countries — **0%**

Guatemala

Donor	Agency	Sector	Amount	Project description	CRS ID
Austria	ADA	31120	74	NGO COFINANCED PROJECT: SUPPORT OF FARMERS ORGANISATIONS	2006000022b
Austria	ADA	31194	122	NGO COFINANCED PROJECT: LEGAL ADVICE F. FARMER ORGANISATIONS	2006000022b
Belgium	DGCD	31162	93	INDUSTRIAL CROPS/EXPORT CROPS	2004008711
Belgium	DGCD	31165	77	AGRICULTURAL ALTERNATIVE DEVELOPMENT	2005000439
Belgium	DGCD	31194	206	AGRICULTURAL CO-OPERATIVES	2005000565
Belgium	DGCD	31220	43	FORESTRY DEVELOPMENT	2004007240
Canada	IDRC	31130	172	AN ALTERNATIVE EVALUATION OF THE AGRARIAN SECTOR IN GUATEMALA	070609i
France	MISC	31182	246	RECHERCHE AGRICULTURE	2007800365
Germany	BMZ	31120	212	AGRICULTURAL DEVELOPMENT	2007006074
Germany	BMZ	31181	164	AGRICULTURAL EDUCATION/TRAINING	2007006289
Ireland	DFA	31120	90	AGRICULTURAL DEVELOPMENT	2007001244
Ireland	DFA	31120	566	SUPPORT TO NGOS - IRISH - MAPS - GENERAL - IMPROVE THE LIVELIHOOD	2007000221
Ireland	DFA	31181	796	SUPPORT TO NGOS - IRISH - CENTRAL AMERICA CAPACITY BUILDING	2007000239
Ireland	DFA	31191	22	AGRICULTURAL SERVICES	2007002012
Ireland	DFA	31194	15	AGRICULTURAL CO-OPERATIVES	2007002415
Italy	DGCS	31120	20	AGRICULTURAL DEVELOPMENT	060708
Japan	JICA	31110	1351	TC AGGREGATED ACTIVITIES	072952T
Japan	JICA	31163	57	TC AGGREGATED ACTIVITIES	073203T
Japan	JICA	31210	248	TC AGGREGATED ACTIVITIES	073386T
Japan	JICA	31310	29	TC AGGREGATED ACTIVITIES	073497T
Luxembourg	MFA	31181	98	AGRICULTURAL EDUCATION/TRAINING	071287
Netherlands	MFA	31110	88	AGRICULTURAL POLICY & ADMIN. MGMT	2007100139
Norway	NORAD	31166	134	School- and Agricultural Program. -Agricultural	2007003678
Spain	MFA	31120	554	AGRICULTURAL DEVELOPMENT	077423
Spain	MFA	31120	1211	LOCAL SUSTAINABLE DEVPT IN THE NARANJO RIVER MANCUERNA	077819
Spain	AG	31120	51	EDUCATION CENTRE FOR THE RURAL DEVELOPMENT OF THE AREA OF SACPUCY	073347
Spain	AG	31120	55	REHAB. OF THE SANITARY CONDITIONS FOOD & THE LIVING CONDITIONS OF RURAL COM.	073459
Spain	AG	31120	276	STRENGTHENING PROCESSES OF SUSTAINABLE MANAGEMENT OF NATURAL RESOURCES	076402
Spain	AG	31140	40	INSTALLATION OF A SPRINKLER IRRIGATION SYSTEM (2PHASE)	072922
Spain	AG	31161	221	PRODUCTION AND MARKETING OF PRODUCTS AND AGROFORESTRY PLANTATIONS	075999
Spain	AG	31161	83	FOOD CROP PRODUCTION: STRENGTHENING TOURISM COMPETITIVENESS IN HONDURAS	074301
Spain	AG	31161	84	IMPROVEMENT OF FOOD SECURITY FOR PEASANT FAMILIES IN 25 INDIGENOUS COM.	074320
Spain	AG	31161	274	IMPROVEMENT OF THE CONDITIONS FOR FOOD SOVEREIGNTY AND SECURITY	076387
Spain	AG	31181	272	STREGTHENING OF LOCAL GUATAMALEAN ORGANISATIONS IN THE FIELD OF FOOD	076296
Spain	Misc	31182	27	STUDIES OF THE STRUCTURE. PATHOGENICITY/ CONTROL OF SPECIES OF PHYTOPHTHORA	072497
Spain	AG	31194	21	AGRICULTURAL CO-OPERATIVES	072031
Spain	MFA	31320	86	EQUIPMENT SUPPORT TO A FISH SELLING CENTRE	074337
Spain	MFA	31320	287	STRENGTHENING THE SYSTEM OF THE ORGANISATIONS OF TRADITIONAL FISHERMEN	076445
United States	AID	31110	1400	AGRICULTURAL ENABLING ENVIRONMENT	2007001651
United States	AID	31110	53	AGRICULTURAL POLICY & ADMIN. MGMT	2007013882
United States	AID	31110	772	PROGRAM SUPPORT (AGRICULTURE)	2007013889

RECIPIENT Donor	Agency	Sector	Amount USD thousand	Project description	CRS ID Number
United States	AID	31120	3550	AGRICULTURAL SECTOR PRODUCTIVITY	2007001750
United States	AID	31120	250	AGRICULTURAL DEVELOPMENT	2007001753
United States	IADF	31194	16	AGRICULTURAL CO-OPERATIVES	2007024955

Total aid for water in 2007 for Guatemala   USD thousand   14506
And as a share of aid to total recipient countries   0.22%

Guinea

Donor	Agency	Sector	Amount	Project description	CRS ID
Belgium	MPRF	31161	25	FOOD CROP PRODUCTION	2005002051
EU institutions	EDF	31310	1916	FISHING POLICY AND ADMIN. MANAGEMENT	2007200172
France	AFD	31120	12320	APPUI FILIERE RIZ BASSE GUINEE	2007155800
France	MISC	31182	6585	RECHERCHE AGRICULTURE	2007800347
IDA		31110	4420	VILLAGE COMMUNITY SUPPORT	2007000255
IFAD		31110	10000	PROJET D'APPUI AUX COMMUNAUTéS VILLAGEOISES – PHASE II	070008
Japan	JICA	31110	405	TC AGGREGATED ACTIVITIES	072922T
Japan	MOFA	31150	2632	GRANT ASSISTANCE FOR UNDERPRIVILEGED FARMERS	070258
Japan	JICA	31150	120	TC AGGREGATED ACTIVITIES	073141T
Japan	JICA	31310	483	TC AGGREGATED ACTIVITIES	073550T
Japan	MOFA	31391	3803	PROJET D AMELIORATION DU PORT DE PECHE ARTISANALE	070257
Korea	KOICA	31130	2123	PROVISION OF EQUIPMENT	2007017809
Spain	MFA	31140	1154	TO IMPROVE THE FOOD & NUTRITION SITUATION OF THE POPULATION	077805
United States	ADF	31163	232	LIVESTOCK	2007000094

Total aid for water in 2007 for Guinea   USD thousand 46217
And as a share of aid to total recipient countries   0.69%

Guinea-Bissau

Donor	Agency	Sector	Amount	Project description	CRS ID
Belgium	DGCD	31120	87	AGRICULTURAL DEVELOPMENT	2004005463
EU institutions	CEC	31192	1369	PLANT/POST-HARVEST PROT. & PEST CTRL	2007100097
Italy	DGCS	31161	206	FOOD CROP PRODUCTION	070356
Japan	JICA	31110	7	TC AGGREGATED ACTIVITIES	072923T
Japan	JICA	31195	6	TC AGGREGATED ACTIVITIES	073246T
Japan	MAFF	31310	37	TC AGGREGATED ACTIVITIES	070034T
Portugal	ICP	31110	15	AGRICULTURAL POLICY & ADMIN. MGMT	070899
Portugal	ICP	31182	37	AGRICULTURAL RESEARCH	070808
Portugal	ICP	31193	14	AGRICULTURAL FINANCIAL SERVICES	070464
Spain	AG	31150	156	IMPROVED THE AGRICULTURAL INFRASTRUCTURE IN THE CANCHUNGA SECTOR	075382
Spain	MFA	31161	684	SUPPORT TO THE AGR. PRODUCERS AFFECTED BY THE CRISIS OF CASHEW NUTS	077579
Spain	MFA	31381	25	FISHERY EDUCATION/TRAINING	072188
Spain	MFA	31381	137	SUPPORT FOR THE FISHERIES TRAINING CENTRE OF BOLAMA	075146
Spain	MFA	31391	34	DELIVERY OF EQUIPMENT FOR SURVEILLANCE AND FISHERIES INSPECTION	072743
UNDP		31120	15	AGRICULTURAL DEVELOPMENT	2007012851

Total aid for water in 2007 for Guinea-Bissau   USD thousand 2829
And as a share of aid to total recipient countries   0.04%

Guyana

Donor	Agency	Sector	Amount	Project description	CRS ID
EU institutions	EDF	31162	37050	GUYANA ANNUAL ACTION PLAN 2007 ON ACCOMPANYING MEASURES FOR SUGAR	2007201277

RECIPIENT Donor	Agency	Sector	Amount USD thousand	Project description	CRS ID Number
Japan	MAFF	31210	542	FORESTRY POLICY & ADMIN. MANAGEMENT	072165
United Kingdom	MISC	31210	120	SUSTAINABLE FOREST MANAGEMENT	2007800827
United States	AID	31120	210	AGRICULTURAL DEVELOPMENT	2007001751

Total aid for water in 2007 for Guyana   **USD thousand**   **37922**
And as a share of aid to total recipient countries   **0.57%**

Haiti

Donor	Agency	Sector	Amount	Project description	CRS ID
Belgium	DGCD	31120	190	APPUI AU DéVELOPPEMENT RURAL EN HAïTI	2004008712
Belgium	DGCD	31161	57	FOOD CROP PRODUCTION	2005000337
Canada	CIDA	31220	156	REBOISEMENT ET DéVELOP. DURABLE HAïTI	070723
Finland	MFA	31220	41	LOCAL COOPERATION FUND (LCF) IN HAITI	2007001049
France	MISC	31182	27	RECHERCHE AGRICULTURE	2007800366
Germany	BMZ	31120	244	AGRICULTURAL DEVELOPMENT	2007006254
Germany	BMZ	31220	315	FORESTRY DEVELOPMENT	2007006189
Japan	MOFA	31110	764	AGRICULTURAL POLICY & ADMIN. MGMT	070058
Japan	JICA	31110	37	TC AGGREGATED ACTIVITIES	073044T
Japan	JICA	31150	89	TC AGGREGATED ACTIVITIES	073155T
Luxembourg	MFA	31120	17	AGRICULTURAL DEVELOPMENT	070216
Luxembourg	MFA	31220	29	FORESTRY DEVELOPMENT	070214
Spain	MFA	31110	342	FOOD SOVEREIGNTY THROUGH SUPPORT TO SMALL PRODUCERS	076739
Spain	MFA	31120	684	SUPPORT FOR AGRICULTURAL DEV. AND THE IMPLEMENTATION OF GOOD AGR. PRACTICES	077581
Spain	MFA	31181	173	EMERGENCY ACTION FOR THE IMPROVEMENT OF FOOD SECURITY IN HAITI	075591
Spain	MFA	31194	274	RURAL DEVELOPMENT	076316
Spain	AG	31194	206	AGRICULTURAL CO-OPERATIVES	075915
Spain	MFA	31320	62	TECHNICAL ASSISTANCE (SUBSIDY) IN SUPPORT OF THE MINISTRY OF AGRICULTURE	073643
Spain	MFA	31320	342	PROJECT ON IMPROVING FISHERIES IN THE SOUTH-EASTERN DEPARTMENT OF HAITI	076702
United States	AID	31110	400	AGRICULTURAL POLICY & ADMIN. MGMT	2007013890
United States	AID	31120	6715	AGRICULTURAL SECTOR PRODUCTIVITY	2007001752

Total aid for water in 2007 for Haiti   **USD thousand 11165**
And as a share of aid to total recipient countries   **0.17%**

Honduras

Donor	Agency	Sector	Amount	Project description	CRS ID
Belgium	DGCD	31110	50	AGRICULTURAL POLICY & ADMIN. MGMT	2004003266
Belgium	DGCD	31194	127	AGRICULTURAL CO-OPERATIVES	2004003485
Belgium	DGCD	31194	346	LUTTE CONTRE LA PAUVRETE, DEVELOPPEMENT LOCAL ET POUVOIR LOCAL	2004003465
Germany	BMZ	31120	694	AGRICULTURAL DEVELOPMENT	2007006310
Germany	BMZ	31210	213	FORESTRY POLICY & ADMIN. MANAGEMENT	2007008770
Germany	BMZ	31220	56	FORESTRY DEVELOPMENT	2007007026
Ireland	DFA	31120	1022	SUPPORT TO NGOS - IRISH - MAPS - GENERAL	2007000218
Ireland	DFA	31181	363	AGRICULTURAL EDUCATION/TRAINING	2007000458
Japan	JICA	31110	279	TC AGGREGATED ACTIVITIES	072954T
Japan	JICA	31130	56	TC AGGREGATED ACTIVITIES	073125T
Japan	JICA	31310	7	TC AGGREGATED ACTIVITIES	073498T
Norway	NORAD	31120	83	SUSTAINABLE AGRICULTURE AND INSTITUTIONAL EMPOWERMENT	2007003418
Norway	NORAD	31130	70	RISK DIVERSIFICATION	2007003405

RECIPIENT / Donor	Agency	Sector	Amount USD thousand	Project description	CRS ID Number
Spain	MFA	31140	171	CONSOLIDATION OF THE RISK MANAGEMENT SYSTEM 2ND STEP	075554
Spain	MFA	31161	411	STRENGTHENING OF THE COFFEE'S DENOMINATION OF ORIGIN IN HONDURAS	077017
Spain	AG	31161	55	SUPPORT FOR FOOD SECURITY IN 11 MUNICIPALITIES IN HONDURAS.	073509
Spain	AG	31161	83	FOOD CROP PRODUCTION: STRENGTHENING TOURISM COMPETITIVENESS IN HONDURAS	074300
Spain	MFA	31320	126	FISHERIES SECTOR DEVELOPMENT IN THE GULF OF FONSECA (XUNTA GALICIA)	074978
Spain	MFA	31320	383	FISHERIES SECTOR DEVELOPMENT IN THE GULF OF FONSECA (XUNTA GALICIA) PHASE IV	076910
Spain	EMP	31320	25	FISHERY DEVELOPMENT	072312
Spain	AG	31381	74	STRENGTHENING THE FISHING COOPERATIVES OF THE MARCOVIA MUNICIPALITY	074042
Switzerland	SDC	31110	95	AGRICULTURAL POLICY & ADMIN. MGMT	2006000671

Total aid for water in 2007 for Honduras   **USD thousand 4791**
And as a share of aid to total recipient countries   **0.07%**

India

Recipient	Agency	Sector	Amount	Project description	CRS ID
Australia	AusAID	31110	118	AGRICULTURAL POLICY & ADMIN. MGMT	2007001362
Australia	AusAID	31120	238	AGRICULTURAL DEVELOPMENT	2007001266
Australia	AusAID	31130	191	AGRICULTURAL LAND RESOURCES	2007002169
Australia	AusAID	31140	50	IMPROVING WATER RESOURCE MANAGEMENT IN INDIA'S AGR	2007001361
Australia	AusAID	31140	224	WATER ALL. IN THE KRISHNA RIVER BASIN TO IMPROVE WATER PRODUCTIVITY IN AGR.	2007000948
Australia	AusAID	31150	226	AGRICULTURAL INPUTS	2007000947
Australia	AusAID	31161	13	FOOD CROP PRODUCTION	2007001921
Australia	AusAID	31163	65	LIVESTOCK	2007001639
Australia	AusAID	31182	33	DEVELOPMENT OF WATERLOGGING TOLERANCE IN WHEAT	2007001568
Australia	AusAID	31182	120	AGRICULTURAL RESEARCH	2007002039
Australia	AusAID	31182	31	ECO-GEOGRAPHICAL AND PHYSIOLOGICAL APPROACHES	2007001602
Australia	AusAID	31182	8	HAPPY SEEDER POLICY LINKAGE SCOPING STUDY	2007002058
Australia	AusAID	31182	178	OILSEED BRASSICA IMPROVEMENT	2007001669
Australia	AusAID	31182	217	THIS PROJECT BRINGS TOGETHER PLANT BREEDERS (ICRISAT) AND LIVESTOCK NUTRITION	2007000648
Australia	AusAID	31182	63	ENSURING PRODUCTIVITY AND FOOD SECURITY (CONTROL OF YELLOW RUST OF WHEAT	2007001749
Australia	AusAID	31191	11	INTERNATIONAL FOOD SAFETY REGULATION	2007001975
Australia	AusAID	31192	17	PLANT/POST-HARVEST PROT. & PEST CTRL	2007001829
Australia	AusAID	31310	8	MECHANISMS TO MAXIMISE BENEFITS TO SMALL-HOLDER SHRIMP FARMER GROUPS	2007002067
Australia	AusAID	31320	63	DEVELOPING AQUACULTURE IN INLAND SALINE AFFECTED	2007001247
Belgium	DGCD	31110	44	AGRICULTURAL POLICY & ADMIN. MGMT	2004003275
Belgium	MPRG	31161	16	FOOD CROP PRODUCTION	2007005260
Belgium	DGCD	31194	51	AGRICULTURAL CO-OPERATIVES	2005001102
Belgium	DGCD	31195	42	LIVESTOCK/VETERINARY SERVICES	2007003925
Canada	CIDA	31110	931	URI REVITALIZATION INITIATIVE (URI)	070629
Canada	CIDA	31130	146	20+5 INTEGRATED HEALTH AND DEV. PROJECT	070730
Canada	IDRC	31130	23	STRENGTHENING NRM & FARMER'S LIVELIHOODS IN NAGALAND	070003i
Canada	IDRC	31210	77	KEY TRENDS AND DRIVERS OF INDIAN FORESTRY: FUTURE SCENARIOS AND CHALLENGES	070757i
Canada	IDRC	31210	87	COMMUNITY-BASED ECOLOGICAL MONITORING	070628i
Denmark	MFA	31161	937	FOOD & LIVELIHOOD SECURITY FOR THE POOR	071216
Finland	MFA	31166	94	NGO SUPPORT / CAPASITY BUILDING OF RURAL FARMERS FOR ECO PROTECTION	2007000410
Finland	MFA	31166	85	NGO SUPPORT / ECOLOGICAL AGRICULTURE PROGRAMME,RESCUE OF CHILD LABORERS	2007000166
Finland	MFA	31210	32	NGO SUPPORT / INDIGENOUS RIGHTS AND CULTURE IN ADIVASI FOREST	2007000168

RECIPIENT Donor	Agency	Sector	Amount USD thousand	Project description	CRS ID Number
France	MISC	31110	56	POLITIQUE AGRICOLE & GESTION ADMINISTRATIVE	2007800781
France	MISC	31182	2957	RECHERCHE AGRICULTURE	2007800376
Germany	BMZ	31110	52	AGRICULTURAL POLICY & ADMIN. MGMT	2007007087
Germany	BMZ	31120	684	DMI'S INTEGRATED DEVELOPMENT PROGRAMME FOR WOMEN'S WELFARE IN INDIA	2007006083
Germany	BMZ	31120	467	INTEGRATED RURAL DEVELOPMENT PROGRAMME AGAINST CHILD LABOUR, INDIA	2007005648
Germany	BMZ	31120	939	PEACE ADVOCACY AND SUSTAINABLE DEVELOPMENT IN 4 STATES OF NORTHERN INDIA	2007006557
Germany	BMZ	31120	1155	AGRICULTURAL DEVELOPMENT	2007006569
Germany	L G	31120	59	AGRICULTURAL DEVELOPMENT	2007011578
Germany	BMZ	31130	637	AGRICULTURAL LAND RESOURCES	2007006300
Germany	BMZ	31165	421	AGRICULTURAL ALTERNATIVE DEVELOPMENT	2007006430
Germany	BMZ	31166	274	AGRICULTURAL EXTENSION	2007006401
Germany	BMZ	31181	381	INTEGRATED RURAL DEVELOPMENT, DEEPDI, INDIA	2007006431
Germany	BMZ	31182	632	RURAL DEVELOPMENT PROGRAMME IN THE REGION OF CHHOTA UDEPUR, INDIA	2007006432
Germany	BMZ	31182	118	AGRICULTURAL RESEARCH	2007007086
Germany	Fed Min	31182	1350	PRODUCTION TECHNOLOGIES AND PRODUCTION EQUIPMENT	2007010618
Germany	BMZ	31210	119	FORESTRY POLICY & ADMIN. MANAGEMENT	2007007093
IDA		31110	11340	COMMUNITY TANK MANAGEMENT PROJECT	2007000765
IDA		31110	26000	RURAL POVERTY REDUCTION	2007000774
IDA		31110	3200	KARNATAKA TANKS	2007000792
IDA		31140	90000	AGRICULTURAL WATER RESOURCES	2007000814
IDA		31140	80325	COMMUNITY TANK MANAGEMENT PROJECT	2007000767
IDA		31140	19200	KARNATAKA TANKS	2007000790
IDA		31161	22500	FOOD CROP PRODUCTION	2007000812
IDA		31163	3000	LIVESTOCK	2007000816
IDA		31166	6000	AGRICULTURAL EXTENSION	2007000810
IDA		31191	1890	COMMUNITY TANK MANAGEMENT PROJECT	2007000769
IDA		31191	12600	RURAL LIVELIHOODS	2007000784
Ireland	DFA	31130	106	AGRICULTURAL LAND RESOURCES	2007001141
Ireland	DFA	31150	50	AGRICULTURAL INPUTS	2007001511
Italy	DGCS	31120	5	AGRICULTURAL DEVELOPMENT	060877
Japan	JICA	31110	1488	TC AGGREGATED ACTIVITIES	073064T
Japan	JICA	31130	30	TC AGGREGATED ACTIVITIES	073109T
Japan	JBIC	31140	203514	ANDHRA PRADESH IRRIGATION AND LIVELIHOOD IMPROVEMENT PROJECT	073040
Japan	JICA	31163	369	TC AGGREGATED ACTIVITIES	073240T
Japan	PRF	31182	12	TC AGGREGATED ACTIVITIES	075606T
Japan	JICA	31195	24	TC AGGREGATED ACTIVITIES	073272T
Japan	JBIC	31220	65577	TRIPURA FOREST ENVIRONMENTAL IMPROVEMENT AND POVERTY ALLEVIATION PROJECT	073041
Japan	JBIC	31220	148735	GUJARAT FORESTRY DEVELOPMENT PROJECT PHASE 2	073042
Japan	JICA	31310	83	TC AGGREGATED ACTIVITIES	073571T
Korea	Misc	31181	5	AGRICULTURAL EDUCATION/TRAINING	2007002567
Luxembourg	MFA	31120	130	AGRICULTURAL DEVELOPMENT	071480
Luxembourg	MFA	31140	22	AGRICULTURAL WATER RESOURCES	070125
Luxembourg	MFA	31161	22	FOOD CROP PRODUCTION	070126
Luxembourg	MFA	31192	58	PLANT AND POST-HARVEST PROTECTION AND PEST CONTROL	071465
Luxembourg	MFA	31391	68	FISHERY SERVICES	070137

RECIPIENT Donor	Agency	Sector	Amount USD thousand	Project description	CRS ID Number
Netherlands	MFA	31110	1658	AGRIC.: SUBS. TO PSOM 2003-2009	2007100141
Norway	NORAD	31120	16	RURAL DEVELOPMENT, NELC, AGRICULTURE	2007003594
Norway	MFA	31120	23	AGRICULTURAL DEVELOPMENT	2007002555
Norway	NORAD	31320	37	PRE-FEASIBILITY STUDY FOR POSSIBLE INVESTMENT PRODUCTION OF FISH-FARMING	2007004112
Norway	MFA	31320	14	TRAINING PARTNERS ON TILAPIA HATCHERY/GROW-OUT MANAGEMENT AND OPERATIONS	2007002682
Norway	NORAD	31382	17	PARTICIPATION FROM INDIA TO AQUANOR	2007004122
Norway	NORAD	31391	7	FISHERY SERVICES	2007004097
Spain	AG	31120	151	SUSTAINABLE DEVELOPMENT CENTRE AMRITA BHOOMI	075342
Spain	AG	31120	125	BASIC DRINKING WATER SUPPLY AND BASIC SANITATION	074972
Spain	Misc	31130	6	SUSTAINABILITY AND IMPROVEMENT OF THE CULTIVATION OF MEDICINAL PLANTS	070758
Spain	AG	31140	411	IMPROVEMENT OF THE WATER INFRASTRUCTURE IN ANANTAPUR	077096
Spain	AG	31140	479	INSTALLATION OF IRRIGATION SYSTEMS FOR AGRICULTURAL USE	077295
Spain	AG	31140	259	WATER SUPPLY FOR AGRICULTURAL DEVELOPMENT	076225
Spain	AG	31140	37	WATERSHEDS PROJECT	072853
Spain	AG	31194	250	DEV. & STRENGTHENING OF PRODUCTIVE & ORGANIZATIONAL CAPACITIES OF WOMEN	076170
Switzerland	seco	31110	167	ORGANIC CERTIFICATION IN AGUACULTURE INDIA	2007001946
Switzerland	SDC	31182	4001	BIOTECHNOLOGIE	1980001018
United States	AID	31110	1035	AGRICULTURAL POLICY & ADMIN. MGMT	2007013899
United States	AID	31110	800	AGRICULTURAL ENABLING ENVIRONMENT	2007001655
United States	AID	31162	100	INDUSTRIAL CROPS/EXPORT CROPS	2007011450
United States	AID	31181	338	AGRICULTURAL EDUCATION/TRAINING	2007001023
United States	AGR	31182	400	AGRICULTURAL RESEARCH	2007000443
United States	AID	31193	400	AGRICULTURAL FINANCIAL SERVICES	2007015783

Total aid for water in 2007 for India
And as a share of aid to total recipient countries

USD thousand 721824
10.76%

Indonesia

Donor	Agency	Sector	Amount	Project description	CRS ID Number
AsDF		31320	35282	SUSTAINABLE AQUACULTURE DEVT FOR FOOD SECURITY & POVERTY REDUC	070016
Australia	AusAID	31110	240	AGRICULTURAL POLICY & ADMIN. MGMT	2007001175
Australia	AusAID	31120	181	AGRICULTURAL DEVELOPMENT	2007001798
Australia	AusAID	31130	74	AGRICULTURAL LAND RESOURCES	2007001984
Australia	AusAID	31161	37	INTEGRATING FORAGE LEGUMES INTO THE MAIZE CROPPING SYSTEMS OF WEST TIMOR	2007001529
Australia	AusAID	31161	174	INTEGRATED SOIL AND CROP MANAGEMENT FOR REHABILITATION OF VEGETABLE	2007000749
Australia	AusAID	31161	177	FOOD CROP PRODUCTION	2007001225
Australia	AusAID	31161	88	ENHANCING THE ADOPTION OF IMPROVED CASSAVA PRODUCTION	2007001056
Australia	AusAID	31161	31	ACIAR FORAGE LEGUMES IN WEST TIMOR (KX41)	2007001599
Australia	AusAID	31161	377	CP/2005/167	2007000470
Australia	AusAID	31163	124	LIVESTOCK	2007001264
Australia	AusAID	31182	94	POVERTY ALLEVIATION & FOOD SECURITY THROUGH IMPROVING THE SWEET POTATO-PIG	2007001506
Australia	AusAID	31182	94	ASSESSMENT OF ZOONOTIC DISEASES IN INDONESIA	2007001030
Australia	AusAID	31182	26	EVALUATING STRATEGIES TO IMPROVE CALF SURVIVAL IN WEST TIMOR VILLAGES	2007001663
Australia	AusAID	31182	64	ANIMAL AND PLANT HEALTH SURVEY	2007001760
Australia	AusAID	31182	287	AGRICULTURAL RESEARCH: Avian influenza	2007000836
Australia	AusAID	31182	726	AGRICULTURAL RESEARCH	2007002023
Australia	AusAID	31182	128	ACIAR - PHASE 2 CROP-LIVESTOCK SYSTEMS	2007001260

RECIPIENT Donor	Agency	Sector	Amount USD thousand	Project description	CRS ID Number
Australia	AusAID	31182	258	AS1/2004/020	2007000577
Australia	AusAID	31182	21	HUANGLONGBING MANAGEMENT	2007001746
Australia	AusAID	31191	36	AGRICULTURAL SERVICES	2007001545
Australia	AusAID	31192	19	THE PROJECT AIMS TO DEVELOP INTEGRATED MANAG. STRATEGIES FOR CHILLI PEPPER	2007002024
Australia	AusAID	31192	39	ESTABLISHMENT OF FRUIT FLY PEST FREE AREAS	2007001501
Australia	AusAID	31192	46	INTEGRATED DISEASE MANAGEMENT (IDM) FOR ANTHRACNOSE,	2007001403
Australia	AusAID	31192	75	MITIGATING THE THREAT OF BANANA FUSARIUM WILT	2007001154
Australia	AusAID	31192	157	PLANT/POST-HARVEST PROT. & PEST CTRL	2007001691
Australia	AusAID	31195	259	RISK MANAG. FOR PANDEMIC INFLUENZA STRENGTHENING THE CAPACITY	2007000573
Australia	AusAID	31195	84	FUTURE DIRECTIONS FOR ANIMAL HEALTH SERVICES IN INDONESIA	2007001083
Australia	AusAID	31195	25	COMMERCIALISATION OF GUMBORO VACCINE IN INDONESIA	2007001694
Australia	AusAID	31195	8	CONTROL OF HPAI IN DUCKS IN INDO. AND VIETNAM. LB8	2007002044
Australia	AusAID	31210	174	FORESTRY POLICY & ADMIN. MANAGEMENT: TEAK PRODUCTION	2007001063
Australia	AusAID	31220	341	FORESTRY DEVELOPMENT	2007000854
Australia	AusAID	31282	213	FORESTRY RESEARCH	2007001759
Australia	AusAID	31310	42	REGIONAL MINISTERIAL INITIATIVE ON IUU FISHING	2007001758
Australia	AusAID	31310	8	STRENGTHENING REGIONAL MECHANISMS : SHRIMP FARMERS	2007002068
Australia	AusAID	31320	38	CAPACITY DEVELOPMENT FOR INDONESIAN TUNA FISHERIES	2007001512
Australia	AusAID	31320	222	FIS/2005/009	2007000638
Australia	AusAID	31320	662	FISHERY DEVELOPMENT	2007001915
Australia	AusAID	31381	142	FISHERY EDUCATION/TRAINING	2007001830
Australia	AusAID	31382	521	FISHERY RESEARCH	2007001045
Australia	AusAID	31382	40	SUPPORT FOR ANTIBIOTIC RESIDUE TESTING IN FISHERIES PRODUCTS	2007002115
Australia	AusAID	31382	82	MONITORING THE LONGLINE CATCH OF SOUTHERN BLUEFIN	2007002147
Belgium	DGCD	31120	1117	RENFORCEMENT DES COMMUNAUTéS VILLAGEOISES PAR L'AGRICULTURE DURABLE	2006003978
Finland	MFA	31210	945	NGO SUPPORT / PROTECTION OF BIODIVERSITY, LIVELIHOOD AND ECOLOGY	2007000175
France	MISC	31182	25749	RECHERCHE AGRICULTURE	2007800379
Germany	BMZ	31120	342	AGRICULTURAL DEVELOPMENT	2007006622
Germany	Fed Min	31182	776	AGRICULTURAL RESEARCH	2007010612
Germany	Fed Min	31182	2702	SUSTAINABLE BIOPRODUCTION	2007010613
Germany	BMZ	31210	64	FORESTRY POLICY & ADMIN. MANAGEMENT	2007007055
Germany	Fed Min	31320	292	FISHERY DEVELOPMENT	2007010317
IDA		31110	9000	FARMER EMPOWERMENT	2007000929
IDA		31140	18450	AGRICULTURAL WATER RESOURCES	2007000915
IDA		31150	15000	FARMER EMPOWERMENT	2007000933
IDA		31166	36000	FARMER EMPOWERMENT	2007000927
Japan	MAFF	31110	140	TC AGGREGATED ACTIVITIES	070010T
Japan	JICA	31110	2678	TC AGGREGATED ACTIVITIES	073001T
Japan	JICA	31130	64	TC AGGREGATED ACTIVITIES	073114T
Japan	MOFA	31150	1698	THE ASSISTANCE FOR UNDERPRIVILEGED FARMERS	070091
Japan	JICA	31163	793	TC AGGREGATED ACTIVITIES	073226T
Japan	MAFF	31181	157	TC AGGREGATED ACTIVITIES	070019T
Japan	PRF	31181	28	TC AGGREGATED ACTIVITIES	075602T
Japan	MOFA	31195	15119	THE PROJECT FOR IMPROVEMENT OF ANIMAL HEALTH LABORATORIES	070231
Japan	JICA	31195	620	TC AGGREGATED ACTIVITIES	073276T

RECIPIENT Donor	Agency	Sector	Amount USD thousand	Project description	CRS ID Number
Japan	JICA	31210	1546	TC AGGREGATED ACTIVITIES	073438T
Japan	JICA	31310	1872	TC AGGREGATED ACTIVITIES	073577T
Japan	MAFF	31381	116	TC AGGREGATED ACTIVITIES	070073T
Japan	MOFA	31391	9083	THE PROJECT FOR PROMOTION OF SUSTAINABLE COASTAL FISHERIES	070180
Korea	Misc	31110	9	AGRICULTURAL POLICY & ADMIN. MGMT	2007002559
Korea	Misc	31120	10	AGRICULTURAL DEVELOPMENT	2007001932
Korea	KOICA	31130	20	AGRICULTURAL LAND RESOURCES	2007016085
Korea	KOICA	31150	20	AGRICULTURAL INPUTS	2007015955
Korea	KOICA	31181	105	AGRICULTURAL EDUCATION/TRAINING	2007017900
Korea	Misc	31181	21	AGRICULTURAL EDUCATION/TRAINING	2007002568
Netherlands	MFA	31110	1789	AGRIC.: SUBS. TO PSOM 2003-2009	2007100135
Netherlands	MFA	31220	555	KALIMANTAN PEATLANDS PROJECT	2003015576
Netherlands	MFA	31220	3933	JAK KALIMANTAN PEATLANDS- CKPP	2007000704
Norway	MFA	31210	527	ELSDA/FIGHTING ILLEGAL LOGGING	2007004154
Norway	MFA	31210	5	CONSULTANCY SUPPORT PEATLAND WORKSHOP	2007004177
Norway	MFA	31210	589	CIFOR/FACING AN ILLEGAL LOGGING CRISIS	2007004158
Norway	MFA	31210	52	SUPPORT TO PREPARATORY MEETING FOR UNFF7	2007001260
Spain	MFA	31320	411	INFORMATION OF THE FISH MARKET IN ACEH	077048
United States	AID	31110	86	AGRICULTURAL POLICY & ADMIN. MGMT	2007013911
United States	AID	31120	3691	AGRICULTURAL SECTOR PRODUCTIVITY	2007001768

Total aid for water in 2007 for Indonesia   **USD thousand 197885**
And as a share of aid to total recipient countries   **2.95%**

Iran

Donor	Agency	Sector	Amount USD thousand	Project description	CRS ID Number
Japan	JICA	31110	140	TC AGGREGATED ACTIVITIES	073055T
Japan	JICA	31130	566	TC AGGREGATED ACTIVITIES	073096T

Total aid for water in 2007 for Iran   **USD thousand 706**
And as a share of aid to total recipient countries   **0.01%**

Iraq

Donor	Agency	Sector	Amount USD thousand	Project description	CRS ID Number
Japan	JICA	31110	286	TC AGGREGATED ACTIVITIES	072974T
Japan	JICA	31130	215	TC AGGREGATED ACTIVITIES	073127T
Korea	KOICA	31140	98	TRAINING PROGRAM	2007017510
Spain	AG	31166	8	AGRICULTURAL EXTENSION	070915
United States	AID	31110	30000	AGRICULTURAL ENABLING ENVIRONMENT	2007001662
United States	DOD	31110	1771	AGRICULTURAL POLICY & ADMIN. MGMT	2007022927
United States	STATE	31110	148	AGRICULTURAL POLICY & ADMIN. MGMT	2007025738
United States	AID	31120	62500	AGRICULTURAL SECTOR PRODUCTIVITY	2007001775
United States	DOD	31120	732	AGRICULTURAL DEVELOPMENT	2007022934
United States	DOD	31130	35	AGRICULTURAL LAND RESOURCES	2007022936
United States	DOD	31140	8191	AGRICULTURAL WATER RESOURCES	2007022952
United States	DOD	31140	3006	WATER RESOURCES PROJECTS MAJOR IRRIGATION PROJECTS	2007019869
United States	DOD	31140	11748	DOD-CERP - IRRIGATION	2007022947
United States	DOD	31150	3946	AGRICULTURAL INPUTS	2007022961
United States	DOD	31150	995	DOD-CERP - AGRICULTURE	2007020601

RECIPIENT Donor	Agency	Sector	Amount USD thousand	Project description	CRS ID Number
United States	DOD	31161	474	DOD-CERP - AGRICULTURE	2007021496
United States	DOD	31161	1728	DOD-CERP - FOOD PRODUCTION AND DISTRIB.	2007022965
United States	DOD	31161	3175	FOOD CROP PRODUCTION	2007022975
United States	DOD	31163	489	LIVESTOCK	2007022977
United States	DOD	31181	57	AGRICULTURAL EDUCATION/TRAINING	2007022983
United States	DOD	31182	47	AGRICULTURAL RESEARCH	2007022984
United States	DOD	31191	1432	DOD-CERP - ECON. AND RELATED INITIATIVES	2007021921
United States	DOD	31191	225	AGRICULTURAL SERVICES	2007022989
United States	DOD	31191	1778	DOD-CERP - AGRICULTURE	2007020693
United States	DOD	31192	349	PLANT/POST-HARVEST PROT. & PEST CTRL	2007022991
United States	DOD	31192	880	DOD-CERP - AGRICULTURE	2007022992
United States	DOD	31193	71	AGRICULTURAL FINANCIAL SERVICES	2007022993
United States	DOD	31194	4238	AGRICULTURAL CO-OPERATIVES	2007022999
United States	DOD	31194	4017	DOD-CERP - AGRICULTURE	2007022519
United States	DOD	31195	725	LIVESTOCK/VETERINARY SERVICES	2007023003
United States	DOD	31195	495	DOD-CERP - EDUCATION	2007022169
United States	DOD	31320	15	FISHERY DEVELOPMENT	2007023004

Total aid for water in 2007 for Iraq — **USD thousand 143871**
And as a share of aid to total recipient countries — **2.14%**

Jamaica

EU institutions	EDF	31162	4066	EUROPEAN UNION BANANA SUPPORT PROGRAMME (EUBSP)-SFA 2007	2007201284
EU institutions	EDF	31164	17112	ACCOMPANYING MEASURES 2007 FOR SUGAR PROTOCOL COUNTRIES - JAM	2007201262
Japan	JICA	31110	27	TC AGGREGATED ACTIVITIES	072956T
Japan	JICA	31195	40	TC AGGREGATED ACTIVITIES	073258T
Japan	JICA	31310	67	TC AGGREGATED ACTIVITIES	073499T

Total aid for water in 2007 for Jamaica — **USD thousand 21311**
And as a share of aid to total recipient countries — **0.32%**

Jordan

France	AFD	31140	3696	IRRIGATION VALLEE DU JOURDAIN	2007170300
Japan	JICA	31110	270	TC AGGREGATED ACTIVITIES	073056T
Japan	JICA	31210	33	TC AGGREGATED ACTIVITIES	073338T
Korea	KOICA	31120	18	AGRICULTURAL DEVELOPMENT	2007015882
United States	AID	31110	200	AGRICULTURAL POLICY & ADMIN. MGMT	2007001663
United States	AID	31120	300	AGRICULTURAL DEVELOPMENT	2007001778
United States	TDA	31140	319	AGRICULTURAL WATER RESOURCES	2007028238

Total aid for water in 2007 for Jordan — **USD thousand 4836**
And as a share of aid to total recipient countries — **0.07%**

Kazakhstan

Japan	JICA	31110	77	TC AGGREGATED ACTIVITIES	072985T
United States	AID	31110	77	AGRICULTURAL POLICY & ADMIN. MGMT	2007015291
United States	STATE	31110	18	AGRICULTURAL POLICY & ADMIN. MGMT	2007026758
United States	AID	31162	182	INDUSTRIAL CROPS/EXPORT CROPS	2007011451

RECIPIENT Donor	Agency	Sector	Amount USD thousand	Project description	CRS ID Number
United States	AID	31182	182	AGRICULTURAL RESEARCH	2007015634

Total aid for water in 2007 for Kazakhstan   USD thousand 536
And as a share of aid to total recipient countries   0.01%

Kenya

Donor	Agency	Sector	Amount	Project description	CRS ID
AfDF		31161	26014	SMALLSCALE HORTICULTURE DEVELOPMENT PRO	070027
Austria	ADA	31150	274	BUSINESS PARTNERSHIP PROGRAM: SILICATEC TECHNOLOGY F. RURAL FARMING	2007500109d
Austria	ADA	31161	85	NGO COFINANCING: AGRICULTURE ELDORET, SUST. AGRICULTURE & FOOD SECURITY	2006000022ci
Belgium	DGCD	31161	82	FOOD CROP PRODUCTION	2006003924
Belgium	DGCD	31163	212	TURKANA LIVESTOCK DEVELOPMENT PROGRAMME	2005000617
Canada	CIDA	31120	465	STRENGTHENING KENYAN FARM FAMILIES	070727
Canada	IDRC	31120	98	MANAGING RISK, REDUCING VULNERABILITY AND ENHANCING PRODUCTIVITY	070312i
Canada	IDRC	31162	42	DIVERSIFICATION OF HOUSEHOLD LIVELIHOOD STRATEGIES FOR TOBACCO	070180i
Canada	IDRC	31163	74	ENHANCING PASTORALISTS ADAPTIVE CAPACITY TO CLIMATE CHANGE	070696i
France	MISC	31182	9925	RECHERCHE AGRICULTURE	2007800349
Germany	BMZ	31110	5476	PROMOTION OF PRIVATE SECTOR DEVELOPMENT IN AGRICULTURE	2007005892
Germany	BMZ	31120	541	AGRICULTURAL DEVELOPMENT	2007009274
Germany	BMZ	31130	344	AGRICULTURAL LAND RESOURCES	2007009058
Germany	BMZ	31140	78	AGRICULTURAL WATER RESOURCES	2007009059
Germany	Fed Min	31165	684	SUSTAINABLE AGRICULTURE, CONSERVATION AGRICULTURE FOR SARD	2007010311
Germany	Fed Min	31165	630	SUPPORTING FOOD SECURITY AND REDUCING POVERTY	2007010315
Germany	BMZ	31182	66	AGRICULTURAL RESEARCH	2007007104
Germany	BMZ	31195	147	LIVESTOCK/VETERINARY SERVICES	2007007106
IDA		31110	4110	AGRICULTURE	2007000267
IDA		31110	8600	FLOOD MITIGATION	2007000276
IDA		31140	39730	AGRICULTURE	2007000265
IDA		31210	21235	AGRICULTURE	2007000266
IFAD		31161	500	FOOD CROP PRODUCTION	070018a
IFAD		31161	23430	SMALLHOLDER HORTICULTURE MARKETING PROGRAMME	070018
Ireland	DFA	31120	421	AGRICULTURAL DEVELOPMENT	2007002445
Ireland	DFA	31140	90	AGRICULTURAL WATER RESOURCES	2007001246
Ireland	DFA	31161	538	FOOD CROP PRODUCTION	2007000891
Ireland	DFA	31165	22	AGRICULTURAL ALTERNATIVE DEVELOPMENT	2007001951
Ireland	DFA	31166	72	AGRICULTURAL EXTENSION	2007001332
Ireland	DFA	31194	87	AGRICULTURAL CO-OPERATIVES	2007001252
Italy	CA	31120	164	AGRICULTURAL DEVELOPMENT	071306
Italy	DGCS	31163	7	LIVESTOCK	060717
Italy	DGCS	31166	30	AGRICULTURAL EXTENSION	060223
Japan	JICA	31110	2544	TC AGGREGATED ACTIVITIES	073023T
Japan	JICA	31130	281	TC AGGREGATED ACTIVITIES	073123T
Japan	JICA	31163	187	TC AGGREGATED ACTIVITIES	073239T
Japan	JICA	31195	13	TC AGGREGATED ACTIVITIES	073247T
Japan	JICA	31210	663	TC AGGREGATED ACTIVITIES	073368T
Japan	MAFF	31310	40	TC AGGREGATED ACTIVITIES	070035T
Japan	JICA	31310	150	TC AGGREGATED ACTIVITIES	073551T

RECIPIENT Donor	Agency	Sector	Amount USD thousand	Project description	CRS ID Number
Netherlands	MFA	31110	1268	AGRIC.: SUBS. TO PSOM 2003-2009	2007100142
Netherlands	MFA	31181	414	AGRICULTURAL EDUCATION/TRAINING	2007000040
Netherlands	MFA	31182	179	AGRICULTURAL RESEARCH	2007000686
New Zealand	NZAid	31181	16	CHIGA CATHOLIC CHURCH: NUTRITIONAL PROJECT: KENYA	070403
Norway	NORAD	31120	54	POKOT DEVELOPMENT PROGR. (PDP - PIP), AGRICULTURE	2007003708
Spain	AG	31120	27	SUPPORT FOR THE SPREADING OF AGRICULTURE	072416
Spain	MFA	31310	69	BOOSTING THE FISHERIES SECTOR THROUGH THE CREATION & EQPT OF FISHERMEN	073953
Switzerland	SDC	31120	17	AGRICULTURAL DEVELOPMENT	1980001070
United Kingdom	DFID	31163	1890	EARLY DETECTION OF AVIAN FLU	2007000141
United States	AID	31110	17	AGRICULTURAL POLICY & ADMIN. MGMT	2007013853
United States	AID	31110	1233	PROGRAM SUPPORT (AGRICULTURE)	2007013865
United States	AID	31110	1425	AGRICULTURAL ENABLING ENVIRONMENT	2007001644
United States	AID	31120	5750	AGRICULTURAL SECTOR PRODUCTIVITY	2007001730

Total aid for water in 2007 for Kenya — **USD thousand 160509**
And as a share of aid to total recipient countries — **2.39%**

Kiribati

Recipient	Agency	Sector	Amount	Project description	CRS ID
Australia	AusAID	31320	6	AQUACULTURE RESEARCH PRIORITIES	2007002141
Japan	JICA	31110	25	TC AGGREGATED ACTIVITIES	073010T
Japan	JICA	31310	25	TC AGGREGATED ACTIVITIES	073533T
Japan	MAFF	31320	521	TC AGGREGATED ACTIVITIES	070059T
Japan	MOFA	31391	10908	THE PROJECT FOR IMPROVEMENT OF FISHERIES-RELATED ROADS IN SOUTH TARAWA	070013
Korea	KOICA	31320	189	TRAINING PROGRAM	2007017334

Total aid for water in 2007 for Kiribati — **USD thousand 11674**
And as a share of aid to total recipient countries — **0.17%**

Korea, Dem. Rep.

Recipient	Agency	Sector	Amount	Project description	CRS ID
Finland	MFA	31161	561	FRAME AGREEMENT WITH NGO (FIDA INT.)	2007000258
Germany	BMZ	31182	132	AGRICULTURAL RESEARCH	2007007652
Switzerland	SDC	31120	1250	AGRICULTURAL SUPPORT PROGRAMME	2002002096

Total aid for water in 2007 for Korea, Dem. Rep. — **USD thousand 1943**
And as a share of aid to total recipient countries — **0.03%**

Kyrgyz Republic

Recipient	Agency	Sector	Amount	Project description	CRS ID
AsDF		31120	15668	SOUTHERN AGRICULTURE AREA DEVELOPMENT PROJECT	070041
Germany	BMZ	31166	78	AGRICULTURAL EXTENSION	2007007114
Germany	BMZ	31181	233	AGRICULTURAL EDUCATION/TRAINING	2007007110
IDA		31140	12640	AGRICULTURAL WATER RESOURCES	2007000716
Japan	JICA	31110	917	TC AGGREGATED ACTIVITIES	073058T
Japan	JICA	31130	50	TC AGGREGATED ACTIVITIES	073102T
Japan	JICA	31210	36	TC AGGREGATED ACTIVITIES	073413T
Norway	MFA	31210	170	FOREST AND ENVIRONMENTAL SECTOR PROGRAMME	2007001839
Sweden	Sida	31120	3359	AGRICULTURAL DEVELOPMENT	2007005533
Switzerland	seco	31110	1542	ORGANIC COTTON PRODUCTION AND TRADE PROMOTION PHASE II	2007001863
Switzerland	FA	31120	33	AGRICULTURAL DEVELOPMENT	2007001405

RECIPIENT Donor	Agency	Sector	Amount USD thousand	Project description	CRS ID Number
Switzerland	SDC	31120	1084	RADS/KSAP RURAL ADVISORY SERVICES	1994000597
Switzerland	SDC	31220	1750	SUPPORT TO FORESTRY SECTOR	1995000363
United States	AID	31110	696	AGRICULTURAL ENABLING ENVIRONMENT	2007001666
United States	AID	31110	294	AGRICULTURAL POLICY & ADMIN. MGMT	2007015292
United States	AID	31110	894	AGRICULTURAL RESOURCE POLICY	2007001695
United States	AID	31140	717	LAND AND WATER MANAGEMENT	2007010927
United States	AID	31162	841	MARKETS AND TRADE CAPACITY	2007011452
United States	AID	31181	63	AGRICULTURAL EDUCATION/TRAINING	2007001027
United States	AID	31182	841	RESEARCH AND TECHNOLOGY DISSEMINATION	2007015635
United States	AID	31193	102	AGRICULTURAL FINANCIAL SERVICES	2007015784

Total aid for water in 2007 for Kyrgyz Republic   **USD thousand 42009**
And as a share of aid to total recipient countries   **0.63%**

Laos

Donor	Agency	Sector	Amount	Project description	CRS ID
AsDF		31163	9848	NORTHERN REGION SUSTAINABLE LIVELIHOODS THRU LIVESTOCK DEVT PR	070006
Australia	AusAID	31110	101	AGRICULTURAL POLICY & ADMIN. MGMT	2007001564
Australia	AusAID	31120	99	AGRICULTURAL DEVELOPMENT	2007001383
Australia	AusAID	31150	6	DEVELOPMENT OF CONSERVATION FARMING IMPLEMENTS FOR TWO-WHEELER	2007002131
Australia	AusAID	31161	91	FOOD CROP PRODUCTION	2007001042
Australia	AusAID	31163	119	LIVESTOCK	2007001513
Australia	AusAID	31181	21	AGRICULTURAL EDUCATION/TRAINING	2007001748
Australia	AusAID	31182	26	IMPROVED RISK ASSESSMENT FOR TRANSBOUNDARY DISEASE IN LAOS AND CAMBODIA	2007001668
Australia	AusAID	31182	18	DEVELOPING BEST PRACTICE CATTLE AND BUFFALO HEALTH AND HUSBANDRY SYSTEMS	2007001804
Australia	AusAID	31182	63	VACCINE BUSINESS DEVELOPMENT IN LAO PDR	2007001243
Australia	AusAID	31182	102	AGRICULTURAL RESEARCH: RICE	2007000982
Australia	AusAID	31182	38	ADAPTATION OF LOW-CHILL TEMPERATE FRUITS	2007001509
Australia	AusAID	31182	33	AGRICULTURAL RESEARCH	2007001582
Australia	AusAID	31195	25	IDENTIFYING RESEARCH PRIORITIES FOR THE DEVELOPMENT OF THE BEEF INDUSTRY	2007001681
Australia	AusAID	31195	28	DIAGNOSIS AND EPIDEMIOLOGY OF FOOT AND MOUTH DISEASE IN LAO PDR 1997-2006	2007001647
Australia	AusAID	31195	75	LIVESTOCK/VETERINARY SERVICES	2007001853
Australia	AusAID	31195	13	Livestock health and vaccines in Cambodia and Laos: scoping study and economic assessment	2007001925
Australia	AusAID	31195	60	FORAGE LEGUMES FOR SUPPLEMENTING VILLAGE PIGS IN LAO PDR	2007001268
Australia	AusAID	31282	266	FORESTRY RESEARCH	2007001989
Australia	AusAID	31320	100	FISHERY DEVELOPMENT	2007001630
Australia	AusAID	31382	10	FISHERY RESEARCH	2007002007
Belgium	DGCD	31110	28	AGRICULTURAL POLICY & ADMIN. MGMT	2004003276
Belgium	DGCD	31120	222	SECURITé ALIMENTAIRE POUR DES GROUPES ETHNIQUES PAUVRES	2006003977
France	AFD	31140	1369	ASSISTANCE TECHNIQUE SECT. IRRIGATION	2007153000
France	MISC	31182	18125	RECHERCHE AGRICULTURE	2007800380
Germany	BMZ	31165	1882	PHONGSALY ALTERNATIVE DEVELOPMENT PROJECT	2007006690
Germany	BMZ	31181	373	AGRICULTURAL EDUCATION/TRAINING	2007009068
Germany	Fed Min	31195	972	AVIAN INFLUENZA CONTROL (LAOS)	2007010306
Ireland	DFA	31161	464	SUPPORT TO NGOS - IRISH - MAPS - GENERAL - COMMUNITY LIVELIHOODS DEVPT	2007000236
Japan	MAFF	31110	46	TC AGGREGATED ACTIVITIES	070005T
Japan	JICA	31110	1285	TC AGGREGATED ACTIVITIES	073002T

RECIPIENT Donor	Agency	Sector	Amount USD thousand	Project description	CRS ID Number
Japan	JICA	31130	51	TC AGGREGATED ACTIVITIES	073115T
Japan	PRF	31140	12	TC AGGREGATED ACTIVITIES	075593T
Japan	PRF	31163	12	TC AGGREGATED ACTIVITIES	075598T
Japan	JICA	31210	1651	TC AGGREGATED ACTIVITIES	073441T
Japan	MAFF	31310	32	TC AGGREGATED ACTIVITIES	070041T
Japan	JICA	31310	1129	TC AGGREGATED ACTIVITIES	073525T
Korea	Misc	31110	9	AGRICULTURAL POLICY & ADMIN. MGMT	2007002554
Korea	KOICA	31140	23	AGRICULTURAL WATER RESOURCES	2007017123
Korea	KOICA	31150	21	AGRICULTURAL INPUTS	2007017116
Korea	KOICA	31163	52	LIVESTOCK	2007017177
Korea	KOICA	31181	64	AGRICULTURAL EDUCATION/TRAINING	2007017157
Korea	Misc	31181	15	AGRICULTURAL EDUCATION/TRAINING	2007002569
Korea	KOICA	31381	22	FISHERY EDUCATION/TRAINING	2007015944
Luxembourg	MFA	31120	781	AGRICULTURAL DEVELOPMENT	070723
New Zealand	NZAid	31162	426	NEW OPPORTUNITIES FOR COFFEE FARMERS	070410
Sweden	Sida	31182	12958	AGRICULTURAL RESEARCH	2006001379
Switzerland	SDC	31110	667	AGROENTREPRISE DEVELOPMENT LAOS/VIETNAM	2003005522
Switzerland	SDC	31110	679	LAEP LAOS AGRICULTURAL EXTENSION PROJECT	2001002701

Total aid for water in 2007 for Laos   **USD thousand 54510**
And as a share of aid to total recipient countries   **0.81%**

Lebanon

RECIPIENT Donor	Agency	Sector	Amount USD thousand	Project description	CRS ID Number
Austria	Reg	31150	14	OLIVE OIL MILL	2007828525
Austria	MISC	31210	7	SMALL-SCALE COMMITMENTS AGGREGATED BY SECTOR AND RECIPIENT COUNTRY	2007980194
Finland	MFA	31166	158	NGO SUPPORT / IMPROVING MARKET ACCESS FOR SMALL PRODUCERS	2007000178
Italy	DGCS	31120	137	AGRICULTURAL DEVELOPMENT	060071
Italy	LA	31120	36	AGRICULTURAL DEVELOPMENT	071527
Italy	DGCS	31166	735	AGRICULTURAL EXTENSION	070637
Italy	DGCS	31320	328	FISHERY DEVELOPMENT	061101
Spain	MFA	31110	821	IMPROVE PRODUCTIVE CAPACITY OF AGRICULTURAL SECTOR TRAINING FARMERS	077664
Spain	MFA	31110	338	AGRICULTURAL POLICY & ADMIN. MGMT	076692
Spain	MFA	31120	353	STRENGTHENING APICULTURE IN AKKAR NORTHERN LEBANON	076819
Switzerland	SDC	31120	56	AGRICULTURAL DEVELOPMENT	2007000475
United States	AID	31120	2545	AGRICULTURAL SECTOR PRODUCTIVITY	2007001772

Total aid for water in 2007 for Lebanon   **USD thousand 5527**
And as a share of aid to total recipient countries   **0.08%**

Lesotho

RECIPIENT Donor	Agency	Sector	Amount USD thousand	Project description	CRS ID Number
Canada	CIDA	31194	26	éLEVAGE DE POULES PONDEUSES	070520
Germany	BMZ	31120	958	COMMUNITY SELF RELIANCE AND ENVIRONMENTALPROTECTION (CSREP)	2007006649
Germany	BMZ	31220	98	FORESTRY DEVELOPMENT	2007007171
IDA		31161	1701	PRIVATE SECTOR COMPETITIVENESS	2007000284
Japan	JICA	31210	40	TC AGGREGATED ACTIVITIES	073369T
United Kingdom	DFID	31120	110	Potato Culture Project	2007000151

RECIPIENT Donor	Agency	Sector	Amount USD thousand	Project description	CRS ID Number

Total aid for water in 2007 for Lesotho — **USD thousand 2933**
And as a share of aid to total recipient countries — **0.04%**

Liberia

Recipient/Donor	Agency	Sector	Amount	Project description	CRS ID
Germany	BMZ	31181	205	AGRICULTURAL EDUCATION/TRAINING	2007006282
IDA		31110	6660	AGRICULTURE AND INFRASTRUCTURE DEVELOPMENT	2007000297
Ireland	DFA	31161	804	SUPPORT TO NGOS - IRISH - MAPS - GENERAL - FARMER SUPPORT SERVICES	2007000232
United States	AID	31110	331	AGRICULTURAL POLICY & ADMIN. MGMT	2007013848
United States	AID	31110	940	AGRICULTURAL ENABLING ENVIRONMENT	2007001632
United States	AID	31120	7719	AGRICULTURAL SECTOR PRODUCTIVITY	2007001716
United States	ADF	31150	99	AGRICULTURAL INPUTS	2007000241
United States	AID	31181	149	AGRICULTURAL EDUCATION/TRAINING	2007001018
United States	AID	31182	50	AGRICULTURAL RESEARCH	2007015629
United States	AID	31193	38	AGRICULTURAL FINANCIAL SERVICES	2007015780
United States	ADF	31310	100	FISHING POLICY AND ADMIN. MANAGEMENT	2007000129
United States	ADF	31320	90	FISHERY DEVELOPMENT	2007000227

Total aid for water in 2007 for Liberia — **USD thousand 17185**
And as a share of aid to total recipient countries — **0.26%**

Libya

Recipient/Donor	Agency	Sector	Amount	Project description	CRS ID
Italy	DGCS	31120	968	AGRICULTURAL DEVELOPMENT	060103
Italy	DGCS	31182	333	AGRICULTURAL RESEARCH	060100
Spain	MFA	31195	47	VIGILENCE AND CONTROL SYSTEMS OF THE HIGHLY INFECTIOUS BIRD FLU	073194
Switzerland	SDC	31110	14	AGRICULTURAL POLICY & ADMIN. MGMT	2006003190

Total aid for water in 2007 for Libya — **USD thousand 1362**
And as a share of aid to total recipient countries — **0.02%**

Macedonia, FYR

Recipient/Donor	Agency	Sector	Amount	Project description	CRS ID
Germany	BMZ	31120	52	AGRICULTURAL DEVELOPMENT	2007007206
Japan	JICA	31110	52	TC AGGREGATED ACTIVITIES	072903T
Norway	MFA	31110	171	DEVELOPMENT OF LEGAL FRAMEWORK GROUPS	2007001738
Norway	MFA	31194	217	**GROWING POTATOES FOR SEED PRODUCTION**	2007001666
Norway	MFA	31194	260	IMPROVING MILK PRODUCTION IN DELCHEVO	2007001665
Norway	MFA	31320	84	FISHERY DEVELOPMENT	2007001704
Sweden	Sida	31110	3700	AGRICULTURAL POLICY & ADMIN. MGMT	2007002480
Switzerland	SDC	31110	322	PROJECT FOR ORGANIC AGRICULTURE	2003005500
Switzerland	SDC	31110	17	AGRICULTURAL POLICY & ADMIN. MGMT	2003005501
United States	AID	31110	12	AGRICULTURAL POLICY & ADMIN. MGMT	2007013817
United States	AID	31162	525	MARKETS AND TRADE CAPACITY	2007011446
United States	AID	31182	450	AGRICULTURAL RESEARCH	2007015628
United States	AID	31193	300	AGRICULTURAL FINANCIAL SERVICES	2007015779

Total aid for water in 2007 for Macedonia, FYR — **USD thousand 6162**
And as a share of aid to total recipient countries — **0.09%**

RECIPIENT Donor	Agency	Sector	Amount USD thousand	Project description	CRS ID Number
Madagascar					
Belgium	DGCD	31120	65	AGRICULTURAL DEVELOPMENT	2004007813
Belgium	DGCD	31181	48	AGRICULTURAL EDUCATION/TRAINING	2004005364
Belgium	DGCD	31194	321	RENFORCEMENT DESCAPACITéS DES COOPéRATIVES D'éLEVEURS	2004005347
EU institutions	EDF	31120	2092	AGRICULTURAL DEVELOPMENT	2007201264
EU institutions	EDF	31161	684	FOOD CROP PRODUCTION	2007201287
France	AFD	31165	1848	PROTECTION DES BASSINS LAC ALAOTRA	2007163900
France	AFD	31165	205	PROTECTION BASSINS LAC ALAOTRA	2007169700
France	MISC	31182	53333	RECHERCHE AGRICULTURE	2007800350
Germany	L G	31181	90	AGRICULTURAL EDUCATION/TRAINING	2007011810
Germany	BMZ	31210	61	FORESTRY POLICY & ADMIN. MANAGEMENT	2007007195
Italy	LA	31120	11	AGRICULTURAL DEVELOPMENT	070828
Italy	DGCS	31192	353	PLANT AND POST-HARVEST PROTECTION AND PEST CONTROL	070646
Japan	JICA	31110	1159	TC AGGREGATED ACTIVITIES	073024T
Japan	JICA	31130	110	TC AGGREGATED ACTIVITIES	073082T
Japan	JICA	31150	373	TC AGGREGATED ACTIVITIES	073143T
Japan	MOFA	31181	4907	PROJET D EXTENSION ET D AMENAGEMENT	070244
Japan	JICA	31195	38	TC AGGREGATED ACTIVITIES	073248T
Japan	JICA	31210	215	TC AGGREGATED ACTIVITIES	073371T
Japan	MAFF	31310	33	TC AGGREGATED ACTIVITIES	070036T
Japan	JICA	31310	405	TC AGGREGATED ACTIVITIES	073553T
Japan	MAFF	31320	1097	TC AGGREGATED ACTIVITIES	070050T
Norway	NORAD	31120	31	INTEGRATED DEV. PROJECT SOFASPAN FIRST PHASE, AGRICULTURE	2007003610
Norway	NORAD	31120	44	INTEGRATED VILLAGE DEV. PROGRAM BARA AGRICULTURAL DEVELOPMENT	2007003563
Norway	MFA	31120	5121	SUPPORT TO FIFAMANOR	2007004316
Norway	NORAD	31320	415	MARINE-SPIM: Seaweed Production in Madagascar	2007003788
Switzerland	SDC	31220	402	FORECA EMISSION REDUCTION THRU AVOIDED DEFORESTATION	2007000244
United States	AID	31110	123	AGRICULTURAL POLICY & ADMIN. MGMT	2007013866
United States	AID	31120	606	AGRICULTURAL SECTOR PRODUCTIVITY	2007001731

Total aid for water in 2007 for Madagascar   **USD thousand 74190**
And as a share of aid to total recipient countries   **1.11%**

RECIPIENT Donor	Agency	Sector	Amount USD thousand	Project description	CRS ID Number
Malawi					
Belgium	MPRF	31161	5613	CONTRIBUTION POUR SéCURITé ALIMENTAIRE	2007004771
Belgium	DGCD	31182	82	AGRICULTURAL RESEARCH	2002000107
Canada	IDRC	31110	173	STRENGTHENING LOCAL AGRICULTURAL INNOVATION SYSTEMS	070306i
EU institutions	EDF	31162	6845	SUGAR ANNUAL ACTION PLAN FOR 2007	2007201263
Greece	YPEJ	31150	41	FOOD SECURITY	2007000254
Ireland	DFA	31120	456	AGRICULTURAL DEVELOPMENT	2007000936
Ireland	DFA	31140	110	AGRICULTURAL WATER RESOURCES	2007001126
Ireland	DFA	31150	291	AGRICULTURAL INPUTS	2007000555
Ireland	DFA	31182	1421	CGIAR CENTRE RESEARCH AGRIC. / FORRESTRY - WORLD	2007000240
Ireland	DFA	31220	13	FORESTRY DEVELOPMENT	2007002555
Italy	CA	31120	113	AGRICULTURAL DEVELOPMENT	071318
Italy	DGCS	31191	353	AGRICULTURAL SERVICES	070491

RECIPIENT Donor	Agency	Sector	Amount USD thousand	Project description	CRS ID Number
Japan	JICA	31110	3941	TC AGGREGATED ACTIVITIES	073025T
Japan	JICA	31130	150	TC AGGREGATED ACTIVITIES	073083T
Japan	JICA	31150	114	TC AGGREGATED ACTIVITIES	073176T
Japan	JICA	31163	225	TC AGGREGATED ACTIVITIES	073193T
Japan	JICA	31195	54	TC AGGREGATED ACTIVITIES	073286T
Japan	JICA	31210	214	TC AGGREGATED ACTIVITIES	073373T
Japan	JICA	31310	219	TC AGGREGATED ACTIVITIES	073481T
Norway	MFA	31110	26	TECHNICAL SUPPORT TO ADP-ADDENDUM	2007004423
Norway	MFA	31110	171	TECHNICAL SUPPORT TO MALAWI NATIONAL AGRICULTURAL DEVELOPMENT PROGRAMME	2007004402
Norway	MFA	31110	21	SCOPING STUDY ON FAIR TRADE ANALYSIS FOR MALAWI'S AGRICULTURAL PRODUCTS	2007004413
Norway	NORAD	31120	341	FARMERS AGRICULTURE INNOVATION IN RURAL MALAWI (FAIR)	2007003416
Norway	MFA	31120	192	AGRICULTURAL DEVELOPMENT	2007002618
Norway	MFA	31130	768	CHIA LAGOON WATERSHED MANAGEMENT PROJECT	2007004415
Norway	MFA	31150	1707	GOVERNMENT OF MALAWI SEEDS AND FERTILSER PROGRAMME	2007004399
Norway	NORAD	31166	112	IMPROVING FOOD SECURITY THROUGH INTEGRATED CROP PRODUCTION	2007003397
Norway	NORAD	31166	112	EXTENDING GROUNDNUT PRODUCTION TO THE NON-TRADITIONAL AND DRY LAND AREAS	2007003396
Norway	MFA	31194	13656	NASFAM PHASE III - IMPROVING THE LIVELIHOODS OF SMALLHOLDER FARMERS	2007004391
Spain	AG	31161	60	IMPROVEMENT OF AGRICULTURAL PRODUCTION AND DIET DIVERSIFICATION	073616
United Kingdom	MISC	31120	171	UNIVERSITY OF STIRLING – AQUACULTURE PROJECT	2007801098
United Kingdom	MISC	31120	171	EQUAL EXCHANGE – NENO MACADEMIA NUT PROJECT	2007801081
United Kingdom	MISC	31382	61	SCOTTISH CROP RESEARCH INSTITUTE – EXPLORING POTATO FARMING OPPORTUNITIES	2007801120
United States	AID	31110	611	AGRICULTURAL POLICY & ADMIN. MGMT	2007015290
United States	AID	31120	700	AGRICULTURAL SECTOR PRODUCTIVITY	2007001787
United States	AID	31120	280	AGRICULTURAL DEVELOPMENT	2007001756
United States	AID	31162	670	MARKETS AND TRADE CAPACITY	2007011449
United States	AID	31181	90	AGRICULTURAL EDUCATION/TRAINING	2007001020
United States	AID	31182	600	RESEARCH AND TECHNOLOGY DISSEMINATION	2007015632

Total aid for water in 2007 for Malawi   **USD thousand 40948**
And as a share of aid to total recipient countries   **0.61%**

Malaysia

	Agency	Sector	Amount	Project description	CRS ID
France	MISC	31182	1492	RECHERCHE AGRICULTURE	2007800381
Germany	BMZ	31320	71	FISHERY DEVELOPMENT	2007007244
Japan	JICA	31110	688	TC AGGREGATED ACTIVITIES	073003T
Japan	JICA	31163	14	TC AGGREGATED ACTIVITIES	073228T
Japan	MAFF	31181	88	TC AGGREGATED ACTIVITIES	070020T
Japan	JICA	31195	148	TC AGGREGATED ACTIVITIES	073278T
Japan	JICA	31210	36	TC AGGREGATED ACTIVITIES	073444T
Japan	JICA	31310	192	TC AGGREGATED ACTIVITIES	073578T
Korea	Misc	31110	9	AGRICULTURAL POLICY & ADMIN. MGMT	2007002558
Korea	Misc	31181	21	AGRICULTURAL EDUCATION/TRAINING	2007002576
United Kingdom	MISC	31210	30	DEVELOPMENT OF A CREDIBLE HIGH CONSERVATION VALUE FORESTS (HCVFS)	2007801149

Total aid for water in 2007 for Malaysia   **USD thousand 2788**
And as a share of aid to total recipient countries   **0.04%**

RECIPIENT Donor	Agency	Sector	Amount USD thousand	Project description	CRS ID Number
Maldives					
IFAD		31110	3505	PROGRAMME DE DIVERSIFICATION DE LA PÊCHE ET DE L'AGRICULTURE	070012
Japan	JICA	31110	29	TC AGGREGATED ACTIVITIES	072994T
UNDP		31320	8	FISHERY DEVELOPMENT	2007014174

Total aid for water in 2007 for Maldives   **USD thousand 3542**
And as a share of aid to total recipient countries   **0.05%**

RECIPIENT Donor	Agency	Sector	Amount USD thousand	Project description	CRS ID Number
Mali					
AfDF		31163	22953	DVPT PRODUCTIONS ANIMALES KAYES SUD	070034
Belgium	DGCD	31161	296	PRODUIRE DES SEMENCES DE CÉRÉALES EN MILIEU PAYSAN AU MALI	2004003880
Belgium	DGCD	31163	130	LIVESTOCK	2004008414
Belgium	DGCD	31166	46	AGRICULTURAL EXTENSION	2004008422
Belgium	DGCD	31191	89	AGRICULTURAL SERVICES	2004003370
Canada	CIDA	31191	279	PROJET D'APPUI A LA COMMERCIALISATION DES CEREALES	030741
Denmark	MFA	31110	3603	INSTITUTIONAL DEVELOPMENT	071356
Denmark	MFA	31110	1011	UNALLOCATED FUNDS	071361
Denmark	MFA	31120	9271	IMPROVEMENT OF INFRASTRUCTURE AT MUNICIPALITY LEVEL	071357
Denmark	MFA	31120	919	REVIEWS STUDIES AND AUDITS	071360
Denmark	MFA	31120	5112	ADVISORS AND TECHNICAL ASSISTANCE	071359
Denmark	MFA	31120	7644	LOCAL AGRICULTURE DEVELOPMENT	071358
EU institutions	CEC	31162	20534	INDUSTRIAL CROPS/EXPORT CROPS	2007100163
France	MISC	31182	21684	RECHERCHE AGRICULTURE	2007800351
Germany	BMZ	31110	46	AGRICULTURAL POLICY & ADMIN. MGMT	2007008795
Germany	BMZ	31120	441	AGRICULTURAL DEVELOPMENT	2007008796
Germany	KFW	31120	4107	FONDS D'ENTRE-AIDE AU PAYS DOGON	2007000897
Germany	KFW	31140	10951	OFFICE DU NIGER-INTE. MARGIN. LANDNUTZER	2007000904
Germany	BMZ	31181	382	AGRICULTURAL EDUCATION/TRAINING	2007008797
IDA		31110	2400	AGRICULTURAL POLICY & ADMIN. MGMT	2007000344
IDA		31140	1600	AGRICULTURAL WATER RESOURCES	2007000343
IDA		31161	1600	FOOD CROP PRODUCTION	2007000342
IDA		31166	12000	AGRICULTURAL EXTENSION	2007000341
Italy	DGCS	31110	16	AGRICULTURAL POLICY & ADMIN. MGMT	060784
Italy	CA	31181	34	AGRICULTURAL EDUCATION/TRAINING	071313
Japan	JICA	31110	1303	TC AGGREGATED ACTIVITIES	073026T
Japan	MOFA	31150	3311	GRANT ASSISTANCE FOR UNDERPRIVILEGED FARMERS	070246
Luxembourg	MFA	31120	84	AGRICULTURAL DEVELOPMENT	070648
Luxembourg	MFA	31140	1891	AGRICULTURAL WATER RESOURCES	071745
Luxembourg	MFA	31161	548	FOOD CROP PRODUCTION	071706
Netherlands	MFA	31110	866	AGRIC.: SUBS. TO PSOM 2003-2009	2007100130
Netherlands	MFA	31140	1169	AGRIC. WATER RES.: BAM CONTRAT PLAN ON 2005/07	2003013915
Norway	NORAD	31130	279	DCG MALI	2007004749
Norway	NORAD	31161	141	HEALTH GARDEN IN TOMORA	2007003745
Spain	MFA	31140	259	TECHNICAL ASSISTANCE FOR A REVIEW STUDY OF THE SITUATION OF IRRIGATION	076216
Spain	MFA	31140	1154	THIS PROJECT WILL HELP TO IMPROVE THE FOOD & NUTRITION SITUATION	077806

RECIPIENT Donor	Agency	Sector	Amount USD thousand	Project description	CRS ID Number
Spain	AG	31140	21	AGRICULTURAL WATER RESOURCES	072044
Spain	AG	31161	433	AGRICULTURAL AND LIVESTOCK SUPPORT AND AWARENESS OF THE ISSUE OF NUTRITION	077162
Spain	AG	31163	285	IMPROVEMENT OF CATTLE PRODUCTION AND INFRASTRUCTURE SUPPORT	076438
Spain	MFA	31165	306	IMPROVEMENT OF FOOD SOVEREIGNTY	076537
Spain	AG	31181	46	NEYLENI FORUM FOR FOOD SOVEREIGNTY IN MALI NEYLENI 2007.	073189
Spain	MFA	31310	46	CONTINENTAL AQUACULTURE DEVELOPMENT PROGRAM.	073166
Sweden	Sida	31210	52	FORESTRY POLICY & ADMIN. MANAGEMENT	2007002461
Sweden	Sida	31220	1926	FORESTRY DEVELOPMENT	2007002487
Switzerland	SDC	31110	208	AGRICULTURAL POLICY & ADMIN. MGMT	2004001775
United States	AID	31110	565	AGRICULTURAL POLICY & ADMIN. MGMT	2007015289
United States	MCC	31110	150125	AGRICULTURAL POLICY AND ADMINISTRATIVE MANAGEMENT COMPACT ACTIVITY	2007025214
United States	AID	31140	375	AGRICULTURAL WATER RESOURCES	2007010926
United States	MCC	31140	4596	AGRICULTURAL WATER RESOURCES ACTIVITY	2007025150
United States	ADF	31161	69	FOOD CROP PRODUCTION	2007000090
United States	AID	31162	300	INDUSTRIAL CROPS/EXPORT CROPS	2007011448
United States	AID	31181	522	ADMINISTRATION AND OVERSIGHT (AGRICULTURE)	2007001021
United States	AID	31182	1675	RESEARCH AND TECHNOLOGY DISSEMINATION	2007015633
United States	AID	31191	613	AGRICULTURAL SERVICES	2007011989
United States	AID	31192	300	PLANT/POST-HARVEST PROT. & PEST CTRL	2007007371
United States	AID	31193	300	AGRICULTURAL FINANCIAL SERVICES	2007015781

Total aid for water in 2007 for Mali   **USD thousand 300916**
And as a share of aid to total recipient countries   **4.48%**

Marshall Islands

Donor	Agency	Sector	Amount	Project description	CRS ID
Japan	JICA	31310	38	TC AGGREGATED ACTIVITIES	073535T
Japan	MAFF	31320	368	TC AGGREGATED ACTIVITIES	070061T

Total aid for water in 2007 for Marshall Islands   **USD thousand 406**
And as a share of aid to total recipient countries   **0.01%**

Mauritania

Donor	Agency	Sector	Amount	Project description	CRS ID
Japan	JICA	31110	767	TC AGGREGATED ACTIVITIES	072929T
Japan	MOFA	31150	2547	THE ASSISTANCE FOR UNDERPRIVILEGED FARMERS	070077
Japan	JICA	31150	33	TC AGGREGATED ACTIVITIES	073145T
Japan	MAFF	31381	46	TC AGGREGATED ACTIVITIES	070068T
Netherlands	MFA	31382	342	FISHERY RESEARCH	2007000666
Spain	MFA	31120	732	SPECIAL FOOD SECURITY PROGRAM (PESA) IN MAURITANIA	077626
Spain	MFA	31140	862	VISA PROJECT. RENOVATION OF IRRIGATION FOR FOOD SOVEREIGNTY.	077696
Spain	MFA	31161	684	REINFORCEMENT OF AGRICULTURAL PRODUCTION TRANSFORMATION	077553
Spain	EMP	31161	25	FOOD CROP PRODUCTION	072317
Spain	AG	31161	63	SECOND PHASE IN THE CONSOLIDATION OF FOOD SECURITY	073727
Spain	MFA	31195	47	VIGILENCE AND CONTROL SYSTEMS OF THE HIGHLY INFECTIOUS BIRD FLU	073196
Spain	MFA	31210	34	FORESTRY POLICY & ADMIN. MANAGEMENT	072775
Spain	AGR	31310	562	CAMPAIGN OF FISHING INVESTIGATION B/O VIZCONDE DE EZA IN MAURITANIAN WATERS	077431
Spain	MFA	31381	25	FISHERY EDUCATION/TRAINING	072186

Total aid for water in 2007 for Mauritania   **USD thousand 6770**

RECIPIENT Donor	Agency	Sector	Amount USD thousand	Project description	CRS ID Number
			And as a share of aid to total recipient countries	**0.1%**	
Mauritius					
EU institutions	CEC	31162	5636	DRU MAURITIUS SUGAR SECTOR SUPPORT PROGRAMME	2007100154
EU institutions	CEC	31162	602	INDUSTRIAL CROPS/EXPORT CROPS	2007100155
EU institutions	EDF	31310	6649	FISHING POLICY AND ADMIN. MANAGEMENT	2007200174
Japan	MAFF	31320	470	TC AGGREGATED ACTIVITIES	070051T
United States	TDA	31310	405	FISHING POLICY AND ADMIN. MANAGEMENT	2007028274
			Total aid for water in 2007 for Mauritius	**USD thousand 13762**	
			And as a share of aid to total recipient countries	**0.21%**	
Mayotte					
France	MISC	31110	5221	POLITIQUE AGRICOLE & GESTION ADMINISTRATIVE	2007800847
France	MISC	31181	2446	EDUCATION & FORMATION DANS LE DOMAINE AGRICOLE	2007800851
France	MISC	31182	12813	RECHERCHE AGRICULTURE	2007800352
			Total aid for water in 2007 for Mayotte	**USD thousand 20481**	
			And as a share of aid to total recipient countries	**0.31%**	
Mexico					
Australia	AusAID	31182	16	OPTIMIZING MARKER-ASSISTED BREEDING SYSTEMS	2007001849
Australia	AusAID	31182	9	AGRICULTURAL RESEARCH	2007002025
Belgium	DGCD	31194	78	AGRICULTURAL CO-OPERATIVES	2004007911
Canada	CIDA	31120	379	ORGANIC NETWORK OF CHANGE	070304
France	MISC	31110	35	POLITIQUE AGRICOLE & GESTION ADMINISTRATIVE	2007800774
France	MISC	31182	5613	RECHERCHE AGRICULTURE	2007800367
Germany	BMZ	31120	548	AGRICULTURAL DEVELOPMENT	2007006667
Germany	BMZ	31130	226	AGRICULTURAL LAND RESOURCES	2007006356
Germany	BMZ	31210	8	FORESTRY POLICY & ADMIN. MANAGEMENT	2007008768
Ireland	DFA	31163	10	LIVESTOCK	2007002779
Japan	JICA	31110	799	TC AGGREGATED ACTIVITIES	073046T
Japan	JICA	31150	18	TC AGGREGATED ACTIVITIES	073156T
Japan	JICA	31163	7	TC AGGREGATED ACTIVITIES	073204T
Japan	JICA	31210	123	TC AGGREGATED ACTIVITIES	073388T
Japan	JICA	31310	90	TC AGGREGATED ACTIVITIES	073561T
Japan	PRF	31382	15	TC AGGREGATED ACTIVITIES	075614T
Spain	AG	31120	170	STRENGTHENING OF THE TECHNICAL-ORGANIZATIONAL CAPACITIES (COOP.)	075550
Spain	AG	31161	103	SUPPORT FOR FOOD SOVEREIGNTY FOR THE INDIGENOUS	074636
Spain	AG	31162	77	ACTIVITIES TO SUPPORT ORGANIC PRODUCTION PROCESSES AND COFFEE MARKETING	074116
Spain	MFA	31181	63	HUMANITARIAN ACTION AND RISK MANAGEMENT IN REGIONAL MICRO DEV. PROCESSES	073707
Spain	AGR	31320	41	FISHERY DEVELOPMENT	072190
United Kingdom	MISC	31210	61	FRAMING THE FUTURE: A SUSTAINABLE DEVELOPMENT POLICY FRAMEWORK FOR MEXICO	2007801140
United Kingdom	MISC	31210	49	GOOD WOOD	2007801141
United States	IADF	31161	65	FOOD CROP PRODUCTION	2007024913

RECIPIENT Donor	Agency	Sector	Amount USD thousand	Project description	CRS ID Number
United States	AID	31310	200	FISHING POLICY AND ADMIN. MANAGEMENT	2007011378
United States	Misc	31382	7	FISHERY RESEARCH	2007024291

Total aid for water in 2007 for Mexico — **USD thousand 8810**
And as a share of aid to total recipient countries — **0.13%**

Micronesia, Fed. States

Donor	Agency	Sector	Amount	Project description	CRS ID
Germany	BMZ	31381	89	FISHERY EDUCATION/TRAINING	2007006369
Japan	JICA	31110	78	TC AGGREGATED ACTIVITIES	073012T
Japan	MAFF	31310	32	TC AGGREGATED ACTIVITIES	070045T
Japan	JICA	31310	484	TC AGGREGATED ACTIVITIES	073536T
Japan	MAFF	31320	1105	TC AGGREGATED ACTIVITIES	070062T
Japan	MAFF	31381	194	TC AGGREGATED ACTIVITIES	070075T

Total aid for water in 2007 for Micronesia, Fed. States — **USD thousand 1982**
And as a share of aid to total recipient countries — **0.03%**

Middle East, regional

Donor	Agency	Sector	Amount	Project description	CRS ID
Japan	JICA	31110	20	TC AGGREGATED ACTIVITIES	072981T
Japan	JICA	31130	25	TC AGGREGATED ACTIVITIES	073100T

Total aid for water in 2007 for Middle East, regional — **USD thousand 45**
And as a share of aid to total recipient countries — **0%**

Moldova

Donor	Agency	Sector	Amount	Project description	CRS ID
Austria	ADA	31120	159	BUSINESS PARTNERSHIP PROGRAM: BEAN CULTIVATION FOR SMALL-SCALE FARMERS	2007500108b
Austria	ADA	31181	616	AGRICULTURAL SCHOOLS AS CENTERS FOR RURAL EDUCATION AND TRAINING	2007500202
Austria	ADA	31181	14	CONTRACTED SERVICES FOR AGRICULTURAL SCHOOLS PROJECT	2007500900al
Austria	BM/BWK	31181	33	COOPERATION F. EDUCATION: AGRICULTURAL EDUCATION SYSTEM & SCHOOLS	2007818288
Greece	YPEJ	31191	55	PROVISION OF KNOW-HOW IN ORDER TO ORGANIZE AND DEVELOP ECOLOGICAL ARGICULTURE	2007000250
Japan	JICA	31110	29	TC AGGREGATED ACTIVITIES	072906T
Japan	MOFA	31150	1952	THE ASSISTANCE FOR UNDERPRIVILEGED FARMERS	070001
Japan	JICA	31150	322	TC AGGREGATED ACTIVITIES	073136T
Japan	MOFA	31181	4499	THE PROJECT FOR IMPROVEMENT OF EQUIPMENT FOR TRAINING CENTER	070241
Norway	MFA	31150	512	RELIEF/TECHNICAL ASSISTANCE RESPONSE TO DROUGHT	2007002036
Sweden	Sida	31120	1480	AGRICULTURAL DEVELOPMENT	2007005511
United States	AID	31120	2063	AGRICULTURAL SECTOR PRODUCTIVITY	2007001782

Total aid for water in 2007 for Moldova — **USD thousand 11734**
And as a share of aid to total recipient countries — **0.17%**

Mongolia

Donor	Agency	Sector	Amount	Project description	CRS ID
Canada	IDRC	31110	463	COLLABORATIVE LEARNING FOR THE CO-MANAGEMENT OF NATURAL RESOURCES	070632i
Canada	IDRC	31130	23	SUSTAINABLE MANAGEMENT OF COMMON NATURAL RESOURCES IN MONGOLIA (PH III)	070028i
France	MISC	31110	17	POLITIQUE AGRICOLE & GESTION ADMINISTRATIVE	2007800784
Germany	BMZ	31220	56	FORESTRY DEVELOPMENT	2007009089
IDA		31163	6600	SUSTAINABLE LIVELIHOODS	2007000945
Italy	LA	31120	82	AGRICULTURAL DEVELOPMENT	071572

RECIPIENT Donor	Agency	Sector	Amount USD thousand	Project description	CRS ID Number
Japan	JICA	31110	380	TC AGGREGATED ACTIVITIES	073004T
Japan	JICA	31163	411	TC AGGREGATED ACTIVITIES	073229T
Japan	PRF	31181	14	TC AGGREGATED ACTIVITIES	075604T
Japan	JICA	31195	284	TC AGGREGATED ACTIVITIES	073279T
Japan	JICA	31210	45	TC AGGREGATED ACTIVITIES	073353T
Japan	JICA	31310	12	TC AGGREGATED ACTIVITIES	073527T
Korea	Misc	31120	108	TRAINING PROGRAMME ON TECHNOLOGY SUPPORT FOR AGRICULTURE DEVELOPMENT	2007002561
Korea	KOICA	31163	1300	PROJECT FOR ESTABLISHMENT OF VIRAL ANIMAL DISEASE DIAGNOSTIC CENTER AT T	2007017683
Korea	KOICA	31163	95	TRAINING PROGRAM	2007017380
Korea	Misc	31163	129	SUPPORT FOR THE ENHANCEMENT OF LLIVESTOCK SANITATION TECHNOLOGY	2007002548
Korea	Misc	31181	51	AGRICULTURAL EDUCATION/TRAINING	2007002570
Korea	Misc	31210	9500	THE GREENBELT PLANTATION PROJECT IN MONGOLIA	2007000593
Norway	NORAD	31163	158	LIVESTOCK	2007003719
Switzerland	SDC	31110	25	AGRICULTURAL POLICY & ADMIN. MGMT	2005003858
Switzerland	SDC	31120	3584	PATATO SEED IMPROVEMENT	2004004628

Total aid for water in 2007 for Mongolia   **USD thousand 23337**
And as a share of aid to total recipient countries   **0.35%**

Montenegro

EU institutions	EDF	31195	1916	LIVESTOCK/VETERINARY SERVICES	2007201411
Italy	DGCS	31120	667	AGRICULTURAL DEVELOPMENT	060968
Luxembourg	MFA	31120	15	AGRICULTURAL DEVELOPMENT	070619
Luxembourg	MFA	31210	533	FORESTRY POLICY & ADMIN. MANAGEMENT	070621

Total aid for water in 2007 for Montenegro   **USD thousand 3131**
And as a share of aid to total recipient countries   **0.05%**

Morocco

Belgium	DGCD	31130	41	AGRICULTURAL LAND RESOURCES	2005008486
Belgium	DGCD	31140	112	AGRICULTURAL WATER RESOURCES	2002000173
Belgium	DGCD	31150	189	AMELIORATION QUALITE INTRANTS AGRICOLES	2003001475
Belgium	DGCD	31282	94	FORESTRY RESEARCH	2006003918
France	AFD	31130	2053	LUTTE CONTRE LA DEGRADATION DES TERRES	2007168700
France	MISC	31182	8556	RECHERCHE AGRICULTURE	2007800335
Germany	BMZ	31182	84	AGRICULTURAL RESEARCH	2007007178
IFAD		31110	494	AGRICULTURAL POLICY & ADMIN. MGMT	070032a
IFAD		31110	18263	RURAL DEVELOPMENT PROJECT IN THE MOUNTAIN ZONES OF ERRACHIDIA PROVINCE	070032
Italy	LA	31110	68	AGRICULTURAL POLICY & ADMIN. MGMT	071007
Italy	DGCS	31120	395	AGRICULTURAL DEVELOPMENT	060765
Japan	JICA	31110	308	TC AGGREGATED ACTIVITIES	072909T
Japan	JICA	31130	47	TC AGGREGATED ACTIVITIES	073075T
Japan	JICA	31150	154	TC AGGREGATED ACTIVITIES	073175T
Japan	JICA	31163	48	TC AGGREGATED ACTIVITIES	073187T
Japan	JICA	31310	610	TC AGGREGATED ACTIVITIES	073543T
Japan	MAFF	31320	481	TC AGGREGATED ACTIVITIES	070048T
Japan	MOFA	31382	8217	PROJET DE CONSTRUCTION DES LABORATOIRES CENTRAUX	070200
Korea	KOICA	31163	77	LIVESTOCK	2007016487

RECIPIENT Donor	Agency	Sector	Amount USD thousand	Project description	CRS ID Number
Korea	KOICA	31320	84	FISHERY DEVELOPMENT	2007017381
Korea	KOICA	31382	9	FISHERY RESEARCH	2007017737
Spain	MFA	31120	30	TEC. ASSISTANCE PROJECT TO SUPPORT THE NATIONAL PLAN TO COMBAT DESERTIF.	072570
Spain	AG	31120	32	AGRICULTURAL DEVELOPMENT	072653
Spain	AGR	31140	252	IRRIGATION PROJECT MOROCCO	076182
Spain	AG	31140	42	AGRICULTURAL AND ENVIRONMENTAL DEVELOPMENT IN THE REGION OF AIT KAMRA	073081
Spain	AG	31140	50	AGRICULTURAL AND ENVIRONMENTAL DEVELOPMENT IN TAZAGHINE - ₄2 PHASE	073331
Spain	AG	31163	323	DEVELOPMENT OF THE CAPRINE RANCHING	076626
Spain	AG	31166	55	SUPPORT TO PROVIDE 48 PRODUCING WOMEN ACCESS TO THE LABOUR MARKET	073460
Spain	MFA	31181	131	TRANSFERENCE OF TECHNOLOGY AND AGRARIAN FORMATION	075033
Spain	AG	31181	35	VOCATIONAL TRAINING	072785
Spain	MFA	31182	15	AGRICULTURAL RESEARCH	070893
Spain	MFA	31182	7	ETHYLENE AS AN ALTERNATIVE FOR THE USE OF INSECTICIDES	070798
Spain	MFA	31195	47	VIGILENCE AND CONTROL SYSTEMS OF THE HIGHLY INFECTIOUS BIRD FLU	073195
Spain	MFA	31210	34	FORESTRY POLICY & ADMIN. MANAGEMENT	072779
Spain	AG	31220	144	DIVERSIF. OF THE SOURCES OF INCOME AND STABILITY OF THE OASIANAS COMMUNITIES	075276
Spain	AGR	31320	82	MANAGEMENT AND OPERATION UNDER SUITABLE HYGIENIC-SANITARY CONDITIONS	074222
Spain	MFA	31320	103	SUSTAINABLE DEVELOPMENT OF TRADITIONAL FISHERIES IN THE MEDITERRANEAN SEA	074614
Spain	AG	31320	84	PRODUCTION MARKET. & SUSTAINABILITY CONDITIONS OF THE ARTISANAL FISHERIES	074311
Spain	AGR	31381	41	SEMINAR ON COMMERCIALIZATION AND TRACEABILITY OF FISH AND AGRIC. PRODUCTS	072968
Spain	MFA	31381	342	CONSTRUCTION AND EQUIPMENT OF THE CENTRE FOR MARITIME SECURITY IN LARACHE	076759
Spain	MFA	31381	342	DEVELOPING TEACHING CAPACITIES OF THE ITPM DE ALHUCEMAS	076760
Spain	MFA	31381	25	FISHERY EDUCATION/TRAINING	072185
Spain	AG	31382	137	LABORATORY FOR RESEARCH TECHNOLOGICAL DEV. AND TRAINING IN AQUACULTURE	075179
United States	AID	31110	167	AGRICULTURAL POLICY & ADMIN. MGMT	2007013900
United States	AID	31110	633	AGRICULTURAL ENABLING ENVIRONMENT	2007001657
United States	AID	31120	969	AGRICULTURAL SECTOR PRODUCTIVITY	2007001759
United States	TDA	31140	350	AGRICULTURAL WATER RESOURCES	2007028865
United States	MCC	31161	4937	FOOD CROP PRODUCTION COMPACT ACTIVITY	2007025222
United States	MCC	31191	2023	AGRICULTURAL SERVICES COMPACT ACTIVITY	2007025228
United States	MCC	31320	3936	FISHERY DEVELOPMENT COMPACT ACTIVITY	2007025232
United States	MCC	31391	3070	FISHERY SERVICES COMPACT ACTIVITY	2007025233

Total aid for water in 2007 for Morocco   **USD thousand 58822**
And as a share of aid to total recipient countries   **0.88%**

Mozambique

	Agency	Sector	Amount	Project description	CRS ID
AfDF		31140	26014	MASSINGIR DAM AND SMALLHOLDER AGRICULTUR	070040
Austria	Reg	31130	27	NGO/EU COFINANCE - SECURING OF LAND RIGHTS IN SOFALA	2007828628
Belgium	DGCD	31110	145	AGRICULTURAL POLICY & ADMIN. MGMT	2004003270
Belgium	DGCD	31120	201	SOUTIEN D'ORGANISATIONS PAYSANNES ET COMMUNAUTAIRES EN MOZAMBIQUE	2004003206
Belgium	DGCD	31181	89	AGRICULTURAL EDUCATION/TRAINING	2006002681
EU institutions	EDF	31162	8214	ACCOMPANYING MEASURES 2007/10 FOR SUGAR PROTOCOL COUNTRIES - MOZAMBIQUE	2007201269
Germany	BMZ	31120	183	AGRICULTURAL DEVELOPMENT	2007006166
Germany	BMZ	31164	719	ACCESS TO LAND AND SUSTAINABLE USE OF NATURAL RESOURCES	2007006650
Ireland	DFA	31110	2423	AGRICULTURE FORESTRY AND FISH - PROAGRI - MINISTERIO DE AGRICULTURE	2007000215
Ireland	DFA	31110	378	AGRICULTURAL POLICY & ADMIN. MGMT	2007000447

RECIPIENT Donor	Agency	Sector	Amount USD thousand	Project description	CRS ID Number
Ireland	DFA	31120	472	AGRICULTURE FORESTRY AND FISH - PRIVATE SECTOR - TECHNOSERVE	2007000223
Ireland	DFA	31120	911	AGRICULTURAL DEVELOPMENT	2007003048
Ireland	DFA	31161	6	FOOD CROP PRODUCTION	2007003126
Ireland	DFA	31320	242	FISHERY DEVELOPMENT	2007000701
Italy	DGCS	31110	101	AGRICULTURAL POLICY & ADMIN. MGMT	070223
Italy	DGCS	31120	1082	AGRICULTURAL DEVELOPMENT	070409
Italy	DGCS	31162	285	INDUSTRIAL CROPS/EXPORT CROPS	070343
Italy	DGCS	31181	12	AGRICULTURAL EDUCATION/TRAINING	060626
Italy	LA	31181	27	AGRICULTURAL EDUCATION/TRAINING	070886
Italy	DGCS	31220	23	FORESTRY DEVELOPMENT	070184
Italy	DGCS	31320	2078	FISHERY DEVELOPMENT	070274
Japan	MOFA	31110	1053	AGRICULTURAL POLICY & ADMIN. MGMT	070114
Japan	JICA	31110	1152	TC AGGREGATED ACTIVITIES	073028T
Japan	JICA	31130	76	TC AGGREGATED ACTIVITIES	073084T
Japan	JICA	31163	43	TC AGGREGATED ACTIVITIES	073194T
Japan	JICA	31195	6	TC AGGREGATED ACTIVITIES	073250T
Japan	JICA	31210	16	TC AGGREGATED ACTIVITIES	073304T
Japan	JICA	31310	25	TC AGGREGATED ACTIVITIES	073483T
Japan	MAFF	31320	648	TC AGGREGATED ACTIVITIES	070052T
Japan	MAFF	31381	23	TC AGGREGATED ACTIVITIES	070069T
Korea	KOICA	31110	109	PROVISION OF EQUIPMENT	2007017769
Netherlands	MFA	31110	834	AGRIC.: SUBS. TO PSOM 2003-2009	2007100131
Netherlands	MFA	31110	25	AGRICULTURAL POLICY & ADMIN. MGMT	2007100146
Norway	NORAD	31150	401	NATURE: BUILDING CAPACITY FOR AGRICULTURE INPUT SUPPLY TO SMALL FARMERS	2007003790
Norway	MFA	31162	3841	SOYA BEAN PRODUCTION AND MARKETING IN NORTHERN MOZAMBIQUE	2007004361
Portugal	ICP	31182	20	AGRICULTURAL RESEARCH	070807
Portugal	ICP	31193	98	SUSTAINABLE EXTENSION OF GAPI- MOZ -	070511
Spain	MFA	31140	767	BASIC COMPREHENSIVE HOUSING IN THE DISTRICT OF CATUANE	077646
Spain	MFA	31161	411	ACCESS TO FOOD INCOME GENERATION GUARANTEES	077027
Spain	AG	31161	320	CONTRIBUTION TO THE ACHIEVEMENT OF FOOD SOVEREIGNTY FOR THE RURAL POP.	076611
Spain	AG	31161	411	INCREASE AND IMPROVEMENT OF THE SOCIO-PRODUCTIVE AND ORGANIZATION CAPACITY	077010
Spain	AG	31161	37	CONSOLIDATION OF THE PRODUCTIVE AND SOCIAL STRUCTURES	072848
Spain	AG	31181	82	SUPPORT FOR THE TRAINING AND DEVELOPMENT CENTRE OF SALELA (INHAMBANE)	074281
Spain	Misc	31282	8	STRATEGY TO IMPROVE THE MANAGEMENT OF WATER AND FOREST	070873
Spain	AGR	31310	548	CAMPAIGN OF FISHING INVESTIGATION B/O VIZCONDE DE EZA IN MOZAMBICAN WATERS	077386
Spain	AGR	31320	21	HYGIENIC AND SANITARY INSPECTION AS WELL AS PROPOSALS FOR IMPROVEMENT	071879
Sweden	Sida	31110	11119	AGRICULTURAL POLICY & ADMIN. MGMT	2007001237
Switzerland	SDC	31110	1917	RURAL DEVELOPMENT NORTHERN MOZAMBIQUE	2005003898
Switzerland	SDC	31110	67	AGRICULTURAL POLICY & ADMIN. MGMT	2005003899
United Kingdom	DFID	31110	972	SMALLHOLDER FOOD SECURITY PROGRAMME CSCF 0402	2007000836
United States	AID	31110	3398	RURAL INCOMES	2007015766
United States	AID	31110	30	AGRICULTURAL POLICY & ADMIN. MGMT	2007013857
United States	AID	31110	1300	AGRICULTURAL ENABLING ENVIRONMENT	2007001645
United States	AID	31110	918	PROGRAM SUPPORT (AGRICULTURE)	2007013867
United States	AID	31120	3590	AGRICULTURAL SECTOR PRODUCTIVITY	2007001732

Total aid for water in 2007 for Mozambique   **USD thousand 77914**
And as a share of aid to total recipient countries   **1.16%**

RECIPIENT / Donor	Agency	Sector	Amount USD thousand	Project description	CRS ID Number

Myanmar

Donor	Agency	Sector	Amount	Project description	CRS ID
Australia	AusAID	31161	148	Increasing food security and farmer livelihoods through enhanced legume cultivation	2007000814
Australia	AusAID	31163	22	LIVESTOCK	2007001731
Japan	JICA	31110	3591	TC AGGREGATED ACTIVITIES	072991T
Japan	Oth. MIN	31110	9	TC AGGREGATED ACTIVITIES	075147T
Japan	JICA	31130	52	TC AGGREGATED ACTIVITIES	073107T
Japan	JICA	31150	205	TC AGGREGATED ACTIVITIES	073181T
Japan	JICA	31163	39	TC AGGREGATED ACTIVITIES	073218T
Japan	JICA	31195	115	TC AGGREGATED ACTIVITIES	073270T
Japan	JICA	31210	643	TC AGGREGATED ACTIVITIES	073417T
Japan	MOFA	31220	518	FORESTRY DEVELOPMENT	070165
Japan	JICA	31310	287	TC AGGREGATED ACTIVITIES	073569T
Korea	Misc	31110	9	AGRICULTURAL POLICY & ADMIN. MGMT	2007002556
Korea	KOICA	31120	103	TRAINING PROGRAM	2007017376
Korea	KOICA	31162	15	INDUSTRIAL CROPS/EXPORT CROPS	2007016370
Korea	KOICA	31181	37	AGRICULTURAL EDUCATION/TRAINING	2007016955
Korea	Misc	31181	64	AGRICULTURAL EDUCATION/TRAINING	2007002575

Total aid for water in 2007 for Myanmar   **USD thousand 5859**
And as a share of aid to total recipient countries   **0.09%**

Namibia

Donor	Agency	Sector	Amount	Project description	CRS ID
Belgium	DGCD	31194	120	AGRICULTURAL CO-OPERATIVES	2004003211
Germany	BMZ	31120	561	AGRICULTURAL DEVELOPMENT	2007006554
Germany	BMZ	31130	2053	AGRICULTURAL LAND RESOURCES	2007002650
Germany	L G	31163	14	LIVESTOCK	2007011690
Germany	BMZ	31164	21	AGRARIAN REFORM	2007002027
Germany	BMZ	31210	1238	FORESTRY POLICY AND ADMINISTRATIVE MANAGEMENT	2007009102
Japan	JICA	31110	15	TC AGGREGATED ACTIVITIES	072940T
Japan	MOFA	31150	1273	THE ASSISTANCE FOR UNDERPRIVILEGED FARMERS	070093
Japan	JICA	31150	40	TC AGGREGATED ACTIVITIES	073151T
Japan	JICA	31210	61	TC AGGREGATED ACTIVITIES	073378T
Luxembourg	MFA	31120	58	AGRICULTURAL DEVELOPMENT	070680
Norway	NORAD	31163	159	LIVESTOCK	2007004424
Norway	MFA	31320	37	SHARE KNOWLEDGE REGARDING FISHERY WITH EMPHASIS ON SEALS AS A RESOURCE	2007002654
Spain	MFA	31120	452	STRENGTHENING FOOD SOVEREIGNTY AT THE FAMILY COMMUNITY	077215
Spain	AGR	31310	702	CAMPAIGN OF FISHING INVESTIGATION B/O VIZCONDE DE EZA IN NAMIBIAN WATERS	077613
Spain	MFA	31310	274	5- TRAINING AND RESEARCH PROGRAM AT NATMIRC	076336
Spain	MFA	31381	329	EDUCATIONAL ACTIVITIES OF THE NAMIBIAN MARITIME AND FISHERIES INSTITUTE (NAMFI)	076655

Total aid for water in 2007 for Namibia   **USD thousand 7407**
And as a share of aid to total recipient countries   **0.11%**

Nauru

Donor	Agency	Sector	Amount	Project description	CRS ID
Japan	MAFF	31320	325	TC AGGREGATED ACTIVITIES	070060T
Japan	MAFF	31381	41	TC AGGREGATED ACTIVITIES	070074T

RECIPIENT Donor	Agency	Sector	Amount USD thousand	Project description	CRS ID Number
			Total aid for water in 2007 for Nauru And as a share of aid to total recipient countries		**USD thousand 366** 0.01%

Nepal

Donor	Agency	Sector	Amount	Project description	CRS ID
Australia	AusAID	31182	14	LENTIL AND LATHYRUS IN CROPPING SYSTEMS OF NEPAL	2007001888
Austria	BM/BWK	31210	42	IDENTIFICATION OF CRITERIA & INDICATORS F. SUSTAIN. COMM. FOREST MANAG.	2007819002
Belgium	DGCD	31195	42	LIVESTOCK/VETERINARY SERVICES	2006003926
Denmark	MFA	31282	1456	COMMUNITY BASED NATURAL FOREST AND TREE MANAGEMENT IN HIMALAYA PHASEII	071371
Finland	MFA	31181	638	FRAME AGREEMENT WITH NGO (SUOMEN L₃HETYSSEURA RY)	2007000318
Finland	MFA	31182	33	FRAME AGREEMENT WITH NGO (SUOMEN L₃HETYSSEURA RY)	2007000323
Germany	BMZ	31161	49	FOOD CROP PRODUCTION	2007007283
Germany	BMZ	31182	76	AGRICULTURAL RESEARCH	2007007284
IDA		31110	1000	IRRIGATION AND WATER RESOURCE MANAGEMENT	2007000838
IDA		31140	25000	AGRICULTURAL WATER RESOURCES	2007000841
IDA		31140	38000	IRRIGATION AND WATER RESOURCE MANAGEMENT	2007000837
IDA		31163	25000	LIVESTOCK	2007000842
IDA		31163	364	AVIAN FLU	2007000827
IDA		31166	6000	IRRIGATION AND WATER RESOURCE MANAGEMENT	2007000836
IDA		31166	2912	AVIAN FLU	2007000826
Ireland	DFA	31195	27	LIVESTOCK/VETERINARY SERVICES	2007001822
Italy	LA	31166	55	AGRICULTURAL EXTENSION	071588
Japan	JICA	31110	1070	TC AGGREGATED ACTIVITIES	072995T
Japan	MOFA	31150	2547	THE ASSISTANCE FOR UNDERPRIVILEGED FARMERS	070076
Japan	JICA	31150	57	TC AGGREGATED ACTIVITIES	073182T
Japan	JICA	31163	216	TC AGGREGATED ACTIVITIES	073221T
Japan	JICA	31210	11	TC AGGREGATED ACTIVITIES	073419T
Japan	JICA	31310	18	TC AGGREGATED ACTIVITIES	073521T
Korea	KOICA	31130	30	AGRICULTURAL LAND RESOURCES	2007016160
Korea	KOICA	31181	13	AGRICULTURAL EDUCATION/TRAINING	2007017165
Korea	Misc	31181	5	AGRICULTURAL EDUCATION/TRAINING	2007002577
Korea	KOICA	31182	77	AGRICULTURAL RESEARCH	2007017876
New Zealand	NZAid	31120	409	CROP & FOOD: SUSTAINABLE SERIAL AND LEGUME FODDER SYSTEMS	070411
Norway	MFA	31120	226	AGRICULTURAL DEVELOPMENT	2007002573
Norway	NORAD	31164	26	Land Rights for Sustainable Development	2007003411
Norway	MFA	31382	171	BIODIVERSITY IN RIVERSYSTEMS IN NEPAL AND FISHERIES ADDENDUM	2007004527
Spain	Misc	31163	7	FEASIBILITY STUDY FOR THE MANAGEMENT OF BUFFALOES	070829
Switzerland	SDC	31120	3792	SSMP SUSTAINABLE SOIL MANAGEMENT	1998000985
Switzerland	SDC	31120	1500	NARC/CIMMYT HILL MAIZE PROJECT	1998000996
Switzerland	SDC	31120	83	AGRICULTURAL DEVELOPMENT	1998000997
United States	AID	31110	68	AGRICULTURAL POLICY & ADMIN. MGMT	2007013901
United States	AID	31120	3315	AGRICULTURAL SECTOR PRODUCTIVITY	2007001760
United States	AID	31161	44	FOOD CROP PRODUCTION	2007016479

Total aid for water in 2007 for Nepal — **USD thousand 114393**
And as a share of aid to total recipient countries — 1.7%

RECIPIENT Donor	Agency	Sector	Amount USD thousand	Project description	CRS ID Number
Nicaragua					
Austria	ADA	31120	193	NGO COFINANCED PROJECT: AGRICULTURAL PROD. & MARKETING - ATLANTIC REGION	2006000022bv
Austria	ADA	31120	70	NGO COFINANCED PROJECT: EMPOWERMENT OF FEMALE FARMERS IN 13 VILLAGES	2006000022bs
Austria	ADA	31120	506	NGO-COFINANCING: STRENGTHENING AGRICULTURAL PROD. OF INDIGEN. MAYANGA	2007500024b
Austria	ADA	31120	1909	TO DEVELOP FORESTRY AND GRASSLANDS ECONOMIC SYSTEMS INCL COCOA CULTIV.	2007500142
Austria	ADA	31130	43	NGO COFINANCED PROJECT: SUPPORT FOR PEACEFUL SOLUTION OF LAND PROBLEMS	2005000043b
Austria	ADA	31162	237	BUSINESS PARTNERSHIP PROGRAM: MARKET. & EXPORT F. QUALITY CERTIFIED COCO	2007500108d
Austria	ADA	31166	28	NGO-COFINANCING: BUSINESS TRAINING FOR AGRICULTURAL COOPERATIVES	2007500028c
Austria	Reg	31191	55	SUPPORT OF MARKETING & EXPORT OF FAIR TRADE ORGANIC COCOA	2007828548
Austria	ADA	31191	46	NGO COFINANCED PROJECT: CECALLI - PROD. & MARKETING OF MEDICAL HERBS	2006000022b
Austria	ADA	31191	438	NGO COFINANCING:SUPPORT FOR AGRICULT. PRODUCTION & MARKETING, WASLALA	2007000021a
Belgium	DGCD	31110	90	AGRICULTURAL POLICY & ADMIN. MGMT	2004003267
Belgium	DGCD	31194	53	AGRICULTURAL CO-OPERATIVES	2004003227
Denmark	MFA	31161	395	FOOD CROP PRODUCTION	071372
France	MISC	31182	972	RECHERCHE AGRICULTURE	2007800368
Germany	BMZ	31120	420	AGRICULTURAL DEVELOPMENT	2007006342
Germany	BMZ	31120	1540	POVERTY REDUCTION IN PROTECTED AREAS, NICARAGUA ETC.	2007005685
Germany	L G	31120	34	AGRICULTURAL DEVELOPMENT	2007011703
Germany	L G	31163	32	LIVESTOCK	2007010880
Germany	BMZ	31194	117	AGRICULTURAL CO-OPERATIVES	2007007280
Germany	BMZ	31210	65	FORESTRY POLICY & ADMIN. MANAGEMENT	2007007281
Ireland	DFA	31110	35	AGRICULTURAL POLICY & ADMIN. MGMT	2007001673
Ireland	DFA	31150	70	AGRICULTURAL INPUTS	2007001342
Ireland	DFA	31181	329	AGRICULTURAL EDUCATION/TRAINING	2007000528
Italy	DGCS	31120	7	AGRICULTURAL DEVELOPMENT	060420
Italy	CA	31120	215	AGRICULTURAL DEVELOPMENT	071317
Japan	JICA	31110	1118	TC AGGREGATED ACTIVITIES	073047T
Japan	JICA	31163	1140	TC AGGREGATED ACTIVITIES	073205T
Japan	JICA	31195	104	TC AGGREGATED ACTIVITIES	073260T
Japan	JICA	31210	422	TC AGGREGATED ACTIVITIES	073390T
Japan	JICA	31310	11	TC AGGREGATED ACTIVITIES	073501T
Netherlands	MFA	31110	65	AGRICULTURAL POLICY & ADMIN. MGMT	2007100140
New Zealand	NZAid	31120	12	SMALL FARMERS PRODUCTIVE CAPACITY	070339
Norway	MFA	31110	14	ASSISTANCE TO PRORURAL PROGRAM	2007004504
Norway	NORAD	31120	183	IMPROVED PEASANTS ECONOMIES	2007003394
Norway	NORAD	31120	50	FOOD SECURITY AND PEASANT'S EMPOWERMENT	2007003392
Norway	MFA	31150	239	SUPPORT REHABILITATION TO RAAN COMMUNITIES AFFECTED BY HURRICANE FELIX	2007004506
Norway	NORFUND	31310	80	FISHING POLICY AND ADMIN. MANAGEMENT	2007004463
Norway	NORAD	31320	162	NICAFISH - PILOT PROJECT - CATCH OF SMALL CRAYFISH OFF NICARAGUAN COAST	2007004500
Spain	MFA	31110	1369	TRUST FUND - NICARAGUA	077866
Spain	AG	31120	390	DIVERSIFICATION OF PRODUCTION IN THE RURAL ECONOMY	076934
Spain	AG	31120	27	ENCOURAGEMENT TO THE LOCAL ECONOMIES OF 10 INDIGENOUS CHOROTEGO FAMILIES	072457
Spain	Misc	31120	8	AGRICULTURAL DEVELOPMENT	070981
Spain	AG	31150	49	SUPPORT FOR AGR. DEV. IN THE MUNICIPALITY OF MATAGALPA NICARAGUA. PHASE II	073269
Spain	EMP	31161	24	FOOD CROP PRODUCTION	072154

RECIPIENT Donor	Agency	Sector	Amount USD thousand	Project description	CRS ID Number
Spain	AG	31161	33	FOOD CROP PRODUCTION	072688
Spain	AG	31161	346	IMPROVEMENT CONDITIONS TO ENSURE FOOD SOVEREIGNTY IN THE DEP. OF MATAGALPA	076805
Spain	AG	31161	233	PROGRAMME OF FOOD SECURITY AND SOVEREIGNTY IN NICARAGUA	076073
Spain	AG	31162	29	SMALL HUMID BENEFIT FOR THE IMPR. OF THE PROCESS AND QUALITY OF THE COFFEE	072529
Spain	AG	31163	185	IMPROVEMENT OF THE QUALITY OF LIFE AND CHILD NUTRITION	075666
Spain	AG	31163	25	AGRICULTURAL ACTIVITY WITH WOMEN'S GROUPS	072203
Spain	AG	31166	82	DEV. ACTIVITIES FOR RURAL WOMEN OF FIVE COMMUNITIES OF THE MUNICIPALITY	074238
Spain	AG	31181	45	AGRICULTURAL EDUCATION/TRAINING	072135
Spain	AG	31194	123	STRENGTHENING OF THE SELFGOVERNING CAPACITIES OF 19 SOCIAL COOPERATIONS	074902
Spain	AG	31194	243	METHODS & TECHNICAL CAPABILITIES IN PRODUCTION ORGANISATIONS AND LOCAL GOV	076127
Spain	AG	31194	74	LIVING CONDITIONS OF THE MEMBERS OF THE FARMERS COOPERATIVE	074047
Spain	AG	31194	274	IMPROVEMENT OF PRODUCTION CAPACITIES AND MARKETING OF 29 COOPERATIVES	076386
Spain	AG	31194	399	DEVELOPMENT OF NEW MARKETING ALTERNATIVES FOR RURAL PRODUCTION	076954
Spain	AG	31194	63	SUPPORT TO COOPERATIVES FOR THE TRADE OF THEIR PRODUCTS: RECYCLING CENTRE	073725
Spain	AG	31220	11	FOREST GENETICS CONSERVATION AND SUSTAINABLE USE OF WOOD AND FOREST	071226
Spain	ICO	31310	14983	SUPPLY OF FOUR CRAFTS FOR ALERTNESS OF FISHING RESOURCES	070004
Spain	AG	31310	50	FISHING POLICY AND ADMIN. MANAGEMENT	073313
Spain	MFA	31320	223	SUPPORT FOR FISHING ACTIVITIES OF THE ARTISANAL FISHING COMMUNITIES	076003
Spain	AG	31320	135	IMPROVEMENT OF THE INSHORE FISHERMEN'S CAPACITIES	075069
Sweden	Sida	31182	148	AGRICULTURAL RESEARCH	2007001813
United States	AID	31110	77	AGRICULTURAL POLICY & ADMIN. MGMT	2007013888
United States	IADF	31191	219	AGRICULTURAL SERVICES	2007025072
United States	IADF	31194	34	AGRICULTURAL CO-OPERATIVES	2007025135

Total aid for water in 2007 for Nicaragua   **USD thousand 31397**
And as a share of aid to total recipient countries   **0.47%**

Niger

Donor	Agency	Sector	Amount	Project description	CRS ID
Belgium	DGCD	31120	203	AGRICULTURAL DEVELOPMENT	2006003801
Belgium	DGCD	31140	5476	DéVELOPPEMENT DE L'IRRIGATION DANS LA RéGION DE TILLABERI	2007004174
Belgium	DGCD	31163	588	APPUI A LA SELECTION, LA PROMOTION ET A LA DIFFUSION DE LA CHEVRE ROUSSE	2004001113
Belgium	DGCD	31163	410	RESEAU DE SANTE ET CONSEILS EN ELEVAGE - PROXEL	2002000571
Belgium	DGCD	31163	772	ELEVAGE DES BOVINS DE LA RACE AZWAK	2002000180
Belgium	DGCD	31163	121	LIVESTOCK	2004008411
Belgium	DGCD	31166	46	AGRICULTURAL EXTENSION	2004008419
Belgium	DGCD	31193	463	INTRANTS - PROGRAMME 2001-2007	2006004208
Belgium	DGCD	31195	120	LIVESTOCK/VETERINARY SERVICES	2005001051
EU institutions	CEC	31194	5219	AGRICULTURAL CO-OPERATIVES	2007100184
France	MISC	31182	1574	RECHERCHE AGRICULTURE	2007800353
Germany	BMZ	31110	68	AGRICULTURAL POLICY & ADMIN. MGMT	2007008821
Germany	BMZ	31120	39	AGRICULTURAL DEVELOPMENT	2007008822
Germany	BMZ	31181	56	AGRICULTURAL EDUCATION/TRAINING	2007008823
IDA		31163	1350	AVIAN FLU	2007000364
IDA		31166	1350	AVIAN FLU	2007000363
Italy	DGCS	31120	2343	AGRICULTURAL DEVELOPMENT	060186
Italy	LA	31163	66	LIVESTOCK	061422

RECIPIENT	Agency	Sector	Amount USD thousand	Project description	CRS ID Number
Donor					
Japan	JICA	31110	1991	TC AGGREGATED ACTIVITIES	073029T
Japan	MOFA	31150	2547	THE ASSISTANCE FOR UNDERPRIVILEGED FARMERS	070113
Japan	JICA	31150	35	TC AGGREGATED ACTIVITIES	073146T
Japan	JICA	31210	116	TC AGGREGATED ACTIVITIES	073374T
Switzerland	SDC	31110	167	APPUI AUX ORGANISATIONS RURALES DU NIGER	2007000295
United States	ADF	31161	91	FOOD CROP PRODUCTION	2007000152

Total aid for water in 2007 for Niger   **USD thousand 25211**
And as a share of aid to total recipient countries   **0.38%**

Nigeria

RECIPIENT	Agency	Sector	Amount USD thousand	Project description	CRS ID Number
Austria	Reg	31150	14	TRANSPORT COSTS FOR AGRICULTURAL EQUIPMENT	2007828530
Canada	IDRC	31166	94	STRENGTHEN THE CAPACITY OF SMALLHOLDER FARMERS TO ADAPT TO CLIMATE CHANGE	070304i
France	MISC	31182	1259	RECHERCHE AGRICULTURE	2007800354
Germany	BMZ	31120	253	AGRICULTURAL DEVELOPMENT	2007006184
Germany	BMZ	31130	253	AGRICULTURAL LAND RESOURCES	2007006142
IFAD		31120	400	AGRICULTURAL DEVELOPMENT	070013a
IFAD		31120	42758	RURAL MICROENTERPRISE DEVELOPMENT PROGRAMME (RUMEDP)	070013
Ireland	DFA	31110	15	AGRICULTURAL POLICY & ADMIN. MGMT	2007002404
Ireland	DFA	31140	23	AGRICULTURAL WATER RESOURCES	2007001895
Italy	LA	31140	12	AGRICULTURAL WATER RESOURCES	071731
Japan	JICA	31110	38	TC AGGREGATED ACTIVITIES	072932T
Japan	JICA	31150	217	TC AGGREGATED ACTIVITIES	073177T
Japan	JICA	31163	45	TC AGGREGATED ACTIVITIES	073195T
Japan	JICA	31195	42	TC AGGREGATED ACTIVITIES	073251T
Japan	JICA	31310	48	TC AGGREGATED ACTIVITIES	073484T
Korea	KOICA	31161	1600	THE PROJECT FOR ESTABLISHMENT OF AGRICULTURAL PRODUCT PROCESSING CENTER	2007017685
Norway	MFA	31381	8	FISHERY EDUCATION/TRAINING	2007002599
United Kingdom	MISC	31210	65	COMMUNITY MANAGEMENT PLANNING FOR SUSTAINABLE FOREST LIVELIHOODS	2007801178
United States	AID	31110	612	AGRICULTURAL POLICY & ADMIN. MGMT	2007013874
United States	AID	31110	500	AGRICULTURAL ENABLING ENVIRONMENT	2007001647
United States	AID	31120	2882	AGRICULTURAL SECTOR PRODUCTIVITY	2007001739
United States	AID	31120	380	AGRICULTURAL DEVELOPMENT	2007001754

Total aid for water in 2007 for Nigeria   **USD thousand 51521**
And as a share of aid to total recipient countries   **0.77%**

Niue

RECIPIENT	Agency	Sector	Amount USD thousand	Project description	CRS ID Number
New Zealand	NZAid	31120	26	ORGANIC FARMING	070369
New Zealand	NZAid	31162	74	NIUE VANILLA AND NONI	070379

Total aid for water in 2007 for Niue   **USD thousand 100**
And as a share of aid to total recipient countries   **0%**

North & Central America, regional

RECIPIENT	Agency	Sector	Amount USD thousand	Project description	CRS ID Number
Belgium	MPRF	31161	34	FOOD CROP PRODUCTION	2006003465
Canada	IDRC	31310	519	COASTAL AND MARINE RESOURCE GOVERNANCE PROGRAMME (CERMES)	070230i

RECIPIENT	Agency	Sector	Amount USD thousand	Project description	CRS ID Number
Donor					
Canada	IDRC	31310	279	SUSTAINABLE AQUATIC RESOURCE MANAGEMENT INITIATIVE	070472i
Norway	MFA	31162	5121	CENTRAL AMERICA CACAO PROJECT	2007002313
Norway	MFA	31181	702	EARTH Scholarships, agriculture	2007002304
Spain	MFA	31110	205	DIVERSIFICATION OF CULTURES AGRICULTURAL TECHNOLOGY EXCHANGE	075824
Spain	MFA	31120	2124	FOOD SOVEREIGNTY AND LOCAL DEVELOPMENT DIRECTED	077971
Spain	MFA	31120	137	TRANSFORMATION OF RESIDUES OF THE BANANA AND ALTERNATIVE CROPS	075100
Spain	MFA	31161	318	FINANCING REGIONAL PROGRAMME OF COFFEE QUALITY ACCORDING TO ITS ORIGIN	076597
Spain	AG	31181	55	AGRICULTURAL TRAINING FOR GIRLS BOYS AND ADOLESCENTS	073507
Spain	MFA	31320	137	FISHERY DEVELOPMENT	075101
Spain	AG	31381	1153	PLAN TO THE PROFESSIONAL TRAINING OF CENTRO AMERICAN INSHORE FISHERMEN	077803
Switzerland	SDC	31120	615	MIP-EAP PROTECTION INTEGREE VEGETAUX	1998000656
United States	AID	31110	56	AGRICULTURAL POLICY & ADMIN. MGMT	2007013884
United States	AID	31110	1034	AGRICULTURAL ENABLING ENVIRONMENT	2007001649
United States	STATE	31162	100	INDUSTRIAL CROPS/EXPORT CROPS	2007026984
United States	AGR	31182	80	AGRICULTURAL RESEARCH	2007000442

Total aid for water in 2007 for North & Central America, regional   **USD thousand 12667**
And as a share of aid to total recipient countries   **0.19%**

North of Sahara, regional

RECIPIENT	Agency	Sector	Amount USD thousand	Project description	CRS ID Number
France	MISC	31110	68	POLITIQUE AGRICOLE & GESTION ADMINISTRATIVE	2007800769
France	MISC	31181	1756	ÉDUCATION & FORMATION DANS LE DOMAINE AGRICOLE	2007800790
Germany	BMZ	31391	2738	SUPPORT FOR FISHERIES MANAGEMENT IN WESTAFRICA	2007005951
Spain	MFA	31161	1232	FOOD SOVEREIGNTY IN W. SAHARIAN REFUGEE CAMPS OF TINDOUF THRU	077826
Spain	MFA	31181	260	TRAINING ON SUSTAINABLE AGRICULTURE AND	076230
Spain	MFA	31381	72	ADVANCED NAUTICAL SEMINAR ON THE SYSTEMS OF MANAG. OF FISHERY STATISTICS.	074011
Spain	MFA	31381	68	ADVANCED NAUTICAL SEMINAR ON MARITIME SECURITY EN THE FISHING SECTOR	073866

Total aid for water in 2007 for North of Sahara, regional   **USD thousand 6194**
And as a share of aid to total recipient countries   **0.09%**

Oceania, regional

RECIPIENT	Agency	Sector	Amount USD thousand	Project description	CRS ID Number
Australia	AusAID	31210	8	FORESTRY POLICY & ADMIN. MANAGEMENT	2007002077
Australia	AusAID	31210	16	DEVELOPMENT OF A PACIFIC ISLAND REGIONAL RESEARCH	2007001851
Australia	AusAID	31310	149	PACIFIC FISHERIES FRAMEWORK AND PIPELINE	2007000810
Australia	AusAID	31320	335	RESEARCH: CC IMPACT ON COASTAL FISHERIES	2007000504
EU institutions	CEC	31162	5476	FACILITATING AGRICULTURAL COMMODITY TRADE (FACT)	2007100203
EU institutions	CEC	31310	9049	SCIENTIFIC SUPPORT FOR OCEANIC FISHERIES MANAGEMENT	2007100375
Japan	JICA	31110	10	TC AGGREGATED ACTIVITIES	073017T
Japan	MAFF	31381	190	TC AGGREGATED ACTIVITIES	070077T
New Zealand	NZAid	31182	59	SOUTH PACIFIC AGRICULTURAL CHEMISTRY LAB NETWORK	070388
New Zealand	NZAid	31192	294	CAPACITY BUILDING FOR BIOCONTROL OF FOOD CROP PESTS IN THE SOUTH PACIFIC	070389

Total aid for water in 2007 for Oceania, regional   **USD thousand 15585**
And as a share of aid to total recipient countries   **0.23%**

RECIPIENT Donor	Agency	Sector	Amount USD thousand	Project description	CRS ID Number

Oman

Japan	JICA	31110	131	TC AGGREGATED ACTIVITIES	072978T
Japan	JICA	31210	32	TC AGGREGATED ACTIVITIES	073339T

Total aid for water in 2007 for Oman — **USD thousand 163**
And as a share of aid to total recipient countries — **0%**

Pakistan

AsDF		31140	10644	MFF-PUNJAB IRRIGATED AGRICULTURE INVESTMENT PROGRAM - PROJECT	070029
Australia	AusAID	31120	19	REFINEMENT OF TECHNOLOGY FOR DIRECT DRILLING	2007001792
Australia	AusAID	31120	99	EVALUATION, REFINEMENT & PROMOTION OF TECHN. FOR DIRECT DRILLING INTO RICE	2007001005
Australia	AusAID	31130	68	Integrated on-farm management of saline drainage	2007001190
Australia	AusAID	31130	18	AGRICULTURAL LAND RESOURCES	2007002094
Australia	AusAID	31161	94	FOOD CROP PRODUCTION	2007001539
Australia	AusAID	31182	63	FOOD SECURITY THROUGH SUSTAINABLE CONTROL OF YELLOW RUST OF WHEAT IN ASIA	2007001750
Australia	AusAID	31195	64	LIVESTOCK/VETERINARY SERVICES	2007001222
Canada	CIDA	31110	2275	OXFAM EARTHQUAKE RECONSTRUCTION PROGRAM	070627
Canada	CIDA	31110	2327	RESTORE AND IMPROVE SUST. LIVELIHOODS	070808
IDA		31130	45650	LAND RECORDS MANAGEMENT AND INFORMATION SYSTEMS	2007000852
IDA		31140	120160	WATER SECTOR IMPROVEMENT PROJECT	2007000859
Ireland	DFA	31130	182	AGRICULTURAL LAND RESOURCES	2007000839
Italy	LA	31181	260	RECONSTRUCTION OF ESMA	070912
Japan	JICA	31110	367	TC AGGREGATED ACTIVITIES	073065T
Japan	JICA	31130	305	TC AGGREGATED ACTIVITIES	073110T
Japan	JICA	31163	19	TC AGGREGATED ACTIVITIES	073222T
Netherlands	MFA	31110	402	AGRICULTURAL POLICY & ADMIN. MGMT	2007100143
United States	AID	31110	2700	AGRICULTURE GROWTH AND EMPLOYMENT	2007001860
United States	AID	31110	146	AGRICULTURAL POLICY & ADMIN. MGMT	2007013922
United States	AID	31120	7884	AGRICULTURAL SECTOR PRODUCTIVITY	2007001779

Total aid for water in 2007 for Pakistan — **USD thousand 193745**
And as a share of aid to total recipient countries — **2.89%**

Palau

Japan	JICA	31163	52	TC AGGREGATED ACTIVITIES	073236T
Japan	MAFF	31310	42	TC AGGREGATED ACTIVITIES	070046T
Japan	MAFF	31320	252	TC AGGREGATED ACTIVITIES	070063T

Total aid for water in 2007 for Palau — **USD thousand 346**
And as a share of aid to total recipient countries — **0.01%**

Palestinian Adm. Areas

Austria	ADA	31140	444	NGO-COFINANCING: INTEGRATED WATER RESOURCE MANAGEMENT IN PALESTINE	2007500024c
Belgium	DGCD	31110	62	AGRICULTURAL POLICY & ADMIN. MGMT	2004003278
Belgium	DGCD	31140	136	AGRICULTURAL WATER RESOURCES	2005001845
Belgium	DGCD	31194	128	AGRICULTURAL CO-OPERATIVES	2005001847

RECIPIENT Donor	Agency	Sector	Amount USD thousand	Project description	CRS ID Number
France	AFD	31162	1711	RENFORCEMENT CAPACITE HUILE OLIVE	2007150100
Germany	BMZ	31140	155	AGRICULTURAL WATER RESOURCES	2007009131
Italy	DGCS	31110	1600	AGRICULTURAL POLICY & ADMIN. MGMT	070155
Italy	DGCS	31161	16	FOOD CROP PRODUCTION	060560
Italy	DGCS	31181	447	AGRICULTURAL EDUCATION/TRAINING	060034
Italy	LA	31195	38	LIVESTOCK/VETERINARY SERVICES	070900
Japan	MOFA	31110	849	AGRICULTURAL POLICY & ADMIN. MGMT	070116
Japan	MOFA	31110	1613	ASSISTANCE FOR UNDERPRIVILEGED FARMERS	070197
Japan	JICA	31110	746	TC AGGREGATED ACTIVITIES	073057T
Japan	JICA	31130	1809	TC AGGREGATED ACTIVITIES	073128T
Netherlands	MFA	31181	2433	AGRIC. TRAINING: RAM EUREPGAP PROJECT I	2007000914
Netherlands	MFA	31193	3761	RAM FINANCIAL INST. (REEF)	2007000915
Norway	NORAD	31130	235	AGRICULTURAL LAND RESOURCES	2007002910
Norway	MFA	31161	2152	PSE/EMERGENCY SUPPORT TO AGRICULTURAL COMMUNITIES	2007000788
Spain	MFA	31110	274	SUPPORT TO THE FAO COORDINATION PROGRAM FOR THE IMPR. THE AGR. SECTOR	076349
Spain	MFA	31120	137	VILLAGE PROFILES AND AZAHAR NEEDS ASSESSMENT IN HEBRON GOBERNORATE	075150
Spain	AG	31161	70	DEVELOPMENT OF IRRIGATED FAMILY FARMS IN THE RURAL ZONES OF JENIN	073980
Spain	MFA	31195	722	BRUCELLOSIS CONTROL PROGRAM IN THE PALESTINIAN TERRITORIES	077619
Spain	MFA	31210	34	FORESTRY POLICY & ADMIN. MANAGEMENT	072776
Switzerland	SDC	31110	325	PARC REHABILITATION DAMAGED LAND GAZA	2007000426
Switzerland	SDC	31120	1150	SUPPORT TO OLIVE OIL PRODUCERS	2005003923
United States	AID	31120	2000	AGRICULTURAL SECTOR PRODUCTIVITY	2007001771

Total aid for water in 2007 for Palestinian Adm. Areas   **USD thousand 23047**
And as a share of aid to total recipient countries   **0.34%**

Panama

RECIPIENT Donor	Agency	Sector	Amount USD thousand	Project description	CRS ID Number
Japan	JICA	31110	1157	TC AGGREGATED ACTIVITIES	072959T
Japan	JICA	31163	54	TC AGGREGATED ACTIVITIES	073206T
Japan	JICA	31210	712	TC AGGREGATED ACTIVITIES	073393T
Japan	JICA	31310	97	TC AGGREGATED ACTIVITIES	073502T
Japan	MAFF	31320	186	TC AGGREGATED ACTIVITIES	070055T
Spain	MIE	31110	82	AGRICULTURAL POLICY & ADMIN. MGMT	074227
Spain	AG	31161	83	FOOD CROP PRODUCTION: STRENGTHENING TOURISM COMPETITIVENESS IN HONDURAS	074303
Spain	AG	31320	1916	DEVELOPMENT PROGRAM FOR THE COMMUNITY OF FARALLON	077940
Spain	AG	31391	411	PROGRAM TO SUPPORT THE ARTISANAL FISHERMEN OF THE MUNICIPALITY OF RIO HATO	077073
United States	IADF	31162	59	INDUSTRIAL CROPS/EXPORT CROPS	2007024960

Total aid for water in 2007 for Panama   **USD thousand 4758**
And as a share of aid to total recipient countries   **0.07%**

Papua New Guinea

RECIPIENT Donor	Agency	Sector	Amount USD thousand	Project description	CRS ID Number
Australia	AusAID	31110	9	PNG COFFEE POLICY ANALYSIS	2007002032
Australia	AusAID	31120	144	AGRICULTURAL DEVELOPMENT	2007000823
Australia	AusAID	31130	61	OVERCOMING MAGNESIUM DEFICIENCY IN OIL PALM CROPS	2007001261
Australia	AusAID	31130	289	AGRICULTURAL LAND RESOURCES	2007001037
Australia	AusAID	31161	31	VITAMIN A SUPPLEMENTATION OF INFANTS IN PNG TO REDUCE THE EFFECTS OF MALARIA	2007002017
Australia	AusAID	31161	146	FOOD CROP PRODUCTION	2007001627

RECIPIENT Donor	Agency	Sector	Amount USD thousand	Project description	CRS ID Number
Australia	AusAID	31162	31	OVERCOMING MAGNESIUM DEFICIENCY IN OIL PALM CROPS	2007001596
Australia	AusAID	31162	63	INDUSTRIAL CROPS/EXPORT CROPS: cocoa	2007001245
Australia	AusAID	31162	67	INDUSTRIAL CROPS/EXPORT CROPS	2007001967
Australia	AusAID	31163	115	LIVESTOCK	2007001336
Australia	AusAID	31163	465	AQIS / NAQIA TWINNING AGREEMENT	2007000388
Australia	AusAID	31181	116	AGRICULTURAL EDUCATION/TRAINING	2007002051
Australia	AusAID	31181	125	ANALYTICAL EQUIPMENT IN SUPPORT OF THE ACIAR/UNITECH	2007000888
Australia	AusAID	31181	102	AQIS DOMESTIC PLANT AND ANIMAL HEALTH SURVEYS	2007001658
Australia	AusAID	31182	38	DEVELOPING THE ORNAMENTALS INDUSTRY IN THE PACIFIC	2007001520
Australia	AusAID	31182	58	MANAG OF EUMETOPINA FLAVIPES:THE VECTOR OF RAMU STUNT DISEASE OF SUGARCANE	2007001302
Australia	AusAID	31182	121	SUSTAINABLE MANAGEMENT OF COFFEE GREEN SCALES IN PAPUA NEW GUINEA	2007000910
Australia	AusAID	31182	66	THE USE OF PATHOGEN TESTED PLANTING MATERIALS	2007001697
Australia	AusAID	31191	86	IMPROVING THE MARKETING SYSTEM FOR FRESH PRODUCE O	2007001066
Australia	AusAID	31191	17	MICROBIAL CONTAMINANTS ASSOCIATED WITH SAGO PROCESS	2007001819
Australia	AusAID	31191	268	RE-COMMERC. OF THE PNG PYRETHRUM INDUSTRY MPROVING HARVESTED YIELDS	2007001109
Australia	AusAID	31192	43	TARO BEETLE MANAGEMENT	2007001426
Australia	AusAID	31192	63	MANAGEMENT OF POTATO BLIGHT IN PAPUA NEW GUINEA	2007001723
Australia	AusAID	31192	134	PLANT/POST-HARVEST PROT. & PEST CTRL	2007002059
Australia	AusAID	31195	22	LIVESTOCK/VETERINARY SERVICES	2007001729
Australia	AusAID	31220	19	FORESTRY DEVELOPMENT	2007001782
Australia	AusAID	31282	237	FORESTRY RESEARCH	2007000942
Australia	AusAID	31310	71	FISHING POLICY AND ADMIN. MANAGEMENT	2007001609
Australia	AusAID	31320	6	AQUACULTURE RESEARCH PRIORITIES	2007002155
Australia	AusAID	31320	355	FISHERY DEVELOPMENT	2007001600
Germany	BMZ	31120	178	AGRICULTURAL DEVELOPMENT	2007006570
IDA		31120	1375	SMALLHOLDER AGRICULTURE DEVELOPMENT	2007001002
IDA		31161	5775	SMALLHOLDER AGRICULTURE DEVELOPMENT	2007001000
IDA		31166	2200	SMALLHOLDER AGRICULTURE DEVELOPMENT	2007000999
Japan	JICA	31110	1874	TC AGGREGATED ACTIVITIES	073013T
Japan	JICA	31130	31	TC AGGREGATED ACTIVITIES	073120T
Japan	JICA	31150	105	TC AGGREGATED ACTIVITIES	073184T
Japan	JICA	31163	182	TC AGGREGATED ACTIVITIES	073237T
Japan	MAFF	31210	350	FORESTRY POLICY & ADMIN. MANAGEMENT	072167
Japan	JICA	31210	39	TC AGGREGATED ACTIVITIES	073460T
Japan	JICA	31310	433	TC AGGREGATED ACTIVITIES	073537T
Japan	MAFF	31320	33	TC AGGREGATED ACTIVITIES	070064T
Korea	KOICA	31195	134	TRAINING PROGRAM	2007017390
New Zealand	NZAid	31191	1793	FRESH PRODUCE DEVELOPMENT :PNG	070068
UNDP		31310	49	FISHING POLICY AND ADMIN. MANAGEMENT	2007013878

Total aid for water in 2007 for Papua New Guinea   **USD thousand 17918**
And as a share of aid to total recipient countries   **0.27%**

Paraguay

France	MISC	31182	2382	RECHERCHE AGRICULTURE	2007800374
Germany	BMZ	31120	1243	AGRICULTURAL DEVELOPMENT	2007006668
Germany	BMZ	31164	339	AGRARIAN REFORM	2007006133

RECIPIENT Donor	Agency	Sector	Amount USD thousand	Project description	CRS ID Number
Germany	BMZ	31166	246	AGRICULTURAL EXTENSION	2007006071
Japan	JICA	31110	2488	TC AGGREGATED ACTIVITIES	073053T
Japan	MAFF	31130	438	TC AGGREGATED ACTIVITIES	070014T
Japan	JICA	31150	86	TC AGGREGATED ACTIVITIES	073160T
Japan	JICA	31163	266	TC AGGREGATED ACTIVITIES	073213T
Japan	JICA	31195	67	TC AGGREGATED ACTIVITIES	073290T
Japan	PRF	31195	22	TC AGGREGATED ACTIVITIES	075610T
Korea	KOICA	31163	14	LIVESTOCK	2007016989
Korea	KOICA	31181	71	AGRICULTURAL EDUCATION/TRAINING	2007017158
Korea	KOICA	31381	16	FISHERY EDUCATION/TRAINING	2007016372
Spain	MFA	31110	544	FOOD SOVEREIGNTY	077377
Spain	MFA	31120	411	COOPERATION WITH THE AIM OF REDUCING POVERTY IN THE INDIGENOUS COMMUNITIES	077028
Spain	AG	31120	82	COMPREHENSIVE RURAL DEVELOPMENT OF SMALL FARMERS	074240
Spain	AG	31161	137	SUPPORT FOR THE REACTIVATION OF AGRICULTURAL PRODUCTION	075189
Spain	AG	31320	78	FISHERY DEVELOPMENT	074123

Total aid for water in 2007 for Paraguay — **USD thousand 8932**
And as a share of aid to total recipient countries — **0.13%**

Peru

RECIPIENT Donor	Agency	Sector	Amount USD thousand	Project description	CRS ID Number
Australia	AusAID	31195	8	ALPACA EMBRYO TRANSFER	2007002037
Belgium	DGCD	31110	26	AGRICULTURAL POLICY & ADMIN. MGMT	2005000998
Belgium	DGCD	31110	188	PROMOTION D'UN COULOIR ECONOMIQUE SUR L'ALTIPLANO-ETUDE PREPARATOIRE	2005007118
Belgium	DGCD	31110	196	ARTICULATION DES FILIERES AGRICOLES DANS LES ANDES PERUVIENNES	2005007117
Belgium	DGCD	31120	153	AGRICULTURAL DEVELOPMENT	2005000570
Belgium	DGCD	31120	643	DEVELOPPEMENT PRODUCTIF ET PROMOTION DE LA FEMME RURALE	2002000206
Belgium	DGCD	31161	189	FOOD CROP PRODUCTION	2005000996
Belgium	DGCD	31181	254	CRéATION D'UN RéSEAU DE CFRAS (CENTRE FAMILAL DE FORMATION RURALE)	2004007228
Belgium	DGCD	31193	53	AGRICULTURAL FINANCIAL SERVICES	2006008735
Belgium	DGCD	31194	494	CONSOLIDATION ECONOMIQUE ET INSTITUT. D'ORGANISATIONS DE PRODUCTEURS	2004003374
Belgium	DGCD	31194	316	PLAN DE CONSOLIDATION DE LA CENTRALE DES COOPéRATIVES CAFE PERU	2004003375
Belgium	DGCD	31194	15	AGRICULTURAL CO-OPERATIVES	2005000997
Belgium	DGCD	31210	530	ADEFOR/GESTION FORESTIERE + INDUSTRIE BOIS FORMULATION	2002000202
Belgium	MPRF	31210	102	FORESTRY POLICY & ADMIN. MANAGEMENT	2007004813
Canada	IDRC	31110	109	ENHANCING CAPACITY FOR INNOVATION, INCREASING PRODUCTIVITY	070450i
Finland	MFA	31165	130	NGO SUPPORT / USE OF APPR. TECHNOLOGIES IN PROCESSING OF ANDEAN CROPS	2004003441
Germany	BMZ	31110	80	AGRICULTURAL POLICY & ADMIN. MGMT	2007007285
Germany	BMZ	31120	1058	AGRICULTURAL DEVELOPMENT	2007008727
Germany	BMZ	31120	1506	DECENTRALIZED ECONOMIC STRATEGIES TOWARDS POVERTY REDUCTION	2007005587
Germany	BMZ	31161	207	FOOD CROP PRODUCTION	2007007303
Germany	BMZ	31166	342	AGRICULTURAL EXTENSION	2007006295
Germany	BMZ	31181	96	AGRICULTURAL EDUCATION/TRAINING	2007006358
Germany	BMZ	31182	334	AGRICULTURAL RESEARCH	2007007641
Germany	BMZ	31191	52	AGRICULTURAL SERVICES	2007007291
Germany	BMZ	31193	89	AGRICULTURAL FINANCIAL SERVICES	2007007289
Germany	BMZ	31210	99	FORESTRY POLICY & ADMIN. MANAGEMENT	2007008777
Ireland	DFA	31120	213	AGRICULTURAL DEVELOPMENT	2007000754

RECIPIENT Donor	Agency	Sector	Amount USD thousand	Project description	CRS ID Number
Italy	LA	31140	41	AGRICULTURAL WATER RESOURCES	071325
Italy	DGCS	31181	17	AGRICULTURAL EDUCATION/TRAINING	060819
Japan	JICA	31110	110	TC AGGREGATED ACTIVITIES	072969T
Japan	JICA	31130	88	TC AGGREGATED ACTIVITIES	073126T
Japan	PRF	31182	14	TC AGGREGATED ACTIVITIES	075605T
Japan	JICA	31195	183	TC AGGREGATED ACTIVITIES	073265T
Japan	MAFF	31210	80	FORESTRY POLICY & ADMIN. MANAGEMENT	072166
Japan	JICA	31310	335	TC AGGREGATED ACTIVITIES	073566T
Japan	MAFF	31381	37	TC AGGREGATED ACTIVITIES	070072T
Luxembourg	MFA	31140	88	AGRICULTURAL WATER RESOURCES	070401
Luxembourg	MFA	31161	251	FOOD CROP PRODUCTION	070403
Luxembourg	MFA	31181	313	AGRICULTURAL EDUCATION/TRAINING	070405
Netherlands	MFA	31110	567	AGRIC.: SUBS. TO PSOM 2003-2009	2007100150
Netherlands	MFA	31220	9	FORESTRY DEVELOPMENT	2007000371
Norway	NORFUND	31140	79	GALLITO CIEGO WATER HARVESTING STUDY	2007004469
Spain	MFA	31110	753	PROMOTE FOOD SOVEREIGNTY: GUARANTEE ACCESS TO ENOUGH FOOD NUTRITIVE	077635
Spain	MIE	31110	92	EST. IMPROVEMENT OF THE PORT FOR ARTISANAL FISHERIES PERU	074430
Spain	MFA	31120	137	IMPROVEMENT OF NATURAL RESOURCES MANAGEMENT	075091
Spain	MFA	31120	205	SUPPORT TO ENABLE THE MINISTRY OF AGRICULTURE	075848
Spain	AG	31120	98	PROCESSING AND PRODUCTION OF PEARS AND DERIVATES IN CH÷PARRA PERU.	074536
Spain	AG	31120	27	AGRICULTURAL DEVELOPMENT	072424
Spain	AG	31120	328	TECHNICAL & MANAGEMENT CAPACITY BUILDING OF SMALL AGRICULTURAL	076640
Spain	AG	31120	107	PRODUCTIVE DIVERSIFICATION AND INCOME GENERATION FOR SMALL-SCALE PRODUCERS	074685
Spain	AG	31120	93	RURAL DEVELOPMENT WITH FARMERS UNDER A POVERTY SITUATION IN VALLE DE LURIN	074437
Spain	AG	31120	280	RURAL DEVELOPMENT	076414
Spain	AG	31120	82	PRODUCTION PROCESSING AND MARKETING OF TUNA AND COCHINEAL	074241
Spain	AG	31120	148	BUILDING IN STRATEGIC ACTORS FOR THE DEVELOPMENT OF AGRICULTURAL ACTIVITY	075314
Spain	AG	31120	334	AGRICULTURAL INTEGRATION PROGRAM	076679
Spain	AG	31120	98	AGRO-INDUSTRIES	074535
Spain	AG	31120	197	ACCESS TO THE REGIONAL MARKET OF AGRICULTURAL PRODUCTS THE VALLEY OF COLCA	075751
Spain	AG	31130	321	MANAGEMENT OF PRODUCTIVE RESOURCES WITH GENDER EQUALITY	076620
Spain	AG	31140	411	CAPACITIES IN THE FIELD OF SUSTAINABLE DEVELOPMENT POLICY IN THE	077076
Spain	AG	31140	68	CONSTRUCTION AND FUNCTIONING OF A MULTI-FAMILY IRRIGATION SYSTEM	073843
Spain	AG	31140	195	MANAG. OF WATER IRRIGATION AND CREATION OF SUSTAINABLE INCOME WITH SMALL	075738
Spain	AG	31140	13	INSTALLATION OF A SPRINKLER IRRIGATION SYSTEM	071361
Spain	AG	31150	329	RESTRUCTURING OF THE AGRICULTURAL PRODUCTION IN THE COMMUNITIES	076660
Spain	MFA	31161	439	SUPPORT TO REPAIR THE DAMAGE RESULTED BY THE EARTHQUAKE IN THE SUBURBAN AG.	077186
Spain	AG	31161	285	PRODUCTIVE AND SOCIAL CAPACITIES OF THE ALPAQUEROS PRODUCERS	076437
Spain	AG	31161	27	COMMUNITY FARM SCHOOL FOR THE IMPROVEMENT OF CHILD AND FAMILY NUTRITION	072410
Spain	AG	31161	410	FOOD SECURITY AND AGROFISHERY DEVELOPMENT	077007
Spain	AG	31163	411	LIVESTOCK	077079
Spain	AG	31163	80	PRODUCTION OF REPRODUCTIVE ANIMALS IN ORDER TO IMPROVE CAPRINES BREEDING	074168
Spain	AG	31163	410	AGRICULTURAL DEVELOPMENT AND IRRIGATION WATER MANAGEMENT	077001
Spain	AG	31165	110	GRANULATED SUGAR PRODUCTION FOR SUSTAINABLE DEVELOPMENT OF SMALL FARMERS	074745
Spain	AG	31166	209	PILOT CENTRE OF FARMING EDUCATION TRAINING AND RESEARCH	075933
Spain	MFA	31181	393	NETWORK OF CENTRES OF TECHNICAL FORMATION IN THE RURAL AREA	076940
Spain	AG	31181	222	AGRICULTURAL EDUCATION/TRAINING	074950

RECIPIENT Donor	Agency	Sector	Amount USD thousand	Project description	CRS ID Number
Spain	Misc	31181	8	AGRICULTURAL EDUCATION/TRAINING	070974
Spain	AG	31182	136	IMPROVEMENT OF INCOME AND LIVING CONDITIONS OF POOR FAMILIES	075079
Spain	Misc	31182	31	PHD COURSE FROM THE AGRO FOREST ECOSYSTEMS AND VEGETABLE PRODUCTION DEP.	072644
Spain	Misc	31220	12	IMPROVEMENT OF THE VEGETAL COVERAGE	071334
Spain	MFA	31310	513	SUPPORT PROGRAM FOR ARTISANAL FISHERIES AQUACULTURE AND SUSTAINABLE MANAG	077340
Spain	AGR	31320	16	FISHERY DEVELOPMENT	071619
Spain	AG	31320	17	COMPREHENSIVE DEVELOPMENT OF THE BAHIA DE SECHURA COVES	071702
Switzerland	seco	31110	2084	BIOTRADE PERU	2007001874
United States	TDA	31110	35	AGRICULTURAL POLICY & ADMIN. MGMT	2007028204
United States	AID	31165	42864	ALTERNATIVE DEVELOPMENT AND ALTERNATIVE LIVELIHOODS	2007001900
United States	AGR	31182	100	AGRICULTURAL RESEARCH	2007000441
United States	IADF	31191	118	AGRICULTURAL SERVICES	2007025095
United States	IADF	31220	17	FORESTRY DEVELOPMENT	2007024991

Total aid for water in 2007 for Peru   **USD thousand 62960**
And as a share of aid to total recipient countries   **0.94%**

Philippines

Recipient	Agency	Sector	Amount	Project description	CRS ID
Australia	AusAID	31110	15	PHILIPPINE POLICY LINKAGE SCOPING STUDY	2007001863
Australia	AusAID	31110	393	ACIAR RESEARCH & DEVELOPMENT ACTIVITIES	2007000454
Australia	AusAID	31110	190	AGRICULTURAL POLICY & ADMIN. MGMT	2007000711
Australia	AusAID	31120	16	PSLP PHILIPPINES, MINDANAO TREES IN FARMS	2007001854
Australia	AusAID	31130	245	AGRICULTURAL LAND RESOURCES	2007001251
Australia	AusAID	31140	124	AGRICULTURAL WATER RESOURCES	2007001589
Australia	AusAID	31140	41	ENHANCING AGRICULTURAL PRODUCTION BY USE OF SHALLO	2007001864
Australia	AusAID	31161	77	RICE AND WHEAT GENES (KN67)	2007001555
Australia	AusAID	31162	58	INDUSTRIAL CROPS/EXPORT CROPS	2007001297
Australia	AusAID	31181	44	AGRICULTURAL EDUCATION/TRAINING	2007001831
Australia	AusAID	31182	75	DEVELOPMENT OF PRSV-P RESISTANT PAPAYA GENOTYPES	2007001753
Australia	AusAID	31182	19	DISEASE CONTROL IN MINDANO VEGETABLE FARMING SYSTEMS	2007001789
Australia	AusAID	31182	105	DEVELOPING MOLECULAR MARKERS TO ENABLE SELECTION AGAINST CHALK IN RICE	2007000969
Australia	AusAID	31182	118	ACIAR PROJECT : FERTILIZATION-INDEPENDENT FORMATION	2007001402
Australia	AusAID	31182	242	FORMATION OF EMBRYO, ENDOSPERM PERICARP FOR APOMICTIC HYBRID RICE	2007000912
Australia	AusAID	31191	34	BRIDGING THE GAP BETWEEN SEASONAL CLIMATE FORECASTS AND DECISION MAKERS	2007001561
Australia	AusAID	31191	153	COMMUNITY AGRICULTURAL TECHNOLOGY PROGRAM	2007001541
Australia	AusAID	31191	194	AGRICULTURAL SERVICES	2007001033
Australia	AusAID	31192	306	PLANT/POST-HARVEST PROT. & PEST CTRL	2007001299
Australia	AusAID	31192	224	PHT/2003/071	2007000637
Australia	AusAID	31220	46	FORESTRY DEVELOPMENT	2007001877
Australia	AusAID	31282	48	MINDANAO OIL PROJECT	2007001371
Australia	AusAID	31282	71	PSLP ASSESSING THE ECONOMIC POTENTIAL OF KEY TREE	2007001182
Australia	AusAID	31291	360	FORESTRY SERVICES	2007001137
Australia	AusAID	31320	332	FISHERY DEVELOPMENT	2007001248
Australia	AusAID	31382	29	QUALITY AND MARKET POTENTIAL FOR THE DEV. OF BIVALVE MOLLUSC AQUACULTURE	2007001628
Austria	ADA	31120	240	NGO COFINANCING: ORGANIC FARMING MINDANAO II	2007000021ai
Austria	ADA	31120	146	NGO COFINANCED PROJECT: ORGANIC FARMING MINDANAO	2006000022a

RECIPIENT Donor	Agency	Sector	Amount USD thousand	Project description	CRS ID Number
Belgium	DGCD	31120	154	AGRICULTURAL DEVELOPMENT	2004005207
Belgium	DGCD	31161	129	FOOD CROP PRODUCTION	2005001247
Belgium	DGCD	31166	45	AGRICULTURAL EXTENSION	2004007247
Belgium	DGCD	31181	97	AGRICULTURAL EDUCATION/TRAINING	2005000900
Belgium	DGCD	31192	69	PLANT/POST-HARVEST PROT. & PEST CTRL	2002000132
Belgium	DGCD	31194	187	AGRICULTURAL CO-OPERATIVES	2005000932
Canada	IDRC	31310	19	STUDY TOUR TO NEWFOUNDLAND BY FISHERIES RESEARCHERS FROM THE PHILIPPINES	070700i
France	MISC	31182	3012	RECHERCHE AGRICULTURE	2007800382
Germany	BMZ	31120	581	INTEGRATED AGRICULTURAL DEVELOPMENT PROGRAMME, NEGROS, PHILIPPINES	2007005720
Germany	BMZ	31120	718	AGRICULTURAL DEVELOPMENT	2007007313
Germany	BMZ	31194	77	AGRICULTURAL CO-OPERATIVES	2007006069
Germany	BMZ	31220	15	FORESTRY DEVELOPMENT	2007007318
Germany	BMZ	31310	685	FISHING POLICY AND ADMINISTRATIVE MANAGEMENT	2007009144
Germany	BMZ	31320	170	FISHERY DEVELOPMENT	2007009231
Ireland	DFA	31120	13	AGRICULTURAL DEVELOPMENT	2007002566
Japan	JICA	31110	3773	TC AGGREGATED ACTIVITIES	073071T
Japan	MOFA	31120	84	TC AGGREGATED ACTIVITIES	070378T
Japan	JICA	31130	60	TC AGGREGATED ACTIVITIES	073116T
Japan	MOFA	31150	2547	THE ASSISTANCE FOR UNDERPRIVILEGED FARMERS	070112
Japan	JICA	31150	39	TC AGGREGATED ACTIVITIES	073169T
Japan	JICA	31163	552	TC AGGREGATED ACTIVITIES	073230T
Japan	JBIC	31164	84482	AGRARIAN REFORM INFRASTRUCTURE SUPPORT PROJECT (PHASE III)	073084
Japan	JBIC	31164	15705	AGRARIAN REFORM	073086
Japan	MAFF	31181	197	TC AGGREGATED ACTIVITIES	070021T
Japan	PRF	31192	15	TC AGGREGATED ACTIVITIES	075609T
Japan	JICA	31195	194	TC AGGREGATED ACTIVITIES	073280T
Japan	JICA	31210	1039	TC AGGREGATED ACTIVITIES	073447T
Japan	MAFF	31310	39	TC AGGREGATED ACTIVITIES	070042T
Japan	JICA	31310	209	TC AGGREGATED ACTIVITIES	073579T
Korea	KOICA	31110	28	AGRICULTURAL POLICY & ADMIN. MGMT	2007017734
Korea	Misc	31110	9	AGRICULTURAL POLICY & ADMIN. MGMT	2007002557
Korea	KOICA	31120	19	AGRICULTURAL DEVELOPMENT	2007016421
Korea	Misc	31120	10	AGRICULTURAL DEVELOPMENT	2007001931
Korea	KOICA	31140	1000	F.S. AND DETAILED DESIGN ON THE IMPROVEMENT OF MALINAO DAM, BOHOL PROVIN	2007017602
Korea	KOICA	31162	42	INDUSTRIAL CROPS/EXPORT CROPS	2007017008
Korea	KOICA	31163	90	LIVESTOCK	2007017207
Korea	KOICA	31181	136	AGRICULTURAL EDUCATION/TRAINING	2007017225
Korea	Misc	31181	21	AGRICULTURAL EDUCATION/TRAINING	2007002571
Korea	KOICA	31381	26	FISHERY EDUCATION/TRAINING	2007015772
Luxembourg	MFA	31194	18	AGRICULTURAL CO-OPERATIVES	070188
Netherlands	MFA	31110	405	AGRICULTURAL POLICY & ADMIN. MGMT	2007100155
New Zealand	NZAid	31140	132	PHILIPPINES INTEGRATED COMMUNITY ENVIRONMENT	070256
New Zealand	NZAid	31291	11	SUSTAINABLE TREE FARMING	070352
New Zealand	NZAid	31320	37	COASTAL RESOURCE LIVELIHOODS	070255
Norway	MFA	31320	35	TRAINING PARTNERS ON TILAPIA HATCHERY/GROW-OUT MANAGEMENT AND OPERATIONS	2007002683

RECIPIENT Donor	Agency	Sector	Amount USD thousand	Project description	CRS ID Number
Norway	MFA	31320	54	DEV. A HIGH STANDARD TILAPIA HATCHERY AND GROW-OUT MANAGEMENT PROTOCOL	2007002572
Norway	NORAD	31381	165	PILOT PROJECT - GENOMAR SUPREME PHILIPPINES LTD	2007004735
Spain	MFA	31120	1369	SUSTAINABLE HUMAN DEV. AND THE FIGHT AGAINST POVERTY	077850
Spain	MFA	31165	758	DEVPT OF RURAL COMMUNITIES OF BICOL AND CARAGA W/ SPECIAL FOCUS ON WOMEN	077640
Spain	MFA	31194	342	POVERTY REDUCTION THROUGH SUSTAINABLE OPPORTUNITIES IN THE AGROFORESTAL	076736
United States	AID	31110	596	AGRICULTURAL ENABLING ENVIRONMENT	2007001661
United States	AID	31110	45	AGRICULTURAL POLICY & ADMIN. MGMT	2007013919
United States	AID	31120	1005	AGRICULTURAL SECTOR PRODUCTIVITY	2007001780

Total aid for water in 2007 for Philippines   **USD thousand 125204**
And as a share of aid to total recipient countries   **1.87%**

Rwanda

RECIPIENT Donor	Agency	Sector	Amount USD thousand	Project description	CRS ID Number
Belgium	DGCD	31120	82	AGRICULTURAL DEVELOPMENT	2007004649
Belgium	DGCD	31150	1557	APPUI AU SECTEUR SEMENCIER - (ASSR)	2007900002
Belgium	DGCD	31161	5476	APPUI FILI-RE HORTICOLE	2007004421
Belgium	DGCD	31161	206	PETITES EXPLOITATIONS AGRICOLES : AMéLIORATION DE LA PRODUCTIVITé	2005001052
Belgium	MPRF	31161	41	FOOD CROP PRODUCTION	2005001987
Belgium	DGCD	31166	340	SYSTèME NATIONAL DE VULGARISATION AGRICOLE DéCENTRALISé	2005000673
Belgium	DGCD	31181	255	AGRICULTURAL EDUCATION/TRAINING	2004008415
Belgium	DGCD	31192	102	PLANT/POST-HARVEST PROT. & PEST CTRL	2007004607
Belgium	MPRF	31194	31	AGRICULTURAL CO-OPERATIVES	2005001975
Belgium	DGCD	31195	48	LIVESTOCK/VETERINARY SERVICES	2002000129
Belgium	MPRW	31195	257	SANTé ANIMALE	2007004972
France	MISC	31182	27	RECHERCHE AGRICULTURE	2007800356
Germany	BMZ	31120	205	AGRICULTURAL DEVELOPMENT	2007006376
Germany	L G	31130	53	AGRICULTURAL LAND RESOURCES	2007011638
Italy	DGCS	31120	24	AGRICULTURAL DEVELOPMENT	060492
Italy	DGCS	31161	883	FOOD CROP PRODUCTION	060176
Japan	JICA	31110	1409	TC AGGREGATED ACTIVITIES	072934T
Japan	JICA	31130	25	TC AGGREGATED ACTIVITIES	073086T
Japan	MOFA	31150	1104	AGRICULTURAL INPUTS	070078
Japan	JICA	31150	37	TC AGGREGATED ACTIVITIES	073149T
Japan	JICA	31195	37	TC AGGREGATED ACTIVITIES	073253T
Japan	JICA	31310	22	TC AGGREGATED ACTIVITIES	073485T
Luxembourg	MFA	31120	2489	AGRICULTURAL DEVELOPMENT	070666
Netherlands	MFA	31182	519	AGRIC. RESEARCH: KIG ISAR/PROGRAMME FILIERES	2003013558
Switzerland	SDC	31220	22	FORESTRY DEVELOPMENT	1980000968
United Kingdom	DFID	31110	3943	MINAGRI INSTITUTIONAL DEVELOPMENT	2007000225
United States	AID	31110	60	AGRICULTURAL POLICY & ADMIN. MGMT	2007013868
United States	AID	31120	500	AGRICULTURAL SECTOR PRODUCTIVITY	2007001733
United States	ADF	31120	145	AGRICULTURAL DEVELOPMENT	2007000148
United States	ADF	31161	199	FOOD CROP PRODUCTION	2007000064
United States	ADF	31162	94	INDUSTRIAL CROPS/EXPORT CROPS	2007000069

RECIPIENT Donor	Agency	Sector	Amount USD thousand	Project description	CRS ID Number
		Total aid for water in 2007 for Rwanda			USD thousand 20193
		And as a share of aid to total recipient countries			0.3%

Samoa

Australia	AusAID	31110	10	A REVIEW OF THE POLICY AND ECONOMIC ENVIRONMENT IN THE SOUTH	2007001995
Australia	AusAID	31161	84	FOOD CROP PRODUCTION	2007001548
Australia	AusAID	31181	9	AGRICULTURAL EDUCATION/TRAINING	2007002034
Australia	AusAID	31192	28	PLANT/POST-HARVEST PROT. & PEST CTRL	2007002028
Australia	AusAID	31192	41	ACIAR-FUNDED PROJECTS	2007001812
Australia	AusAID	31220	25	FORESTRY DEVELOPMENT	2007002085
Australia	AusAID	31320	6	INTEGRATION OF BROODSTOCK REPLENISHMENT	2007002145
Australia	AusAID	31320	6	Aquaculture research priorities	2007002144
Japan	JICA	31110	229	TC AGGREGATED ACTIVITIES	073016T
Japan	JICA	31163	116	TC AGGREGATED ACTIVITIES	073238T
Japan	JICA	31210	32	TC AGGREGATED ACTIVITIES	073360T
Japan	JICA	31310	102	TC AGGREGATED ACTIVITIES	073586T
		Total aid for water in 2007 for Samoa			USD thousand 687
		And as a share of aid to total recipient countries			0.01%

Sao Tome & Principe

Japan	JICA	31110	7	TC AGGREGATED ACTIVITIES	072935T
Japan	MAFF	31310	26	TC AGGREGATED ACTIVITIES	070037T
Japan	MAFF	31381	20	TC AGGREGATED ACTIVITIES	070070T
Portugal	ICP	31192	8	PLANT/POST-HARVEST PROT. & PEST CTRL	070592
Portugal	ICP	31282	38	FORESTRY RESEARCH	070684
Portugal	ICP	31310	5	FISHING POLICY AND ADMIN. MANAGEMENT	070981
Spain	MFA	31320	55	VALUATION OF FISHING PRODUCTS IN THE COMMUNITIES OF THE AREA SOUTH OF S. TAKE	073508
Spain	MFA	31391	150	PROJECT CREATION OF A SMALL INSTALLATION FOR FISH UNLOADING AND COM.	075320
		Total aid for water in 2007 for Sao Tome & Principe			USD thousand 309
		And as a share of aid to total recipient countries			0%

Senegal

Austria	ADA	31120	452	NGO-COFINANCING: AGRICULTURAL DEVELOPMENT & RESOURCE MANAGEMENT, THIES	2007500024a
Austria	ADA	31120	101	NGO COFINANCING: PADORF III - SUSTAINABLE AGRICULTURE REGION FIMELA	2006000022cr
Belgium	DGCD	31110	10	AGRICULTURAL POLICY & ADMIN. MGMT	2004008718
Belgium	DGCD	31120	491	RENFORCEMENT DES CAPACITéS DES GROUPES DE BASE	2006003971
Belgium	DGCD	31120	108	AGRICULTURAL DEVELOPMENT	2005008474
Belgium	DGCD	31120	509	PROGRAMME D'ACTION POUR UN DéVELOPPEMENT RURAL JUSTE ET DURABLE	2005001595
Belgium	DGCD	31161	409	PROMOTION DE L'AGRICULTURE PAYSANNE	2004003365
Belgium	DGCD	31161	52	FOOD CROP PRODUCTION	2005001866
Belgium	DGCD	31281	254	RENFORCEMENT DURABLE DU CENTRE FORET DANS SA MISSION DE FORMATION	2005001625
Canada	CIDA	31120	17639	PDMAS: DEVELOPPEMENT MARCHES AGRICOLES	070445
France	AFD	31110	329	PROMOTION D'UNE AGRICULT. COMPET & DURABLE	2007151600
France	MISC	31110	6	POLITIQUE AGRICOLE & GESTION ADMINISTRATIVE	2007800772

RECIPIENT Donor	Agency	Sector	Amount USD thousand	Project description	CRS ID Number
France	MISC	31182	23395	RECHERCHE AGRICULTURE	2007800357
Germany	BMZ	31192	71	PLANT/POST-HARVEST PROT. & PEST CTRL	2007006171
Italy	LA	31110	132	AGRICULTURAL POLICY & ADMIN. MGMT	071061
Italy	LA	31130	21	AGRICULTURAL LAND RESOURCES	070846
Italy	LA	31150	19	AGRICULTURAL INPUTS	071670
Italy	DGCS	31163	355	LIVESTOCK	070609
Italy	LA	31166	21	AGRICULTURAL EXTENSION	062164
Italy	LA	31181	28	AGRICULTURAL EDUCATION/TRAINING	061615
Italy	LA	31194	7	AGRICULTURAL CO-OPERATIVES	071645
Italy	LA	31320	120	FISHERY DEVELOPMENT	071499
Japan	JICA	31110	1656	TC AGGREGATED ACTIVITIES	072936T
Japan	JICA	31210	2030	TC AGGREGATED ACTIVITIES	073377T
Japan	JICA	31310	507	TC AGGREGATED ACTIVITIES	073555T
Korea	KOICA	31120	1500	THE PROJECT FOR IMPROVEMENT OF AGRICULTURAL PRODUCTIVITY IN TAGARA REGIO	2007017564
Korea	KOICA	31181	49	AGRICULTURAL EDUCATION/TRAINING	2007016576
Luxembourg	MFA	31381	52	FISHERY EDUCATION/TRAINING	070080
Netherlands	MFA	31310	12	FISHING: DAK PRCM	2003013787
Norway	NORAD	31194	91	AGRICULTURAL CO-OPERATIVES	2007003785
Spain	MFA	31110	41	AGRICULTURAL POLICY & ADMIN. MGMT	072965
Spain	MIE	31110	111	AGRICULTURAL POLICY & ADMIN. MGMT	074774
Spain	AG	31120	275	EFFORT AND DIVERSIFICATION OF AGRICULTURAL ACTIVITIES	076396
Spain	AG	31120	194	INCREASE AND DIVERSIFICATION OF AGRICULTURAL ACTIVITIES	075732
Spain	AG	31120	298	INCREASE AND DIVERSIFICATION OF AGRICULTURAL ACTIVITIES IN CPAS DIEMBERING	076503
Spain	AG	31120	479	SUPPORT FOR MODERNIZATION AND THE INCREASE IN AGRICULTURAL PRODUCTION	077274
Spain	AG	31120	77	AGRICULTURAL DEVELOPMENT	073533
Spain	AG	31130	25	AGRICULTURAL LAND RESOURCES	071659
Spain	AG	31130	38	REINFORCEMENT OF THE PRODUCTIVE AND ORGANIZATIONAL CAPACITY	072893
Spain	MFA	31140	1154	WATER IN AFRICA (BURKINA FASOGUINEA SENEGAL & MALI)	077807
Spain	AG	31140	160	PROJECT WATER FOR FOOD SECURITY	075411
Spain	Misc	31140	5	DESIGN AND INSTALLATION OF AN IRRIGATION SYSTEM IN A PLOT OF FRUIT TREES	070679
Spain	AG	31181	51	AGRICULTURAL EDUCATION/TRAINING	072468
Spain	AG	31182	342	MULTISECTOR AID	076792
Spain	MFA	31320	127	VIABILITY OF LOBSTER CULTURE IN COSTAL POPULATIONS IN SENEGAL	074987
Spain	AG	31320	205	INSHORE FISHERY AND LOCAL DEVELOPMENT TOWN OF NGOR SENEGAL	075817
Spain	AG	31320	203	INSHORE FISHERY AND FIGHT AGAINST POVERTY IN MBAO SENEGAL.	075802
Spain	MFA	31381	25	FISHERY EDUCATION/TRAINING	072187
Spain	AG	31382	235	EXPERIMENTAL MUSSLE CULTIVATE IN SENEGAL	076088
Switzerland	SDC	31110	33	AGRICULTURAL POLICY & ADMIN. MGMT	1990000091
Switzerland	SDC	31110	1500	FONGS FéDéRATION DES ONG AU SéNéGAL	1990000090
Switzerland	SDC	31110	500	GESTION DURABLE RESSOURCES HALIEUTIQUES	2007000171
United States	AID	31110	227	AGRICULTURAL POLICY & ADMIN. MGMT	2007013869
United States	AID	31120	1219	AGRICULTURAL SECTOR PRODUCTIVITY	2007001734
United States	ADF	31163	74	LIVESTOCK	2007000087

Total aid for water in 2007 for Senegal   **USD thousand 58027**
And as a share of aid to total recipient countries   **0.86%**

RECIPIENT / Donor	Agency	Sector	Amount USD thousand	Project description	CRS ID Number
Serbia					
Austria	ADA	31120	163	BUSINESS PARTNERSHIP PROGRAM: BIO DIESEL MLADENOVO, VOJVODINA	2007500108e
Austria	Reg	31150	58	DELIVERY OF TRACTORS AND AGRICULTURAL MACHINERY	2007828638
Austria	Reg	31150	39	ESTABLISHMENT OF A REPAIR-SHOP FOR AGRICULATURAL MACHINERY AND EQUIPMENT	2007828641
Austria	Reg	31163	109	LIVESTOCK PROGRAMME, EXPERTS TRAINING AND EQUIPMENT	2007828637
EU institutions	EDF	31195	2738	LIVESTOCK/VETERINARY SERVICES	2007200102
Finland	MFA	31182	20	LOCAL COOPERATION FUND (LCF) IN SERBIA	2007001185
Finland	MFA	31220	82	ENVIRONMENTAL GIS FOR MONTENEGRO THROUGH UNDP	2004003063
IDA		31166	570	AVIAN FLU	2007000019
Ireland	DFA	31120	222	AGRICULTURAL DEVELOPMENT	2007000736
Italy	DGCS	31110	288	AGRICULTURAL POLICY & ADMIN. MGMT	070353
Japan	JICA	31110	28	TC AGGREGATED ACTIVITIES	072901T
Luxembourg	MFA	31120	955	DEVELOPMENT ASSISTANCE TO FARMERS IN REMOTE AREAS	070428
Norway	MFA	31163	79	ANIMAL HUSBANDRY	2007001784
Norway	MFA	31181	1038	LIFELONG LEARNING	2007001599
Norway	MFA	31194	696	BILAT AGRICULTURAL COOPERATIVES' SUPP	2007001761
Norway	MFA	31195	13	PRESTUDY - PIG BREEDING	2007001554
Norway	MFA	31195	904	BILATPIG BREEDING PHASE 2	2007001583
Norway	MFA	31210	660	FOREST MANAGEMENT PLANNING AND GIS	2007001600
Switzerland	SDC	31110	54	AGRICULTURAL POLICY & ADMIN. MGMT	2007000202
Switzerland	SDC	31120	1554	HORTICULTURE	2001002690
Switzerland	SDC	31120	25	AGRICULTURAL DEVELOPMENT	2001002691
United States	AID	31110	122	AGRICULTURAL POLICY & ADMIN. MGMT	2007013818
United States	AID	31120	6508	AGRICULTURAL SECTOR PRODUCTIVITY	2007001704
United States	AID	31181	279	AGRICULTURAL EDUCATION/TRAINING	2007001010

Total aid for water in 2007 for Serbia USD thousand 17203
And as a share of aid to total recipient countries 0.26%

RECIPIENT / Donor	Agency	Sector	Amount USD thousand	Project description	CRS ID Number
Seychelles					
Japan	MAFF	31320	609	TC AGGREGATED ACTIVITIES	070053T

Total aid for water in 2007 for Seychelles USD thousand 609
And as a share of aid to total recipient countries 0.01%

RECIPIENT / Donor	Agency	Sector	Amount USD thousand	Project description	CRS ID Number
Sierra Leone					
Germany	Fed Min	31150	408	DEVELOPMENT OF A SUSTAINABLE SEED PROGRAMME IN SIERRA LEONE	2007010300
IDA		31110	4500	RURAL DEVELOPMENT & PRIVATE SECTOR DEVELOPMENT	2007000403
IDA		31166	1500	RURAL DEVELOPMENT & PRIVATE SECTOR DEVELOPMENT	2007000402
IDA		31191	11400	RURAL DEVELOPMENT & PRIVATE SECTOR DEVELOPMENT	2007000405
Ireland	DFA	31161	178	FOOD CROP PRODUCTION	2007000856
Japan	JICA	31110	1085	TC AGGREGATED ACTIVITIES	072938T
United States	AID	31110	72	AGRICULTURAL POLICY & ADMIN. MGMT	2007013870
United States	AID	31120	780	AGRICULTURAL SECTOR PRODUCTIVITY	2007001735

Total aid for water in 2007 for Sierra Leone USD thousand 19923
And as a share of aid to total recipient countries 0.3%

RECIPIENT Donor	Agency	Sector	Amount USD thousand	Project description	CRS ID Number
Solomon Islands					
Australia	AusAID	31161	198	INTEGRATED CROP MANAGEMENT PACKAGE FOR SUSTAINABLE SMALLHOLDER GARDENS	2007001020
Australia	AusAID	31161	31	VITAMIN A SUPPLEMENTATION OF INFANTS IN PNG TO REDUCE THE EFFECTS OF MALARIA	2007002018
Australia	AusAID	31163	14	FEEDING VILLAGE POULTRY IN THE SOLOMON ISLANDS	2007001883
Australia	AusAID	31182	33	AGRICULTURAL RESEARCH	2007001946
Australia	AusAID	31182	152	THE USE OF PATHOGEN TESTED PLANTING MATERIALS TO IMPROVE SWEET POTATO PROD	2007001274
Australia	AusAID	31195	200	LIVESTOCK/VETERINARY SERVICES	2007001479
Australia	AusAID	31320	6	AQUACULTURE RESEARCH PRIORITIES	2007002142
Australia	AusAID	31320	119	FISHERY DEVELOPMENT	2007001531
Australia	AusAID	31382	10	FISHERY RESEARCH	2007002010
IDA		31166	608	RURAL DEVELOPMENT PROGRAM	2007001004
Japan	JICA	31110	243	TC AGGREGATED ACTIVITIES	073014T
Japan	JICA	31310	59	TC AGGREGATED ACTIVITIES	073585T
Japan	MAFF	31320	618	TC AGGREGATED ACTIVITIES	070065T
Korea	KOICA	31220	141	TRAINING PROGRAM	2007017347
New Zealand	NZAid	31320	735	FISHERIES LIVELIHOODS REHABILITATION	070572

Total aid for water in 2007 for Solomon Islands — **USD thousand 3166**
And as a share of aid to total recipient countries — **0.05%**

Somalia					
EU institutions	EDF	31140	1752	AGRICULTURAL WATER RESOURCES	2007201286
EU institutions	CEC	31195	5476	SOMALI ECOSYSTEM RINDERPEST ERADICATION COORDINATION UNIT	2007100419
Ireland	DFA	31161	324	FOOD CROP PRODUCTION	2007000531

Total aid for water in 2007 for Somalia — **USD thousand 7552**
And as a share of aid to total recipient countries — **0.11%**

South & Central Asia, regional					
Australia	AusAID	31182	63	SUSTAINABLE CONTROL OF YELLOW RUST OF WHEAT IN ASIA	2007001751
Australia	AusAID	31192	6	PLANT/POST-HARVEST PROT. & PEST CTRL	2007002124
Australia	AusAID	31192	37	ACIAR - Rodent Management in Indonesia & Vietnam	2007001537
Australia	AusAID	31282	86	FORESTRY RESEARCH: DOMESTICATION OF MELIACEAE SPECIES	2007001061
Australia	AusAID	31320	466	WSSV MANAGEMENT IN PRAWNS	2007000639
France	MISC	31181	29	¡DUCATION & FORMATION DANS LE DOMAINE AGRICOLE	2007800793

Total aid for water in 2007 for South & Central Asia, regional — **USD thousand 687**
And as a share of aid to total recipient countries — **0.01%**

South Africa					
Australia	AusAID	31120	135	AGRICULTURAL DEVELOPMENT	2007001224
Australia	AusAID	31120	104	DEVELOPING BEEF BUSINESS SYSTEMS FOR FARMERS	2007000970
Australia	AusAID	31130	83	IMPROVED FERTILISER RECOMMENDATIONS AND POLICY FOR DRY REGIONS OF S. AFRICA	2007001106
Australia	AusAID	31150	11	AGRICULTURAL INPUTS	2007001962
Australia	AusAID	31163	121	LIVESTOCK	2007001298
Australia	AusAID	31195	152	LIVESTOCK/VETERINARY SERVICES	2007001322

RECIPIENT Donor	Agency	Sector	Amount USD thousand	Project description	CRS ID Number
Australia	AusAID	31282	36	FORESTRY RESEARCH	2007001538
Austria	ADA	31130	250	NGO COFINANCED PROJECT: FARMWORKERS SECURITY - LAND RIGHTS SECURITY	2006000022ce
Belgium	DGCD	31110	130	AGRICULTURAL POLICY & ADMIN. MGMT	2004003272
Belgium	MPRF	31110	41	AGRICULTURAL POLICY & ADMIN. MGMT	2007008018
Belgium	DGCD	31163	173	INTEGRATION PETITS AGRICULTEURS MARGINALISES DANS LE SECTEUR AGRICOLE	2004003471
Belgium	DGCD	31164	476	ACCELERER LA RESTITUTION DES TERRES	2005000843
Belgium	DGCD	31164	68	AGRARIAN REFORM	2006003350
Belgium	DGCD	31166	102	AGRICULTURAL EXTENSION	2006003349
Belgium	MPRF	31181	28	AGRICULTURAL EDUCATION/TRAINING	2006008021
Belgium	DGCD	31182	441	IMT - DEPARTMENT OF VETERINARY TROPICAL DISEASES	2005008313
Belgium	DGCD	31194	45	AGRICULTURAL CO-OPERATIVES	2004003213
Canada	CIDA	31162	125	VS/ORGANIC FERTILIZER PRODUCT FACILITY	070174
Canada	IDRC	31182	100	IMPACTING AGRICULTURAL RESEARCH IN AFRICA: THE CGIAR PLAN FOR ACTION	070693i
EU institutions	EDF	31120	1369	AGRICULTURAL DEVELOPMENT	2007201210
France	MISC	31182	16194	RECHERCHE AGRICULTURE	2007800337
Germany	BMZ	31120	684	SUPPORTING LAND REFORM PROCESSES IN THE SOUTHERN CAPE	2007005593
Germany	BMZ	31165	294	AGRICULTURAL ALTERNATIVE DEVELOPMENT	2007007683
Germany	BMZ	31166	315	AGRICULTURAL EXTENSION	2007006201
Germany	BMZ	31281	63	FORESTRY EDUCATION/TRAINING	2007007538
Italy	LA	31120	21	AGRICULTURAL DEVELOPMENT	070992
Japan	JICA	31110	310	TC AGGREGATED ACTIVITIES	072912T
Japan	JICA	31210	55	TC AGGREGATED ACTIVITIES	073362T
Netherlands	MFA	31110	316	AGRICULTURAL POLICY & ADMIN. MGMT	2007100136
Norway	NORAD	31150	65	MARKETING SUPPORT FOR GOODS IMPORTED FROM SOUTH AFRICA TO NORWAY	2007004839
Norway	MFA	31166	7	AGRICULTURAL EXTENSION	2007002598
Norway	MFA	31191	123	AGRICULTURAL SERVICES	2007002834
Norway	NORAD	31192	37	PLANT/POST-HARVEST PROT. & PEST CTRL	2007004852
United States	AID	31110	196	AGRICULTURAL POLICY & ADMIN. MGMT	2007013871
United States	AID	31120	467	AGRICULTURAL DEVELOPMENT	2007001736
United States	Misc	31382	25	FISHERY RESEARCH	2007024287

Total aid for water in 2007 for South Africa   **USD thousand 23160**
And as a share of aid to total recipient countries   **0.35%**

South America, regional

Donor	Agency	Sector	Amount	Project description	CRS ID
Belgium	DGCD	31163	106	LIVESTOCK	2004003279
Belgium	DGCD	31193	171	FONDS DE GARANTIE POUR L'AMÉRIQUE LATINE	2004003376
Belgium	DGCD	31194	6	AGRICULTURAL CO-OPERATIVES	2005001103
Finland	MFA	31210	10	STUDY OF FOREST HARVESTING IN ARGENTINA AND URUGUAY	2007006015
France	MISC	31181	38	ÉDUCATION & FORMATION DANS LE DOMAINE AGRICOLE	2007800792
Germany	BMZ	31120	1538	LOCAL ECONOMICAL DEVELOPMENT IN RURAL REGIONS, BOLIVIA, ECUADOR, PERU	2007005721
Germany	BMZ	31182	64	AGRICULTURAL RESEARCH	2007007645
Italy	DGCS	31120	581	AGRICULTURAL DEVELOPMENT	060344
New Zealand	NZAid	31120	25	CIP ANDEAN PROJECT	070390
Spain	AGR	31120	32	AGRICULTURAL DEVELOPMENT	072652
Spain	AG	31161	65	WORKING TOGETHER FOR THE RIGHT TO A DECENT LIFE SMALL PRODUCERS OF COFFEE	073763
Spain	AGR	31166	38	RURAL AND AGRARIAN EXTENSION COURSE	072866

RECIPIENT Donor	Agency	Sector	Amount USD thousand	Project description	CRS ID Number
Spain	AGR	31194	27	COURSE ON AGRARIAN COOPERATION	072417
Switzerland	SDC	31110	183	PROGRAMME D'ACTIVITéS RéSEAU EVALUATION CONTROLLING	2007000401
Switzerland	seco	31110	165	E-WASTE RECYCLING	2007001909
Switzerland	SDC	31120	95	AGRICULTURAL DEVELOPMENT	2007000322
Switzerland	SDC	31120	209	ANNéE INTERNATIONALE DE LA POMME DE TERRE 2008	2007000279

Total aid for water in 2007 for South America, regional   **USD thousand 3355**
And as a share of aid to total recipient countries   **0.05%**

South Asia, regional

Austria	ADA	31120	892	REGIONAL GRASSLANDS PROGRAMME HINDUKUSH-HIMALAYA, PHASE III	2007500030
Belgium	ASPF	31382	219	ACCORDS DE COOPERATION BILATERALE	2007005493
Germany	DEG	31162	1904	EQUITY INVESTMENT	2007001539

Total aid for water in 2007 for South Asia, regional   **USD thousand 3015**
And as a share of aid to total recipient countries   **0.04%**

South of Sahara, regional

Austria	ADA	31181	22	INTERNATIONAL TRAINING COURSE ON ORGANIC AGRICULTURE - SUMMERSCHOOL	2007500121
Austria	ADA	31220	664	BUSINESS PARTNERSHIP PROGR.: ALLANBLACKIA CULTIV. BY SMALL SCALE FARMERS	2007500109c
Belgium	DGCD	31140	1552	INITIATIVES POUR LA SECURITE ALIMENTAIRE	2005007006
Belgium	DGCD	31161	87	FOOD CROP PRODUCTION	2005001036
Belgium	DGCD	31195	182	LIVESTOCK/VETERINARY SERVICES	2005001056
Denmark	MFA	31182	549	AGRICULTURAL RESEARCH	071370
Denmark	MFA	31182	1510	BIOSAFETRAIN: BIOSAFETY OF GENETICALLY MODIFIED CROPS IN EAST AFRICAII	071341
Denmark	MFA	31182	919	WATERMELONS FOR LIFE: POTENTIAL OF INDIGENOUS AFRICAN GENETIC RESOURCES	071387
Denmark	MFA	31182	919	ENHANCEMENT OF CAPACITY TO CONTROL AND MANAGE BACTERIAL PLANT DISEASES	071347
EU institutions	CEC	31182	5065	AGRICULTURAL RESEARCH	2007100359
Finland	MFA	31182	1478	AGRICULTURAL RESEARCH	2007000097
France	MISC	31110	236	POLITIQUE AGRICOLE & GESTION ADMINISTRATIVE	2007800773
France	MISC	31181	1228	éDUCATION & FORMATION DANS LE DOMAINE AGRICOLE	2007800791
Germany	BMZ	31110	35	AGRICULTURAL POLICY & ADMIN. MGMT	2007008839
Germany	BMZ	31120	12	AGRICULTURAL DEVELOPMENT	2007008840
Germany	Fed Min	31161	684	FAIR TRADE AND ECO-CERTIFICATION-PROJECT (PHASE II OF PROJECT 2003-10)	2007010302
Germany	BMZ	31181	57	AGRICULTURAL EDUCATION/TRAINING	2007008841
Germany	L G	31195	23	LIVESTOCK/VETERINARY SERVICES	2007010971
IDA		31110	9300	NIGER BASIN WATER RESOURCES	2007000482
IDA		31110	9000	AGRICULTURE	2007000502
IDA		31140	9300	NIGER BASIN WATER RESOURCES	2007000480
IDA		31166	3900	WEST AFRICA BIOSAFETY	2007000503
IDA		31166	36000	AGRICULTURE	2007000500
Ireland	DFA	31120	277	AGRICULTURAL DEVELOPMENT	2007000602
Italy	LA	31182	130	AGRICULTURAL RESEARCH	071202
Japan	MAFF	31110	540	TC AGGREGATED ACTIVITIES	070003T
Japan	MAFF	31130	160	TC AGGREGATED ACTIVITIES	070013T
Japan	MAFF	31191	108	TC AGGREGATED ACTIVITIES	070024T
Netherlands	MFA	31110	70	AGRIC.: KIG-IFDC/CATALIST	2006000668

RECIPIENT Donor	Agency	Sector	Amount USD thousand	Project description	CRS ID Number
Norway	MFA	31110	341	QME/FOOD SECURITY, HORN OF AFRICA	2007000876
Norway	NORAD	31194	187	NATURE-ESAANET: EAST AND SOUTH AFRICA AGRICULTURAL NETWORK	2007003791
Spain	MFA	31110	1369	IMPROVE FOOD SECURITY IN SAHEL	077852
Spain	AG	31110	171	STRENGTHENING THE PEASANTS/COUNTRYMEN PLATFORM	075576
Spain	AG	31161	66	STRENGTHENING THE ORGANIZATION OF THE AFRICA NETWORK OF SMALL FAIR TRADE	073778
Spain	MFA	31381	70	ADVANCED NAUTICAL SEMINAR ON SECURITY AND HYGIENE OF FISH PRODUCTS.	073984
Switzerland	SDC	31182	42	AGRICULTURAL RESEARCH	1983000135
Switzerland	SDC	31182	875	REGIONAL RESEARCH BEANS & MAIZE EAST AFRICA	1983000121
Switzerland	SDC	31182	163	ICIPE BIOLOGICAL CONTROL OF MAIZE PESTS	2007000174
Switzerland	SDC	31182	1900	CIMMYT DROUGHT TOLERANCE ON MAIZE	1996001455
Switzerland	SDC	31220	121	RIGHTS AND RESOURCES INITIATIVE	2007000272
Switzerland	SDC	31220	104	FORESTRY DEVELOPMENT	2007000271
United Kingdom	DFID	31110	700	FOOD AGRICULTURE & NATURAL RESOURCES POLICY ANALYSIS NETWORK	2007000803
United States	AID	31110	1525	AGRICULTURAL POLICY & ADMIN. MGMT	2007013860
United States	AID	31110	15019	AGRICULTURAL ENABLING ENVIRONMENT	2007001640
United States	AID	31110	3164	PROGRAM SUPPORT (AGRICULTURE)	2007013849
United States	TDA	31110	9	AGRICULTURAL POLICY & ADMIN. MGMT	2007028205
United States	AID	31120	22241	AGRICULTURAL SECTOR PRODUCTIVITY	2007001725
United States	AID	31120	250	AGRICULTURAL DEVELOPMENT	2007001707
United States	AGR	31181	290	AGRICULTURAL EDUCATION/TRAINING	2007000453
United States	AGR	31210	40	FORESTRY POLICY & ADMIN. MANAGEMENT	2007000473
United States	AID	31220	104	FORESTRY DEVELOPMENT	2007015377

Total aid for water in 2007 for South of Sahara, regional   **USD thousand 132759**
And as a share of aid to total recipient countries   **1.98%**

Sri Lanka

RECIPIENT Donor	Agency	Sector	Amount USD thousand	Project description	CRS ID Number
Australia	AusAID	31192	14	MANAGEMENT OF POST HARVEST DISEASE	2007001894
Australia	AusAID	31210	563	SRI LANKA NATURAL RESOURCE MANAGEMENT	2007000346
Austria	BReg	31182	82	TSUNAMI - RESEARCH PROJECT FOR THE RECREATION OF FLOOD DESTROYED SOIL	2007568307
Canada	CIDA	31161	124	VS/MUSHROOM CULTIVATION FACILITY	070151
Canada	CIDA	31310	1117	TSUNAMI NARA CAPACITY ENHANCEMENT	070640
Canada	CIDA	31320	1117	INT'L FUND FOR AGRICULTURAL DEVELOPMENT - TSUNAMI 04 - IFAD	070036
Germany	BMZ	31193	39	AGRICULTURAL FINANCIAL SERVICES	2007001848
Germany	L G	31310	5	FISHING POLICY AND ADMIN. MANAGEMENT	2007011661
Greece	YPEJ	31391	1972	UNOPS-CONSTRUCTION OF FISHING PORTS AT DODANDUWA & NEGOBO IN SRI LANKA	2007000741
Greece	YPEJ	31391	369	REHABILITATION OF ANURADHAPURA BAZZAR AT TRINCONMALEE OF SRI LANKA	2007000802
Greece	YPEJ	31391	371	FISHERY SERVICES	2007000273
Italy	DGCS	31161	997	FOOD CROP PRODUCTION	070544
Japan	JICA	31110	3681	TC AGGREGATED ACTIVITIES	073063T
Japan	JICA	31130	101	TC AGGREGATED ACTIVITIES	073108T
Japan	MOFA	31150	2801	THE ASSISTANCE FOR UNDERPRIVILEGED FARMERS	070089
Japan	JICA	31150	198	TC AGGREGATED ACTIVITIES	073165T
Japan	JICA	31163	198	TC AGGREGATED ACTIVITIES	073219T
Japan	JICA	31195	301	TC AGGREGATED ACTIVITIES	073291T
Japan	JICA	31210	115	TC AGGREGATED ACTIVITIES	073418T
Japan	JICA	31310	181	TC AGGREGATED ACTIVITIES	073570T

RECIPIENT	Agency	Sector	Amount USD thousand	Project description	CRS ID Number
Donor					
Korea	KOICA	31120	24	AGRICULTURAL DEVELOPMENT	2007016591
Korea	Misc	31120	10	AGRICULTURAL DEVELOPMENT	2007001933
Korea	KOICA	31130	63	AGRICULTURAL LAND RESOURCES	2007016231
Korea	KOICA	31181	10	AGRICULTURAL EDUCATION/TRAINING	2007015878
Korea	Misc	31181	5	AGRICULTURAL EDUCATION/TRAINING	2007002572
Netherlands	MFA	31120	1017	AGRICULTURAL DEVELOPMENT	2007000737
Norway	MFA	31120	48	AGRICULTURAL DEVELOPMENT	2007002617
Norway	NORAD	31162	26	INDUSTRIAL CROPS/EXPORT CROPS	2007004260
Norway	NORAD	31194	46	STRENGTHENING FARMERS' INSTITUTIONS	2007003402
Norway	MFA	31382	172	COMMERCIALISATION OF PRODUCTS FROM FISH	2007004281
Norway	NORAD	31391	83	FISHERY SERVICES	2007004298
Spain	MFA	31320	247	DEVELOPMENT OF RURAL AQUACULTURE AND RENT GENERATING AQUACULTURAL ACTI.	076159

Total aid for water in 2007 for Sri Lanka — **USD thousand 16096**
And as a share of aid to total recipient countries — **0.24%**

St. Lucia

Japan	JICA	31163	117	TC AGGREGATED ACTIVITIES	073207T
Japan	JICA	31310	630	TC AGGREGATED ACTIVITIES	073505T

Total aid for water in 2007 for St. Lucia — **USD thousand 747**
And as a share of aid to total recipient countries — **0.01%**

St.Vincent & Grenadines

Japan	JICA	31110	62	TC AGGREGATED ACTIVITIES	072961T
Japan	JICA	31210	10	TC AGGREGATED ACTIVITIES	073327T
Japan	MOFA	31391	7428	THE PROJECT FOR THE CONSTRUCTION OF OWIA FISHERY CENTER	070169

Total aid for water in 2007 for St.Vincent & Grenadines — **USD thousand 7500**
And as a share of aid to total recipient countries — **0.11%**

States Ex-Yugoslavia

Italy	LA	31120	33	AGRICULTURAL DEVELOPMENT	071526
Luxembourg	MFA	31120	90	AGRICULTURAL DEVELOPMENT	070099
United States	AGR	31110	2100	STATE SUPPORT FOR EAST EUROPEAN DEMOCRACY	2007000472

Total aid for water in 2007 for States Ex-Yugoslavia — **USD thousand 2223**
And as a share of aid to total recipient countries — **0.03%**

Sudan

Canada	IDRC	31120	98	MANAGING RISK, REDUCING VULNERABILITY AND ENHANCING PRODUCTIVITY	070313i
Finland	FF	31220	214	ODA EQUITY THROUGH FINNFUND	2007005001
Ireland	DFA	31120	871	SUPPORT TO NGOS - IRISH - MAPS - GENERAL - A LIVELIHOOD SECURITY PROG.	2007000219
Ireland	DFA	31150	586	SUPPORT TO NGOS - IRISH - MAPS - GENERAL - PRIMARY HEALTH CARE SEVICES	2007000231
Ireland	DFA	31161	460	SUPPORT TO NGOS - IRISH - MAPS - GENERAL - SUPPORT TO IMPROVE AGRICULTURE	2007000237
Ireland	DFA	31163	62	LIVESTOCK	2007001429
Ireland	DFA	31166	84	AGRICULTURAL EXTENSION	2007001262

RECIPIENT Donor	Agency	Sector	Amount USD thousand	Project description	CRS ID Number
Ireland	DFA	31181	64	AGRICULTURAL EDUCATION/TRAINING	2007001419
Italy	CA	31120	189	AGRICULTURAL DEVELOPMENT	071312
Italy	DGCS	31181	624	AGRICULTURAL EDUCATION/TRAINING	070665
Japan	MOFA	31110	1613	THE ASSISTANCE FOR UNDERPRIVILEGED FARMERS	070115
Japan	JICA	31163	62	TC AGGREGATED ACTIVITIES	073197T
Japan	JICA	31210	8	TC AGGREGATED ACTIVITIES	073312T
Japan	JICA	31310	47	TC AGGREGATED ACTIVITIES	073556T
Korea	KOICA	31320	339	TRAINING PROGRAM	2007017363
Norway	MFA	31120	1980	NPA FOOD SECURITY AND LIVELIHOOD 2007	2007004875
Norway	NORAD	31130	162	AGRICULTURAL LAND RESOURCES	2007004750
Norway	NORAD	31140	256	UM JAWASIR COMMUNITY DEVELOPMENT PROJECT	2007003797
Norway	MFA	31210	4950	THE SOUTHERN SUDAN FOREST SECTOR PROGRAMME 2007-2009	2007004872
Spain	MFA	31120	548	AGRICULTURAL DEVELOPMENT	077391
Switzerland	SDC	31120	583	VSF - LIVELIHOOD SUPPORT SOUTHERN SUDAN	2007000080
United States	AID	31110	5000	AGRICULTURAL ENABLING ENVIRONMENT	2007001641
United States	AID	31110	406	AGRICULTURAL POLICY & ADMIN. MGMT	2007013840
United States	AID	31120	3000	AGRICULTURAL SECTOR PRODUCTIVITY	2007001723

Total aid for water in 2007 for Sudan   **USD thousand 22206**
And as a share of aid to total recipient countries   **0.33%**

Suriname

RECIPIENT Donor	Agency	Sector	Amount USD thousand	Project description	CRS ID Number
EU institutions	EDF	31140	1888	AGRICULTURAL WATER RESOURCES	2007201290
Japan	JICA	31310	202	TC AGGREGATED ACTIVITIES	073513T
Japan	MOFA	31391	6935	THE PROJECT FOR CONSTRUCTION OF SMALL-SCALE FISHERIES CENTER	070019
Netherlands	MFA	31110	2886	AGRICULTURAL POLICY & ADMIN. MGMT	2007100151

Total aid for water in 2007 for Suriname   **USD thousand 11912**
And as a share of aid to total recipient countries   **0.18%**

Swaziland

RECIPIENT Donor	Agency	Sector	Amount USD thousand	Project description	CRS ID Number
EU institutions	EDF	31162	20390	ANNUAL ACTION PROGRAMME 2007 - ACCOMPANYING MEASURES TO SUGAR PROTOCOL	2007201266
Italy	LA	31120	41	AGRICULTURAL DEVELOPMENT	071554
United States	ADF	31120	54	AGRICULTURAL DEVELOPMENT	2007000196

Total aid for water in 2007 for Swaziland   **USD thousand 20485**
And as a share of aid to total recipient countries   **0.31%**

Syria

RECIPIENT Donor	Agency	Sector	Amount USD thousand	Project description	CRS ID Number
France	MISC	31110	8	POLITIQUE AGRICOLE & GESTION ADMINISTRATIVE	2007800779
France	MISC	31182	1218	RECHERCHE AGRICULTURE	2007800375
IFAD		31120	20135	NORTH-EASTERN REGION RURAL DEVELOPMENT PROJECT	070014
Italy	DGCS	31161	344	FOOD CROP PRODUCTION	061047
Japan	JICA	31110	318	TC AGGREGATED ACTIVITIES	072979T
Japan	JICA	31130	912	TC AGGREGATED ACTIVITIES	073098T
Japan	JICA	31163	42	TC AGGREGATED ACTIVITIES	073214T
Japan	JICA	31195	112	TC AGGREGATED ACTIVITIES	073268T

Total aid for water in 2007 for Syria   **USD thousand 23089**

RECIPIENT Donor	Agency	Sector	Amount USD thousand	Project description	CRS ID Number
			And as a share of aid to total recipient countries	**0.34%**	

Tajikistan

AsDF		31120	9345	RURAL DEVELOPMENT PROJECT	070040
AsDF		31162	5852	SUSTAINABLE COTTON SUBSECTOR PROJECT	070008
Canada	CIDA	31164	1117	ENHANCING AGRIC. GOVERNANCE IN TAJIK	070081
Canada	CIDA	31191	707	TUGCE & EXPORT MARKET FACILITATION	070082
Germany	BMZ	31130	255	AGRICULTURAL LAND RESOURCES	2007009185
Germany	BMZ	31166	115	AGRICULTURAL EXTENSION	2007009183
IDA		31161	1050	COTTON SECTOR	2007000720
IDA		31162	5850	COTTON SECTOR	2007000722
IDA		31166	300	COTTON SECTOR	2007000719
Japan	JICA	31110	290	TC AGGREGATED ACTIVITIES	073059T
Japan	JICA	31130	23	TC AGGREGATED ACTIVITIES	073103T
Japan	JICA	31150	39	TC AGGREGATED ACTIVITIES	073180T
Japan	JICA	31163	36	TC AGGREGATED ACTIVITIES	073215T
Norway	MFA	31120	11	AGRICULTURAL DEVELOPMENT	2007002574
Norway	MFA	31130	82	Reduce ongoing degradation of agricultural land	2007002041
Norway	MFA	31220	266	VOCATIONAL TRAINING, FORESTRY/ENVIRONMENTAL SECTOR	2007002026
Sweden	Sida	31120	3848	AGRICULTURAL DEVELOPMENT	2007005531
Sweden	Sida	31195	3108	LIVESTOCK/VETERINARY SERVICES	2007005532
United States	AID	31110	271	AGRICULTURAL POLICY & ADMIN. MGMT	2007015293
United States	AID	31110	1156	AGRICULTURAL ENABLING ENVIRONMENT	2007001667
United States	AID	31120	196	AGRICULTURAL DEVELOPMENT	2007001788
United States	AID	31140	953	LAND AND WATER MANAGEMENT	2007010928
United States	AID	31162	428	INDUSTRIAL CROPS/EXPORT CROPS	2007011453
United States	AID	31181	11	AGRICULTURAL EDUCATION/TRAINING	2007001033
United States	AID	31182	428	AGRICULTURAL RESEARCH	2007015636

Total aid for water in 2007 for Tajikistan   **USD thousand 35736**
And as a share of aid to total recipient countries   **0.53%**

Tanzania

AfDF		31110	61209	AGRICULTURE SECTOR DEVELOPMENT PROGRAMME	070047
Austria	MISC	31110	14	SMALL-SCALE COMMITMENTS AGGREGATED BY SECTOR AND RECIPIENT COUNTRY	2007980120
Austria	Reg	31120	14	PROJECT FOR AGRICULTURAL DEVELOPMENT IN SONGEA	2007828578
Austria	Reg	31120	16	PROJECT OF AGRICULTURAL DEVELOPMENT IN IRINGA AND KILOLO MAFINGA	2007828575
Austria	ADA	31120	73	NGO COFINANCED PROJECT: RIIP - PROGR. FOR AGRICULTURAL DEVELOPMENT	2006000022bi
Austria	Reg	31150	25	AGRICULTURAL EQUIPMENT FOR FARMERS IN MUSOMA	2007828542
Austria	ADA	31161	203	NGO COFINANCING: VIFAFI - VICTORIA FARMING AND FISHING PROJECT	2006000022cn
Austria	ADA	31162	274	BUSINESS PARTNERSHIP PROGRAM: SUSTAIN. COFFEE PRODUCTION & PROCESSING	2007500109e
Austria	ADA	31166	113	NGO COFINANCING: RURAL INCOME IMPROVEMENT BIHARAMULO & KARAGWE DISTRICTS	2007000021ax
Austria	ADA	31191	165	NGO COFINANCED: MARA - MSHP II - MARKETING FOR SMALL FARMERS	2006000022cj
Austria	ADA	31191	16	NGO-VOLUNTEERS: MARKETING AND SALES OF AGRICULTURAL PRODUCTS - LAKE ZONE	2007500160dc
Belgium	DGCD	31110	624	RENFORCEMENT ET SOUTIENT DES INITIATIVES SOCIAUX-ECONOMIQUES	2004003470
Belgium	DGCD	31120	274	INITIATIVES FOR IMPROVEMENT OF SOCIO-ECONOMICAL ENVIRONMENT OF RURAL COM.	2002000624
Belgium	DGCD	31120	471	FINANCEMENT PARTENAIRES TANZANIE	2006003969

RECIPIENT Donor	Agency	Sector	Amount USD thousand	Project description	CRS ID Number
Belgium	DGCD	31161	372	SÉCURITÉ ALIMENTAIRE DURABLE POUR LE DISTRICT DE CHUNYA	2005007007
Belgium	DGCD	31163	226	POVERTY REDUCTION & SOCIAL-ECONOMIC EMPOWERMENT OF PASTORAL WOMEN & MEN	2004003787
Belgium	DGCD	31182	41	AGRICULTURAL RESEARCH	2006006600
Belgium	DGCD	31192	40	PLANT/POST-HARVEST PROT. & PEST CTRL	2005000194
Canada	IDRC	31110	173	LOCAL AGRICULTURAL INNOVATION SYSTEMS IN LESS FAVORED & HIGH POTENTIAL AREAS	070307i
Canada	IDRC	31120	98	MANAGING RISK, REDUCING VULNERABILITY AND ENHANCING PRODUCTIVITY	070314i
Denmark	MFA	31210	3307	VILLAGE BASED FOREST MANAGEMENT	071291
Denmark	MFA	31210	919	IDENPENDENT FOREST MONITORING	071564
Denmark	MFA	31282	918	PARTICIPATORY FOREST MANAGEMENT (PFM) FOR RURAL LIVELIHOODS FOREST	071389
Finland	FF	31220	1326	ODA LOAN THROUGH FINNFUND	2003004006
Finland	MFA	31220	149	NGO SUPPORT / VILLAGE FORESTRY PROMOTION PORJECT, II PHASE	2004003437
Germany	BMZ	31161	48	FOOD CROP PRODUCTION	2007007429
IDA		31110	2275	LOWER KIHANSI ENV. MNGT	2007000423
Ireland	DFA	31110	61	AGRICULTURAL POLICY & ADMIN. MGMT	2007003002
Ireland	DFA	31110	554	TANZANIA COUNTRY PROGRAMME - AGRICULTURE FORESTRY AND FISH	2007000217
Ireland	DFA	31110	3323	TANZANIA COUNTRY PROGRAMME - AGRICULTURE FORESTRY AND FISH	2007000214
Ireland	DFA	31110	692	TANZANIA COUNTRY PROGRAMME - AGRICULTURE FORESTRY AND FISH	2007000216
Italy	DGCS	31120	34	AGRICULTURAL DEVELOPMENT	060677
Italy	LA	31120	66	AGRICULTURAL DEVELOPMENT	070982
Italy	DGCS	31166	518	AGRICULTURAL EXTENSION	070375
Japan	JICA	31110	2301	TC AGGREGATED ACTIVITIES	073031T
Japan	JICA	31130	133	TC AGGREGATED ACTIVITIES	073124T
Japan	JICA	31163	31	TC AGGREGATED ACTIVITIES	073198T
Japan	JICA	31195	26	TC AGGREGATED ACTIVITIES	073288T
Japan	JICA	31210	8	TC AGGREGATED ACTIVITIES	073314T
Japan	JICA	31310	22	TC AGGREGATED ACTIVITIES	073489T
Japan	MAFF	31320	374	TC AGGREGATED ACTIVITIES	070054T
Korea	KOICA	31140	1700	PROJECT FOR REHABILITATION OF THE IRRIGATION FACILITIES AND MODERNIZATIO	2007017684
Korea	KOICA	31150	22	AGRICULTURAL INPUTS	2007017071
Korea	KOICA	31181	54	AGRICULTURAL EDUCATION/TRAINING	2007016715
Korea	KOICA	31281	9	FORESTRY EDUCATION/TRAINING	2007016174
Netherlands	MFA	31110	906	AGRIC.: SUBS. TO PSOM 2003-2009	2007100137
Norway	NORAD	31110	20	AGRICULTURAL POLICY & ADMIN. MGMT	2007005014
Norway	MFA	31120	98	FOOD SECURITY, SUSTAINABLE NATURAL RESOURCES MANAGEMENT AND FAIR TRADE	2007002506
Norway	MFA	31120	97	AGRICULTURAL DEVELOPMENT	2007002619
Norway	MFA	31162	29	INDUSTRIAL CROPS/EXPORT CROPS	2007005005
Norway	NORAD	31191	102	SUPPORT TO THE AGRICULTURAL COUNCIL OF TANZANIA - ACTION PLAN	2007005001
Norway	NORAD	31194	307	NATURE-BRITA: BUILDING RURAL INCOME THROUGH FARMERS' ASSOCIATIONS	2007003789
Norway	MFA	31210	15	REPRINT TRAFFIC REPORT	2007005036
Norway	MFA	31261	6	FUELWOOD/CHARCOAL	2007005003
United States	AID	31110	486	AGRICULTURAL POLICY & ADMIN. MGMT	2007013847
United States	AID	31120	825	AGRICULTURAL SECTOR PRODUCTIVITY	2007001726
United States	AID	31120	52	AGRICULTURAL DEVELOPMENT	2007009838

Total aid for water in 2007 for Tanzania   **USD thousand 86253**
And as a share of aid to total recipient countries   **1.29%**

RECIPIENT Donor	Agency	Sector	Amount USD thousand	Project description	CRS ID Number
Thailand					
Australia	AusAID	31161	23	FOOD CROP PRODUCTION	2007001724
Australia	AusAID	31182	113	MIGRATION AND/OR OFF-FARM EMPLOYMENT :WOMEN AND APPROPRIATE TECHNOLOGIES	2007001654
Australia	AusAID	31182	38	ADAPTATION OF LOW-CHILL TEMPERATE FRUITS	2007001510
Australia	AusAID	31182	8	AGRICULTURAL RESEARCH	2007002046
Australia	AusAID	31192	14	PUBLIC SECTOR LINKAGES PROGRAM, THAILAND BUILDING	2007002136
Australia	AusAID	31195	13	BREEDING IMPROVEMENT PROGRAM FOR HIGH QUALITY TROP	2007001916
Australia	AusAID	31195	18	LABORATORY ACT AS THE REFERENCE CENTRE FOR THE SOUTHEAST ASIAN FOOT	2007001811
Australia	AusAID	31210	27	FORESTRY POLICY & ADMIN. MANAGEMENT	2007001661
Australia	AusAID	31220	73	PSLP CAPACITY BUILDING FOR ROYAL FOREST DEPARTMENT	2007001168
Australia	AusAID	31282	32	FORESTRY RESEARCH	2007001926
Australia	AusAID	31310	8	MECHANISMS TO MAXIMISE BENEFITS TO SMALL-HOLDER SHRIMP FARMER GROUPS	2007002069
Australia	AusAID	31320	164	FISHERY DEVELOPMENT	2007001928
Australia	AusAID	31382	77	FISHERY RESEARCH	2007002008
Australia	AusAID	31382	17	GILL-ASSOCIATED VIRUS AND CONSTRUCTION OF A FARMER'S DECISION TREE	2007001843
Canada	IDRC	31210	28	INTERNATIONAL CONFERENCE ON POVERTY REDUCTION AND FORESTS	070675i
EU institutions	EDF	31110	22	AGRICULTURAL POLICY & ADMIN. MGMT	2007201665
France	MISC	31110	13	POLITIQUE AGRICOLE & GESTION ADMINISTRATIVE	2007800785
France	MISC	31182	18412	RECHERCHE AGRICULTURE	2007800383
Germany	BMZ	31195	25	LIVESTOCK/VETERINARY SERVICES	2007007390
Germany	BMZ	31310	101	FISHING POLICY AND ADMIN. MANAGEMENT	2007007382
Italy	DGCS	31320	28	FISHERY DEVELOPMENT	060840
Japan	JICA	31110	2161	TC AGGREGATED ACTIVITIES	073072T
Japan	JICA	31130	51	TC AGGREGATED ACTIVITIES	073134T
Japan	JICA	31150	181	TC AGGREGATED ACTIVITIES	073170T
Japan	JICA	31163	7	TC AGGREGATED ACTIVITIES	073231T
Japan	MAFF	31181	207	TC AGGREGATED ACTIVITIES	070022T
Japan	JICA	31195	242	TC AGGREGATED ACTIVITIES	073281T
Japan	JICA	31210	116	TC AGGREGATED ACTIVITIES	073449T
Japan	JICA	31310	137	TC AGGREGATED ACTIVITIES	073580T
Korea	Misc	31110	9	AGRICULTURAL POLICY & ADMIN. MGMT	2007002553
Korea	Misc	31120	10	AGRICULTURAL DEVELOPMENT	2007001929
Korea	Misc	31181	21	AGRICULTURAL EDUCATION/TRAINING	2007002573
Netherlands	MFA	31110	93	AGRICULTURAL POLICY & ADMIN. MGMT	2007100144
Norway	MFA	31320	1878	MARINE AQUALCULTURE AND FISHERY RESOURCE	2007005054
Norway	NORAD	31391	83	FISHERY SERVICES	2007005052
Spain	MIE	31110	311	IRRIGATION PLAN IN PHITSANULOK	076560

Total aid for water in 2007 for Thailand   **USD thousand 24759**
And as a share of aid to total recipient countries   **0.37%**

Timor-Leste

Australia	AusAID	31120	392	CAPACITY DEVELOPMENT PROGRAM	2007000456
Australia	AusAID	31120	88	AGRICULTURAL DEVELOPMENT	2007001055
Australia	AusAID	31120	8	ANIMAL AND PLANT HEALTH SURVEY	2007002047
Australia	AusAID	31161	38	ADOPTION OF IMPROVED CASSAVA PRODUCTION AND UTILISATION SYSTEMS	2007001523

RECIPIENT Donor	Agency	Sector	Amount USD thousand	Project description	CRS ID Number
Australia	AusAID	31161	107	FOOD CROP PRODUCTION	2007000961
Australia	AusAID	31182	83	AGRICULTURAL RESEARCH	2007001494
Australia	AusAID	31192	23	CAPACITY BUILDING IN CROP PROTECTION IN EASTTIMOR	2007001713
Australia	AusAID	31195	40	LIVESTOCK/VETERINARY SERVICES	2007001488
Germany	BMZ	31120	320	AGRICULTURAL DEVELOPMENT	2007006372
Ireland	DFA	31120	154	AGRICULTURAL DEVELOPMENT	2007000923
Ireland	DFA	31130	494	SUPPORT TO NGOS - IRISH - MAPS - GENERAL - POVERTY REDUCTION OPT & STRATEGY	2007000227
Ireland	DFA	31140	11	AGRICULTURAL WATER RESOURCES	2007002745
Japan	JICA	31110	495	TC AGGREGATED ACTIVITIES	073007T
Japan	JICA	31130	344	TC AGGREGATED ACTIVITIES	073135T
Japan	MOFA	31140	6256	THE PROJECT FOR REHABILITATION AND IMPROVEMENT	070218
Japan	JICA	31163	7	TC AGGREGATED ACTIVITIES	073232T
Japan	JICA	31310	144	TC AGGREGATED ACTIVITIES	073581T
Portugal	ICP	31120	103	AGRICULTURAL DEVELOPMENT IN MANATUTO DISTRICT	070418
Portugal	ICP	31120	198	RURAL DEVELOPMENT PROGRAMME	070320
Spain	MFA	31110	308	POVERTY REDUCTION THROUGH INCREASEMENT OF ECONOMIC CAPACITIES	076547
United States	AID	31110	72	AGRICULTURAL POLICY & ADMIN. MGMT	2007013912
United States	AID	31120	75	AGRICULTURAL DEVELOPMENT	2007001766
United States	AID	31120	2500	AGRICULTURAL SECTOR PRODUCTIVITY	2007001773

Total aid for water in 2007 for Timor-Leste   **USD thousand 12262**
And as a share of aid to total recipient countries   **0.18%**

Togo

Donor	Agency	Sector	Amount	Project description	CRS ID
Belgium	DGCD	31120	417	DéVELOPPEMENT DES CHAINES DES PRODUITS AGRICOLES	2006003974
Belgium	DGCD	31130	179	AMéLIORER LES REVENUS DES AGRICULTEURS ET LA PRODUCTIVITé DES SOLS	2007004695
Belgium	DGCD	31163	85	LIVESTOCK	2006003911
France	MISC	31182	14	RECHERCHE AGRICULTURE	2007800358
Germany	BMZ	31120	616	PROMOTION OF MARKET ACCESS FOR PEASANTS IN TOGO AND BENIN	2007006647
Italy	LA	31120	27	AGRICULTURAL DEVELOPMENT	070904
Japan	JICA	31110	65	TC AGGREGATED ACTIVITIES	073032T
Japan	JICA	31130	57	TC AGGREGATED ACTIVITIES	073088T
Spain	MFA	31310	46	CONTINENTAL AQUACULTURE DEVELOPMENT PROGRAM.	073167

Total aid for water in 2007 for Togo   **USD thousand 1506**
And as a share of aid to total recipient countries   **0.02%**

Tonga

Donor	Agency	Sector	Amount	Project description	CRS ID
Australia	AusAID	31110	10	A REVIEW OF THE POLICY AND ECONOMIC ENVIRONMENT IN THE SOUTH PACIFIC	2007001994
Australia	AusAID	31162	95	INDUSTRIAL CROPS/EXPORT CROPS	2007001427
Australia	AusAID	31163	84	USING LOCAL FEEDS TO REDUCE THE COST OF PIG AND POULTRY PRODUCTION IN TONGA	2007001822
Australia	AusAID	31192	91	PLANT/POST-HARVEST PROT. & PEST CTRL	2007001038
Australia	AusAID	31320	6	AQUACULTURE RESEARCH PRIORITIES	2007002143
Australia	AusAID	31382	78	FISHERY RESEARCH	2007001135
Japan	JICA	31110	13	TC AGGREGATED ACTIVITIES	073015T
Japan	JICA	31210	25	TC AGGREGATED ACTIVITIES	073359T
Japan	JICA	31310	51	TC AGGREGATED ACTIVITIES	073539T

RECIPIENT Donor	Agency	Sector	Amount USD thousand	Project description	CRS ID Number
			Total aid for water in 2007 for Tonga		**USD thousand 453**
			And as a share of aid to total recipient countries		**0.01%**
Trinidad and Tobago					
EU institutions	EDF	31162	8214	INDUSTRIAL CROPS/EXPORT CROPS	2007201268
France	MISC	31182	3258	RECHERCHE AGRICULTURE	2007800369
			Total aid for water in 2007 for Trinidad and Tobago		**USD thousand 11472**
			And as a share of aid to total recipient countries		**0.17%**
Tunisia					
Belgium	DGCD	31140	118	AGRICULTURAL WATER RESOURCES	2006003917
Canada	IDRC	31120	1117	RAINWATER AND GREYWATER HARVESTING IN URBAN AND PERI-URBAN AGRICULTURE	070495i
France	MISC	31110	36	POLITIQUE AGRICOLE & GESTION ADMINISTRATIVE	2007800768
France	MISC	31182	1561	RECHERCHE AGRICULTURE	2007800336
Germany	KFW	31140	2422	AGRICULTURAL WATER RESOURCES	2007001375
Germany	KFW	31140	20998	MOD. DES PPI DE LA BASSE VALLEE DE LA M.	2007001373
Italy	DGCS	31120	1535	AGRICULTURAL DEVELOPMENT	060918
Italy	DGCS	31140	97	AGRICULTURAL WATER RESOURCES	060082
Italy	DGCS	31191	126	AGRICULTURAL SERVICES	060076
Italy	DGCS	31320	342	FISHERY DEVELOPMENT	070333
Japan	JICA	31130	32	TC AGGREGATED ACTIVITIES	073076T
Japan	JBIC	31140	44652	WATER SAVING AGRICULTURE PROJECT IN SOUTHERN OASIS AREA PROJECT	073051
Japan	JICA	31310	1150	TC AGGREGATED ACTIVITIES	073545T
Korea	KOICA	31110	50	AGRICULTURAL POLICY & ADMIN. MGMT	2007017735
Korea	KOICA	31181	24	AGRICULTURAL EDUCATION/TRAINING	2007015988
Spain	EMP	31150	25	AGRICULTURAL INPUTS	072333
Spain	Misc	31182	10	VIRIS AND BACTERIA IN THE TOMATO AND PAPRIKA GROWING IN TUNESIA	071127
Spain	MFA	31195	47	VIGILENCE AND CONTROL SYSTEMS OF THE HIGHLY INFECTIOUS BIRD FLU	073197
Spain	MFA	31210	34	FORESTRY POLICY & ADMIN. MANAGEMENT	072778
Spain	MFA	31320	103	SUSTAINABLE DEVELOPMENT OF TRADITIONAL FISHERIES IN THE MEDITERRANEAN SEA	074613
Switzerland	SDC	31110	450	SEMENCES LÉGUMES BIODYNAMIQUES TUNISIE	2005003884
			Total aid for water in 2007 for Tunisia		**USD thousand 74927**
			And as a share of aid to total recipient countries		**1.12%**
Turkey					
EU institutions	EDF	31110	1513	AGRICULTURAL POLICY & ADMIN. MGMT	2007202462
EU institutions	EDF	31192	7973	PLANT/POST-HARVEST PROT. & PEST CTRL	2007202466
EU institutions	EDF	31310	3012	FISHING POLICY AND ADMIN. MANAGEMENT	2007202463
Japan	JICA	31110	716	TC AGGREGATED ACTIVITIES	073019T
Japan	JICA	31163	42	TC AGGREGATED ACTIVITIES	073185T
Japan	JICA	31310	413	TC AGGREGATED ACTIVITIES	073541T
			Total aid for water in 2007 for Turkey		**USD thousand 13668**
			And as a share of aid to total recipient countries		**0.2%**

RECIPIENT / Donor	Agency	Sector	Amount USD thousand	Project description	CRS ID Number
Turkmenistan					
United States	AID	31110	43	AGRICULTURAL POLICY & ADMIN. MGMT	2007015294
United States	STATE	31161	11	FOOD CROP PRODUCTION	2007026852
United States	AID	31162	117	INDUSTRIAL CROPS/EXPORT CROPS	2007011454
United States	AID	31181	34	AGRICULTURAL EDUCATION/TRAINING	2007001034

Total aid for water in 2007 for Turkmenistan USD thousand 205
And as a share of aid to total recipient countries 0%

Tuvalu					
Japan	MAFF	31320	288	TC AGGREGATED ACTIVITIES	070066T
Japan	MAFF	31381	16	TC AGGREGATED ACTIVITIES	070076T

Total aid for water in 2007 for Tuvalu USD thousand 304
And as a share of aid to total recipient countries 0%

Uganda					
AfDF		31110	45907	COMMUNITY AGRICULTURAL INFRASTRUCTURE IMPROVEMENT PROGRAMME PROJECT 1	070051
Austria	ADA	31120	8	NGO-VOLUNTEERS: FARM DEVELOPMENT PROJECT KITAGWENDA	2007500160a
Austria	ADA	31166	192	NGO COFINANCING: ADP IV - TRAINING F. INTEGR. AGRICULTURAL DEVELOPMENT	2007000021a
Belgium	DGCD	31120	549	AMELIORATION DU STANDARD DE VIE DES PAYSANS	2006003968
Belgium	DGCD	31161	240	SECURITE DE SUBSISTANCE DES MENAGES UGANDA DE L'EST FONDS BELGE SURVIE	2005007008
Belgium	DGCD	31182	100	AGRICULTURAL RESEARCH	2006003903
France	MISC	31182	3847	RECHERCHE AGRICULTURE	2007800359
Germany	BMZ	31120	502	AGRICULTURAL DEVELOPMENT	2007006602
IDA		31166	9600	AGRICULTURAL RESEARCH AND TRAINING	2007000439
IDA		31181	2400	AGRICULTURAL RESEARCH AND TRAINING	2007000440
Ireland	DFA	31120	404	AGRICULTURAL DEVELOPMENT	2007003176
Ireland	DFA	31120	500	SUPPORT TO NGOS - IRISH - MAPS - GENERAL - A LIVELIHOOD SECURITY PROG.	2007000222
Italy	CA	31120	319	AGRICULTURAL DEVELOPMENT	071315
Italy	DGCS	31150	23	AGRICULTURAL INPUTS	060789
Italy	DGCS	31161	411	FOOD CROP PRODUCTION	060947
Italy	DGCS	31191	148	AGRICULTURAL RESEARCH	060788
Italy	DGCS	31191	21	AGRICULTURAL SERVICES	060873
Japan	JICA	31110	2314	TC AGGREGATED ACTIVITIES	073034T
Japan	JICA	31130	54	TC AGGREGATED ACTIVITIES	073089T
Japan	JICA	31163	235	TC AGGREGATED ACTIVITIES	073199T
Japan	JICA	31195	421	TC AGGREGATED ACTIVITIES	073256T
Netherlands	MFA	31110	654	AGRIC.: SUBS. TO PSOM 2003-2009	2007100133
Netherlands	MFA	31120	175	AGRICULTURAL DEVELOPMENT	2007000887
Norway	MFA	31120	2680	FAO - PRODUCTIVE AGRICULTURAL LIVELIHOODS AND INCOME SECURITY	2007005108
Norway	MFA	31120	273	AGRICULTURAL DEVELOPMENT	2007002620
Norway	NORFUND	31150	34	JAMBO ROSES UGANDA RECONSTRUCTION	2007004471
Norway	MFA	31165	172	AGRICULTURAL ALTERNATIVE DEVELOPMENT	2007002642
Norway	NORAD	31194	581	NATURE-EFTAF: EMPOWER FARMERS THROUGH AGRIBUSINESS AND FINANCIAL SERVICE	2007003787
Norway	MFA	31210	16	FORESTRY POLICY & ADMIN. MANAGEMENT	2007005106

RECIPIENT Donor	Agency	Sector	Amount USD thousand	Project description	CRS ID Number
Norway	MFA	31281	26	UGANDA FORESTRY HISTORY	2007005082
Norway	NORAD	31320	256	METSER GReNN - FEASIBILITY STUDY	2007005085
Spain	AG	31163	277	LIFE CHAINS AND STRENGTHENING OF THE ASSOCIATIVE NETWORK OF COMMUNITIES	076406
United States	AID	31110	1032	PROGRAM SUPPORT (AGRICULTURE)	2007013872
United States	AID	31110	59	AGRICULTURAL POLICY & ADMIN. MGMT	2007013861
United States	AID	31120	6909	AGRICULTURAL SECTOR PRODUCTIVITY	2007001737
United States	ADF	31182	84	AGRICULTURAL RESEARCH	2007000192

Total aid for water in 2007 for Uganda   **USD thousand 81424**
And as a share of aid to total recipient countries   **1.21%**

Ukraine

Austria	MISC	31110	13	SMALL-SCALE COMMITMENTS AGGREGATED BY SECTOR AND RECIPIENT COUNTRY	2007980044
Germany	BMZ	31110	173	AGRICULTURAL POLICY & ADMIN. MGMT	2007007458
Germany	Fed Min	31110	274	AGRICULTURAL POLICY & ADMIN. MGMT	2007010324
Germany	BMZ	31120	116	AGRICULTURAL DEVELOPMENT	2007007453
Germany	BMZ	31164	65	AGRARIAN REFORM	2007007454
Germany	BMZ	31166	38	AGRICULTURAL EXTENSION	2007007446
Germany	Fed Min	31166	436	UKRAINE-GERMAN CONSULTANCY POLTAWA -UDOP-	2007010325
Germany	L G	31181	43	AGRICULTURAL EDUCATION/TRAINING	2007011970
Japan	JICA	31110	171	TC AGGREGATED ACTIVITIES	072905T
Switzerland	SDC	31220	876	FORZA FLOOD PREVENTION TRANSCARPATHIA	2002002307
United States	AID	31110	161	AGRICULTURAL POLICY & ADMIN. MGMT	2007013942
United States	AID	31120	100	AGRICULTURAL DEVELOPMENT	2007001786

Total aid for water in 2007 for Ukraine   **USD thousand 2466**
And as a share of aid to total recipient countries   **0.04%**

Uruguay

Australia	AusAID	31110	10	ABARE INTERNSHIP - MARTIN OLIVERA	2007002015
Canada	IDRC	31110	106	ENHANCING CAPACITY FOR INNOVATION, INCREASING PROD. AND ACCESS TO MARKETS	070451i
Germany	BMZ	31130	246	AGRICULTURAL LAND RESOURCES	2007006143
Italy	DGCS	31192	247	PLANT AND POST-HARVEST PROTECTION AND PEST CONTROL	070490
Japan	JICA	31110	321	TC AGGREGATED ACTIVITIES	073054T
Japan	JICA	31195	233	TC AGGREGATED ACTIVITIES	073266T
Japan	JICA	31210	85	TC AGGREGATED ACTIVITIES	073411T
Spain 071244	AG	31182	11	COUNSELLING PROGRAM FOR POST-HARVEST TECHNIQUES IN FRUIT PRODUCTION AND VIDAS	

Total aid for water in 2007 for Uruguay   **USD thousand 1260**
And as a share of aid to total recipient countries   **0.02%**

Uzbekistan

AsDF		31130	29226	LAND IMPROVEMENT PROJECT	070004
France	MISC	31110	19	POLITIQUE AGRICOLE & GESTION ADMINISTRATIVE	2007800782
Germany	BMZ	31166	92	AGRICULTURAL EXTENSION	2007009210
Japan	JICA	31110	320	TC AGGREGATED ACTIVITIES	073061T
Japan	JICA	31130	110	TC AGGREGATED ACTIVITIES	073129T

RECIPIENT Donor	Agency	Sector	Amount USD thousand	Project description	CRS ID Number
United States	AID	31110	362	AGRICULTURAL POLICY & ADMIN. MGMT	2007015295
United States	AID	31140	2297	LAND AND WATER MANAGEMENT	2007010929
United States	AID	31162	512	MARKETS AND TRADE CAPACITY	2007011455
United States	AID	31181	282	AGRICULTURAL EDUCATION/TRAINING	2007001037
United States	AID	31182	512	RESEARCH AND TECHNOLOGY DISSEMINATION	2007015637

Total aid for water in 2007 for Uzbekistan   **USD thousand 33732**
And as a share of aid to total recipient countries   **0.5%**

Vanuatu

Australia	AusAID	31192	73	PLANT/POST-HARVEST PROT. & PEST CTRL	2007001692
Australia	AusAID	31282	314	FORESTRY RESEARCH	2007001234
Australia	AusAID	31320	6	AQUACULTURE RESEARCH PRIORITIES	2007002139
Australia	AusAID	31320	10	INTEGRATION OF BROODSTOCK REPLENISHMENT	2007002013
France	MISC	31182	4709	RECHERCHE AGRICULTURE	2007800385
Japan	JICA	31110	166	TC AGGREGATED ACTIVITIES	073011T
Japan	JICA	31195	24	TC AGGREGATED ACTIVITIES	073293T
Japan	JICA	31310	707	TC AGGREGATED ACTIVITIES	073534T
New Zealand	NZAid	31110	132	AGRICULTURAL CENSUS	070372

Total aid for water in 2007 for Vanuatu   **USD thousand 6141**
And as a share of aid to total recipient countries   **0.09%**

Venezuela

Belgium	DGCD	31161	149	FOOD CROP PRODUCTION	2004008732
Japan	JICA	31110	405	TC AGGREGATED ACTIVITIES	072971T
Japan	JICA	31210	31	TC AGGREGATED ACTIVITIES	073336T
Japan	JICA	31310	125	TC AGGREGATED ACTIVITIES	073515T
Spain	Misc	31130	12	CONSTRUCTION OF AN INDUSTRIAL STOREHOUSE	071289
Spain	AG	31161	21	ENSURING FOOD SECURITY	072016

Total aid for water in 2007 for Venezuela   **USD thousand 744**
And as a share of aid to total recipient countries   **0.01%**

Viet Nam

AsDF		31120	92693	INTEGRATED RURAL DEVELOPMENT SECTOR PROJECT IN THE CENTRAL PRV	070061
AsDF		31182	31853	AGRICULTURE SCIENCE AND TECHNOLOGY PROJECT	070014
Australia	AusAID	31110	125	AGRICULTURAL POLICY & ADMIN. MGMT	2007002049
Australia	AusAID	31120	205	AGRICULTURAL DEVELOPMENT	2007001799
Australia	AusAID	31120	56	IMPROVED BEEF PRODUCTION SYSTEMS IN CENTRAL VIETNAM	2007001315
Australia	AusAID	31130	251	AGRICULTURAL LAND RESOURCES	2007001578
Australia	AusAID	31140	77	AGRICULTURAL WATER RESOURCES	2007001138
Australia	AusAID	31150	118	AGRICULTURAL INPUTS	2007001314
Australia	AusAID	31161	73	FOOD CROP PRODUCTION	2007001169
Australia	AusAID	31163	184	LIVESTOCK	2007001387
Australia	AusAID	31181	43	STRENGTHENING RESEARCH CAPABILITY AND ENHANCING PR	2007001432
Australia	AusAID	31182	66	HUANGLONGBING MANAGEMENT	2007001705

RECIPIENT Donor	Agency	Sector	Amount USD thousand	Project description	CRS ID Number
Australia	AusAID	31182	232	AH/2004/040	2007000627
Australia	AusAID	31182	75	IMPACT OF MIGRATION AND/OR OFF-FARM EMPLOYMENT ON ROLES OF WOMEN	2007001806
Australia	AusAID	31182	14	TAM GIANG LAGOON SYSTEM TO IMPROVE THE LIVELIHOODS OF POOR COMMUNITIES	2007001881
Australia	AusAID	31182	157	AGRICULTURAL RESEARCH	2007001976
Australia	AusAID	31191	38	STRENGTHENING AGRICULTURAL MARKET INFORMATION	2007001517
Australia	AusAID	31192	17	R&D PRACTITIONERS IN HOW TO TAILOR RESEARCH AND DEVELOPMENT AGENDA-SETTING	2007001826
Australia	AusAID	31192	57	VIETNAMESE R&D PRACTITIONERS IN HOW TO TAILOR RESEARCH AND DEVELOPMENT	2007001309
Australia	AusAID	31192	303	PLANT/POST-HARVEST PROT. & PEST CTRL	2007001682
Australia	AusAID	31192	44	MANAGEMENT OF PEST FRUIT FLIES	2007001421
Australia	AusAID	31192	42	EFFECTIVE PHOSPHINE FUMIGATION - TECHNOLOGY TRANSFER	2007001450
Australia	AusAID	31195	16	LIVESTOCK/VETERINARY SERVICES	2007001859
Australia	AusAID	31195	8	CONTROL OF HPAI IN DUCKS IN INDO. AND VIETNAM. LB8	2007002048
Australia	AusAID	31282	144	FORESTRY RESEARCH	2007001886
Australia	AusAID	31310	8	REGIONAL MECHANISMS TO MAXIMISE BENEFITS TO SMALL-HOLDER SHRIMP FARMERS	2007002070
Australia	AusAID	31310	15	POLICY LINKAGE STUDY WITH SOME OF ACIAR'S AQUACULTURE PROJECTS	2007001871
Australia	AusAID	31320	13	CULTURE BASED AND CAPTURE FISHERIES DEVELOPMENT	2007001918
Australia	AusAID	31320	8	DISSEMINATION OF FISHERIES RESERVOIR MANAGEMENT EXTENSION MATERIALS	2007002060
Australia	AusAID	31320	466	FISHERY DEVELOPMENT	2007001554
Australia	AusAID	31382	147	SUSTAINABLE TROPICAL SPINY ROCK LOBSTER AQUACULTURE	2007001219
Australia	AusAID	31382	32	TECHNOLOGY TRANSFER IN APPLIED POPULATION GENETICS OF AQUATIC SPECIES	2007001588
Australia	AusAID	31382	49	FISHERY RESEARCH	2007002009
Belgium	DGCD	31110	74	AGRICULTURAL POLICY & ADMIN. MGMT	2004003268
Belgium	DGCD	31120	327	AGRICULTURE DURABLE ET LUTTE CONTRE LA PAUVRETE	2006003976
Belgium	DGCD	31140	12	AGRICULTURAL WATER RESOURCES	2006002773
Belgium	DGCD	31140	160	IRRIGATION BINH DINH PROVINCE	2005001902
Belgium	DGCD	31163	621	DÉVELOPPEMENT ET EXTENSION DAIRY FARMING ACTIVITIES	2004001109
Belgium	MPRF	31320	21	FISHERY DEVELOPMENT	2005001966
Canada	CIDA	31166	140	SLED- THAN HOA - SLED- THAN HOA-DESIGN PHASE	070030
Canada	CIDA	31220	273	FS/DEVELOPMENT FOREST NURSERY	070662
Denmark	MFA	31110	3675	ADVISERS	071122
Denmark	MFA	31110	5156	CENTRAL COMPONENT	071119
Denmark	MFA	31110	937	PROGRAMME ADMINISTRATION	071121
Denmark	MFA	31110	2719	UNALLOCATED FUNDS	071123
Denmark	MFA	31120	29773	PROVINCIAL COMPONENT	071120
Denmark	MFA	31382	1468	FISHBORNE ZOONOTIC PARASITES	031148
Finland	MFA	31210	17385	FORESTRY TRUST FUND	2004003014
France	MISC	31110	18	POLITIQUE AGRICOLE & GESTION ADMINISTRATIVE	2007800786
France	AFD	31120	54757	DEVLPT INFRASTR. RURALES PROVINC. CENTRE	2007116000
France	AFD	31120	1369	DEVLPT INFRASTR. RURALES PROVINC.CENTRE	2007172500
France	AFD	31140	1095	APPUI A LA GESTION PARTICIP DE L'IRRIGAT	2007158900
France	AFD	31140	15606	INFRASTRUC. CONTRE INONDATIONS RIV. SAIGON	2007001020
France	MISC	31182	22697	RECHERCHE AGRICULTURE	2007800384
Germany	BMZ	31110	55	AGRICULTURAL POLICY & ADMIN. MGMT	2007007508
Germany	BMZ	31120	107	AGRICULTURAL DEVELOPMENT	2007006264
Germany	BMZ	31181	36	AGRICULTURAL EDUCATION/TRAINING	2007008859
Germany	BMZ	31182	7	AGRICULTURAL RESEARCH	2007007648
Germany	BMZ	31210	73	FORESTRY POLICY & ADMIN. MANAGEMENT	2007007495

RECIPIENT Donor	Agency	Sector	Amount USD thousand	Project description	CRS ID Number
Germany	L G	31282	135	FORESTRY RESEARCH	2007011670
IDA		31110	4200	AVIAN AND HUMAN INFLUENZA CONTROL	2007000976
Italy	DGCS	31120	707	AGRICULTURAL DEVELOPMENT	060729
Japan	JICA	31110	3907	TC AGGREGATED ACTIVITIES	073073T
Japan	JICA	31130	836	TC AGGREGATED ACTIVITIES	073119T
Japan	JICA	31150	18	TC AGGREGATED ACTIVITIES	073171T
Japan	JICA	31163	1002	TC AGGREGATED ACTIVITIES	073233T
Japan	JICA	31195	81	TC AGGREGATED ACTIVITIES	073282T
Japan	PRF	31195	17	TC AGGREGATED ACTIVITIES	075611T
Japan	JICA	31210	3461	TC AGGREGATED ACTIVITIES	073457T
Japan	MAFF	31310	64	TC AGGREGATED ACTIVITIES	070043T
Japan	JICA	31310	247	TC AGGREGATED ACTIVITIES	073582T
Korea	Misc	31110	86	CONSULTING FOR RURAL DEVELOPMENT MODEL OF VIETNAM	2007002549
Korea	Misc	31110	9	AGRICULTURAL POLICY & ADMIN. MGMT	2007002560
Korea	Misc	31120	10	AGRICULTURAL DEVELOPMENT	2007001930
Korea	KOICA	31130	41	AGRICULTURAL LAND RESOURCES	2007016651
Korea	KOICA	31140	16	AGRICULTURAL WATER RESOURCES	2007015831
Korea	KOICA	31140	87	TRAINING PROGRAM	2007017367
Korea	KOICA	31161	87	TRAINING PROGRAM	2007017403
Korea	KOICA	31163	43	LIVESTOCK	2007017106
Korea	KOICA	31181	23	AGRICULTURAL EDUCATION/TRAINING	2007016032
Korea	Misc	31181	21	AGRICULTURAL EDUCATION/TRAINING	2007002574
Korea	KOICA	31320	177	TRAINING PROGRAM	2007017370
Korea	KOICA	31381	42	FISHERY EDUCATION/TRAINING	2007016062
Luxembourg	MFA	31140	1069	AGRICULTURAL WATER RESOURCES	070741
Luxembourg	MFA	31181	275	AGRICULTURAL EDUCATION/TRAINING	070738
Netherlands	MFA	31110	902	AGRIC.: SUBS. TO PSOM 2003-2009	2007100128
Netherlands	MFA	31220	6160	HAN_MULTI-D TRUST FUND FORESTS	2003008324
New Zealand	NZAid	31120	96	BINH DINH SUSTAINABLE RURAL LIVELIHOOD	070548
New Zealand	NZAid	31130	417	LAND ADMINISTRATION PROGRAMME	070421
New Zealand	NZAid	31181	22	CECEM VOCATIONAL TRAINING	070288
New Zealand	NZAid	31195	610	AVIAN INFLUENZA SUPPORT PROGRAMME - MARD	070624
Norway	NORAD	31310	17	FISHING POLICY AND ADMIN. MANAGEMENT	2007004927
Norway	NORAD	31320	37	Feasibility Study - gelatine / oil from fisheries	2007004946
Norway	NORAD	31320	80	feasibility study - Aquaculture in Vietnam	2007004953
Spain	MFA	31120	165	SUPPORT TO PRODUCTION DEFENSE AND MARKETING OF ITS PRODUCTS	075506
Switzerland	SDC	31110	7360	SFSP SOCIAL FORESTRY SUPPORT PROGRAMME	2000002315
Switzerland	SDC	31110	58	AGRICULTURAL POLICY & ADMIN. MGMT	2000002316
Switzerland	SDC	31110	133	ISG INTERNATIONAL SUPPORT GROUP	2007000219
Switzerland	seco	31210	625	LINKING TRADE DEMAND MEKONG REGION II	2007001888
United States	AID	31120	7	AGRICULTURAL DEVELOPMENT	2007001774
United States	TDA	31210	35	FORESTRY POLICY & ADMIN. MANAGEMENT	2007028254

Total aid for water in 2007 for Viet Nam   **USD thousand 319570**
And as a share of aid to total recipient countries   **4.76%**

RECIPIENT Donor	Agency	Sector	Amount USD thousand	Project description	CRS ID Number

Wallis & Futuna

France	MISC	31110	1064	POLITIQUE AGRICOLE & GESTION ADMINISTRATIVE	2007800828
France	MISC	31181	370	EDUCATION & FORMATION DANS LE DOMAINE AGRICOLE	200780083

Total aid for water in 2007 for Wallis & Futuna   **USD thousand 1434**
And as a share of aid to total recipient countries   **0.02%**

Yemen

Japan	JICA	31110	68	TC AGGREGATED ACTIVITIES	072980T
Japan	JICA	31130	31	TC AGGREGATED ACTIVITIES	073099T
Japan	MOFA	31150	2801	THE ASSISTANCE FOR UNDERPRIVILEGED FARMERS	070086
Japan	JICA	31150	25	TC AGGREGATED ACTIVITIES	073161T
Japan	MAFF	31310	23	TC AGGREGATED ACTIVITIES	070039T
Japan	JICA	31310	77	TC AGGREGATED ACTIVITIES	073568T
Netherlands	MFA	31110	724	AGRIC.: ORET-PROGRAMME 2007	2007100147
United States	AID	31110	60	AGRICULTURAL POLICY & ADMIN. MGMT	2007013915
United States	AID	31120	150	AGRICULTURAL DEVELOPMENT	2007001755
United States	AID	31120	499	AGRICULTURAL SECTOR PRODUCTIVITY	2007001776

Total aid for water in 2007 for Yemen   **USD thousand 4458**
And as a share of aid to total recipient countries   **0.07%**

Zambia

EU institutions	EDF	31162	8214	ACCOMPANYING MEASURES 2007-2010 FOR SUGAR PROTOCOL COUNTRIES - ZAMBIA	2007201275
Finland	MFA	31161	198	NGO SUPPORT / ZNAPD NGWENA RIVER FARMING PROJECT	2007000150
Finland	MFA	31181	185	NGO SUPPORT / GREEN VILLAGE COMMUNITY LIVELIHOODS	2007000153
Germany	BMZ	31120	249	AGRICULTURAL DEVELOPMENT	2007009164
Germany	BMZ	31181	11	AGRICULTURAL EDUCATION/TRAINING	2007003040
Ireland	DFA	31120	11	AGRICULTURAL DEVELOPMENT	2007003142
Ireland	DFA	31130	127	AGRICULTURAL LAND RESOURCES	2007001067
Ireland	DFA	31150	49	AGRICULTURAL INPUTS	2007001522
Ireland	DFA	31161	443	SUPPORT TO NGOS - IRISH - MAPS - GENERAL - INCREASE DIVERSIFICATION	2007000238
Ireland	DFA	31166	19	AGRICULTURAL EXTENSION	2007002293
Ireland	DFA	31182	145	AGRICULTURAL RESEARCH	2007000962
Japan	JICA	31110	2584	TC AGGREGATED ACTIVITIES	073038T
Japan	JICA	31130	236	TC AGGREGATED ACTIVITIES	073090T
Japan	JICA	31150	281	TC AGGREGATED ACTIVITIES	073153T
Japan	JICA	31163	93	TC AGGREGATED ACTIVITIES	073201T
Japan	JICA	31195	531	TC AGGREGATED ACTIVITIES	073289T
Japan	JICA	31210	95	TC AGGREGATED ACTIVITIES	073382T
Japan	JICA	31310	132	TC AGGREGATED ACTIVITIES	073492T
Korea	KOICA	31140	116	TRAINING PROGRAM	2007017426
Netherlands	MFA	31110	69	AGRICULTURAL POLICY & ADMIN. MGMT	2007100134
Netherlands	MFA	31181	506	AGRIC. TRAINING: LUS LIVESTOCK DEVEL. TRUST	2003014254
Norway	MFA	31120	80	AGRICULTURAL DEVELOPMENT	2007002404
Norway	MFA	31161	13	Conservation Farming Consultancy	2007005118

RECIPIENT Donor	Agency	Sector	Amount USD thousand	Project description	CRS ID Number
Norway	NORAD	31162	15	HONEY CONSULTANCY - ZAMBIA	2007005127
Norway	NORAD	31181	137	CAPACITY BUILDING IN ORGANIC AGRICULTURE - ZAMBIA	2007005137
Norway	MFA	31191	412	EMBASSY DISENGAGEMENT FROM PAM COMACO	2007005152
Sweden	Sida	31110	1480	AGRICULTURAL POLICY & ADMIN. MGMT	2007001323
United Kingdom	DFID	31110	756	RIGHTS TO SEED CSCF 0359	2007000360
United States	AID	31110	18	AGRICULTURAL POLICY & ADMIN. MGMT	2007013862
United States	AID	31110	821	AGRICULTURAL ENABLING ENVIRONMENT	2007001646
United States	AID	31110	644	PROGRAM SUPPORT (AGRICULTURE)	2007013873
United States	AID	31120	100	AGRICULTURAL DEVELOPMENT	2007001727
United States	AID	31120	2681	AGRICULTURAL SECTOR PRODUCTIVITY	2007001738
United States	ADF	31162	23	INDUSTRIAL CROPS/EXPORT CROPS	2007000062
United States	ADF	31320	37	FISHERY DEVELOPMENT	2007000223

Total aid for water in 2007 for Zambia USD thousand 21511
And as a share of aid to total recipient countries 0.32%

Zimbabwe

Recipient Donor	Agency	Sector	Amount	Project description	CRS ID
Belgium	DGCD	31194	90	AGRICULTURAL CO-OPERATIVES	2004003209
Canada	IDRC	31110	319	LACK OF RESILIENCE IN AFRICAN SMALLHOLDER FARMING	070305i
Canada	IDRC	31120	265	ADAPTIVE CAPACITY TO COPE WITH INCREASING VULNERABILITY TO CLIMATIC CHANGE	070310i
EU institutions	CEC	31140	8214	SMALLHOLDER MICRO-IRRIGATION DEVELOPMENT SUPPORT PROGRAMME	2007100294
EU institutions	EDF	31162	3696	ZIMBABWE ANNUAL ACTION PLAN 2007 UNDER AMSP	2007201271
France	MISC	31182	12416	RECHERCHE AGRICULTURE	2007800355
Germany	BMZ	31120	743	AGRICULTURAL DEVELOPMENT	2007006014
Germany	BMZ	31120	958	PROGRAMME ON FOOD SOVEREIGNTY IN ZIMBABWE AND THE SOUTHERN AFRICA REGION	2007006509
Germany	BMZ	31150	513	FURTHER IMPLEMENTATION OF SMALL-SCALE IRRIGATION PROJECTS	2007006016
Germany	BMZ	31161	142	FOOD CROP PRODUCTION	2007003797
Ireland	DFA	31161	632	SUPPORT TO NGOS - IRISH - MAPS - GENERAL	2007000233
Japan	JICA	31110	536	TC AGGREGATED ACTIVITIES	073030T
Japan	JICA	31130	94	TC AGGREGATED ACTIVITIES	073085T
Japan	JICA	31150	48	TC AGGREGATED ACTIVITIES	073148T
Japan	JICA	31163	91	TC AGGREGATED ACTIVITIES	073196T
Japan	JICA	31195	55	TC AGGREGATED ACTIVITIES	073287T
Japan	JICA	31210	52	TC AGGREGATED ACTIVITIES	073306T
Korea	KOICA	31181	120	PROJECT FOR UPGRADING THE AGRICULTURAL ENGINEERING TRAINING CENTER, MINI	2007017686
Norway	MFA	31164	9	AGRARIAN REFORM	2007004778
United Kingdom	DFID	31120	600	T FOOD SECURITY AND PROTECT THE LIVELIHOODS OF ABOUT 2 MILLION PEOPLE	2007000367

Total aid for water in 2007 for Zimbabwe USD thousand 29595
And as a share of aid to total recipient countries 0.44%

OECD PUBLISHING, 2, rue André-Pascal, 75775 PARIS CEDEX 16
PRINTED IN FRANCE
(43 2009 20 3 P) ISBN 978-92-64-07702-7 – No. 57182 2010